ENZYME
IMMUNOASSAYS

ENZYME IMMUNOASSAYS
FROM CONCEPT TO PRODUCT DEVELOPMENT

S.S. DESHPANDE
IDETEK, INC.

CHAPMAN & HALL

 New York • Albany • Bonn • Boston • Cincinnati • Detroit • London • Madrid • Melbourne
Mexico City · Pacific Grove • Paris • San Francisco • Singapore • Tokyo • Toronto • Washington

Library of Congress Cataloging-in-Publication Data

Deshpande, S. S.
   Enzyme immunoassays : from concept to product development / S. S. Deshpande.
      p. cm.
   Includes bibliographical references and index.
   ISBN 0-412-05601-1 (alk. paper)
   1. Enzyme-linked immunosorbant assay. I . Title.
   QP519.9.I44.D47   1996
   616.07'56--dc20

95-32913
CIP

To order this or any other Chapman & Hall book, please contact **International Thomson Publishing, 7625 Empire Drive, Florence, KY 41042.** Phone: (606) 525-6600 or 1-800-842-3636. Fax: (606) 525-7778. e-mail: order@chaphall.com.

For a complete listing of Chapman & Hall's titles, send your request to **Chapman & Hall, Dept. BC, 115 Fifth Avenue, New York, NY 10003.**

In loving memory of my teacher,
Dr. S. P. Manjrekar
Had it not been for his guidance early in my college years,
I would have missed an exciting and eventful career in basic sciences

Contents

Preface

Since the classical work of Rosalyn Yalow and Solomon Berson on the radioimmunoassay of insulin in the late 1950s, which so brilliantly opened new domains in the diagnostics sector, immunoassays have been well received as a diagnostic and research tool in several disciplines of life sciences. The economics, rapidity, sensitivity, specificity, and the easy-to-use approach of immunodiagnostic tests are widely recognized, and have had a tremendous impact on clinical, agricultural, food, veterinary, and environmental diagnostics. At present, in the clinical diagnostic sector alone, immunoassay products command an annual worldwide market value in excess of U.S. $10 billion, which is expanding by about 10% per annum.

The recent trend toward increasing demand for automation of immunoassays and stable reagents clearly shows a definitive shift to nonisotopic immunoassay systems on a large scale. Historically, enzyme immunoassays have been in a prime position for in-house applications, primarily because of the easy access to inexpensive measuring equipment as well as to a vast body of literature on labeling techniques and applications. The recent popularity with the general public of several over-the-counter immunoassay products based on this technology, such as glucose level kits for diabetics, cholesterol screening tests, and pregnancy detection test kits, is quite evident. Because the huge expenditure involved in the health care profession continues to grow at an alarming pace worldwide, immunoassays, especially those based on inexpensive enzyme-based technology, are expected to play an important role in keeping these costs down while providing a high level of service, especially in the underdeveloped markets.

This book is intended to provide useful practical information for the development of a successful commercial immunodiagnostic product based on enzyme immunoassay technology. Nonetheless, the concepts presented here are also quite applicable to products that utilize other detection labels. The reader is taken through all stages of this process—from conceptualization, product design, and manufacture to introduction of the product to the market.

Every diagnostic product commercially available owes its success to a joint team effort, requiring the expertise of scientists from such diverse disciplines as immunochemistry, organic chemistry, protein chemistry, enzymology, statistics, business management, and marketing. Thus, an understanding of the basic concepts and proper manipulation of these techniques is key to a successful product development. For convenience, this book is divided into two parts. Part 1, Chapters 1 through 7, covers the basic concepts including the classification, structure, and function of antibodies; antigen-antibody reactions; general information on various conjugation techniques; antibody development, production, and processing; the properties and characteristics of the most widely used enzymes in immunoassays; and the choice of various solid-phase systems available in the market. The salient features of each of these topics are covered from a product development viewpoint.

Part 2 of the book, Chapters 8 through 12, covers the actual product development process. It includes information on the classification and the various formats of enzyme immunoassays available for product development, reagent formatting, and assay development, data processing, standardization, scale-up, and commercial manufacture of the product. Requirements such as evaluation and clinical or regulatory validation of the product, the importance of good laboratory and manufacturing practices (GLP and GMP), and certain international marketing requirements such as the ISO 9000 certification process are also described. Finally, information is presented for the benefit of young entrepreneurs who would like to venture in this exciting field with their own start-up company.

Acknowledgments

Sincere appreciation is extended to the members of the Chapman & Hall editorial and production staff, especially Dr. Eleanor S. Riemer, Ms. JoAnna Turtletaub, Ms. MaryAnn Cottone, and Ms. Torrey Adams. Their encouragement, support, cooperation, and, above all, tremendous patience during the preparation of this manuscript made this an experience to treasure and cherish. No project of mine has ever been completed without the guidance and encouragement of my teacher, Professor D. K. Salunkhe at Utah State University, Logan. I also wish to thank Dr. Richard M. Rocco, my immediate supervisor at Idetek, Inc., for allowing me to pursue my academic interests and for his invaluable support during my stay with the company.

No words can express my gratitude to my wife, Usha, for her unflagging encouragement, patience, and support; her not so gentle prodding at a time when I wanted to drop out literally gave a second wind to this project. And, finally, to our daughter, Maithili, whose much anticipated and actual arrival in our lives at once delayed, yet provided the necessary impetus to finish this project, this is just the sort of gift I intended for her first birthday.

Part I

Basic Considerations (Concepts)

1
Introduction

ANALYTICAL TECHNIQUES IN LIFE SCIENCES

Qualitative and quantitative analytical techniques are of tremendous importance in all fields and disciplines of life sciences for the detection, identification, and measurement of concentration of a wide variety of biologically important molecules. These analytical techniques can be classified into three general categories: biological techniques or "bioassays," those based on physical and/or chemical methods, and those that depend on noncovalent binding of one reactant to another. Techniques that fall under the third category are also commonly referred to as "binding assays." The salient features of these classes of analytical techniques are briefly described below.

Biological Techniques or Bioassays

Biological techniques or the bioassays measure the response that follows the application of a stimulus to a biological system. The applied stimulus is represented by standard or test samples that contain the biologically active substance or analyte. The biological system that receives the stimulus may be a whole, multicellular organism such as an animal or a plant, isolated organs or tissues from multicellular organisms, and whole cells or microorganisms (Hawcroft et al. 1987; Odell 1983; Clausen 1988). The response, measured as a change in some aspect of the biological system used, may be a positive response associated with an increased activity or a negative response that is inhibitory or even lethal to the biological system. It relates to a biological activity that is normally attributed to the analyte and is often expressed in concentration units, generally relative to an arbitrary although internationally recognized standard.

All bioassays are comparative and require a standard preparation, ideally from the same source, with which each test sample can be compared. The use of "in-house" standards often results in a lack of interlaboratory correlations and, therefore, is not advisable. Better correlations are achieved by the use of International Standards for the Calibration of Bioassays. These standards are determined by the World Health Organization based on the recommendations of the Expert Committee on Biological Standardization (Bangham 1983; Kirkwood 1977; Campbell 1974).

Once the assays are completed, the data can be analyzed in several different ways to establish the relationship between the dose and the intensity of the response. Some representative examples of dose:response curves are shown in Figure 1.1A–D. Although only positive relationships are shown here, it is not uncommon to observe inverse relationships in these types of studies. The intensity of the biological response is dependent on both the slope of the curve as well as its position relative to the abscissa; these determine the sensitivity and detection limits (or range) of the bioassay, respectively. Ideally, the response to an applied stimulus should be sufficiently sensitive to be able to differentiate between small changes in doses but not so great as to restrict the detection limits of the assay.

Some representative examples of this class of analytical techniques include monitoring the potency of pharmaceutical drugs at target sites, tissues, or the organism as a whole, and the measurement of LD_{50} values to determine the toxicity of chemicals that are potentially hazardous to human health.

Bioassays can only be used for substances that produce biological responses in living organisms and tissues. This precludes their widespread use as a general analytical technique in several fields (Odell and Franchimont 1983). Similarly, the use of animals or their tissues in scientific research is increasingly being viewed by the general public as unethical. Advances in cell culture techniques, however, have minimized the necessity for the use of animals in bioassays. At present, most initial screening tests for measuring the biological potency of various drugs involve the use of cells cultured in an artificial medium. By monitoring the obvious morphological or biochemical changes in the cell culture populations, these techniques greatly facilitate the elimination of potentially hazardous drugs in the early stages, and thereby restrict the number of tests to be carried out with actual animal system to a minimum. Clinical drug trials with selected human populations, although they fall under this category, are generally not viewed as bioassays.

Another major disadvantage of bioassays that use animal model systems is the inherent biological variability often found in test animals and their tissues (Odell 1983; Hawcroft et al. 1987). This results in poor precision and reproducibility, and necessitates the extent of replication. Therefore, the use of animal model systems is cost prohibitive and time consuming. Similarly, in some types of bioassays, the animals may die or have to be slaughtered for further tissue analysis.

In this respect, microbiological assays using broth cultures ("tube assays") and assays on semi-solid culture media ("plate assays") offer several advantages. Al-

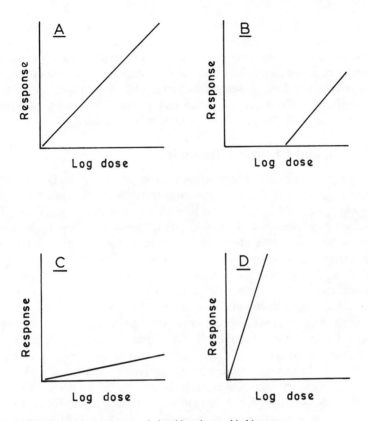

Figure 1.1. Typical dose:response relationships observed in bioassays
A: An ideal response with adequate sensitivity and wide detection limits
B: Although the sensitivity is adequate, the detection limits are restricted to higher doses of drugs
C: Bioassays with wide detection limits but lacking sensitivity
D: The high sensitivity restricts the detection limits of the bioassays to lower doses of drugs

though yeasts, protozoa, and algae can be used for microbiological assays, most methods use bacteria because of the relative ease of use. Microbiological assays do not require highly specialized and expensive equipment. They also provide highly homogeneous cell populations for test and thus result in better assay precision. These methods, therefore, are ideal in situations where the number of tests is likely to be small and, hence, large capital expenditure cannot be justified. Microbiological assays, however, are limited only to those analytes that either promote or inhibit microbiological growth. This fact mandates that no other substance be present in the test sample that either promotes or inhibits growth or modifies the response to the analyte.

In spite of these limitations, the fact that they are generally specific for the biologically active forms of the analyte is a good reason for actually using a bioassay. Bioassays are the preferred analytical methods when the test material contains a mixture of active and inactive forms of the analyte that cannot be separated effectively (Odell 1983; Hawcroft et al. 1987). Alternatively, the analyte may occur in a variety of active forms that affect the same target site but with different biological activities and are present in unknown relative quantities. In addition to high specificity, bioassays are also often very sensitive. Bioassays, therefore, are used when no suitable alternative assay methods are available.

Physical and Chemical Methods

The classical biochemical techniques involving the use of spectrophotometry, gas chromatography (GC), mass spectrometry (MS), high-performance liquid chromatography (HPLC), paper and thin-layer chromatography (PC and TLC), and electrophoretic techniques coupled with UV or fluorescence detection methods are widely used for the identification and quantitation of biologically important compounds. These basic analytical tools that exploit the physical and/or chemical characteristics of the compounds are used both in research and as confirmatory methods in the diagnostic industry. However, the cost of equipment and consumables, sample throughput, and the level of experience and skill required for analysis often preclude their use in rapid diagnosis of medical causes in the clinical fields, in on-line routine monitoring of process variables and quality in the food industry, or in on-site testing of environmental samples. Moreover, such methods are often tedious, laborious, and may require elaborate sample cleanup and concentration procedures. Although not suitable as screening tests, HPLC, GC, and GC-MS methods are often used as confirmatory tools for the presumptive positive samples that are initially screened by some kind of immunoassays.

Binding Assays

Binding assays comprise a variety of methods that utilize the specific reaction between a ligand and a binding protein. The basic principle of a binding assay is a reversible reaction between a ligand (L) and its binding protein (BP) that obeys the law of mass action as follows:

$$L + BP \underset{k_2}{\overset{k_1}{\rightleftharpoons}} L\text{-}BP \tag{1}$$

The ligand combines with the binding protein to form the L-BP complex at rate constant k_1. At equilibrium, the L-BP complex dissociates with a rate constant k_2 to form free ligand and binding protein. In the immunodiagnostic field, the ligand is also often referred to as an antigen, hapten, or analyte.

Binding assays can be further classified into three general categories: immunoassays in which the ligand is an antigen and the binding protein its specific antibody, receptor binding assays in which the receptor protein is usually extracted from the target organ of a hormone or drug, and circulating binding protein assays that employ a naturally occurring plasma protein such as thyroxine binding globulin (TBG) for the assay of thyroxine (T4) and transcortin for the assay of cortisol or progesterone.

Immunoassays are by far the most common form of binding assays. In fact, of all the methods described above, immunodiagnostics has developed into an extremely versatile analytic technique with a diverse range of assay protocols. In comparison to other analytical methods, an immunoassay provides a rapid, economical, highly sensitive, and specific analysis that is relatively simple to perform and interpret. Three factors have been primarily responsible for the rapid development of immunoassays as a powerful analytical tool in many clinical, pharmaceutical, agricultural, veterinary, environmental, and other basic scientific investigations. These include:

1 The generation of antibodies that display marked specificity against an immense range of compounds,

2 The high affinity with which they bind their appropriate antigen, thus enabling great sensitivity to be achieved and indeed defining the ultimate sensitivity that can be attained, and finally

3 The ease of their use in detecting the concentration of an antigen, or in detecting the presence of antibodies, especially in monitoring the onset and spread of infectious diseases such as AIDS, herpes, and viral hepatitis.

As compared to bioassays, which provide the estimates of concentration of an analyte based on its biological activity or function, immunoassays are "structural" assays, which assess only a part of the antigen's structure or its "antigenic determinant." Therefore, the values obtained by a bioassay may be considerably lower than those determined by an immunoassay method. For example, the reduction of disulfide ring of oxytocin or arginine vasopressin causes a complete loss of its biological activity, and thus fails to yield a positive result in a bioassay that determines only the hormone itself. In contrast, such reduction of the disulfide ring does not affect the results of most immunoassays that determine any circulating prohormone, the hormone itself, and any fragment of the hormone that contains the antigenic determinant. Nevertheless, for several clinically important protein or peptide analytes (including hormones, lipoproteins, oncoproteins, pathogen antigens, and specific antibodies), there are as yet no viable alternatives to immunoassays (Gosling 1990). Even where an analyte can feasibly be determined by chromatographic, colorimetric, or other standard analytic procedures, quantitative immunologic methods are often used because of their speed, simplicity, and relatively low cost. This trend has been greatly encouraged by the availability of convenient reliable commercial kits.

Other reasons for the growing popularity of immunoassays include the potential of rapidly measuring a minute quantity of specific analyte from within a complex sample matrix, often with little or no sample cleanup, the development of more sensitive detection systems, the relative ease of performing the assays, as well as the lower cost of the assays relative to most conventional analytic techniques (Deshpande and Sharma 1993).

IMMUNODIAGNOSTICS: A HISTORICAL PERSPECTIVE

Although the classical works of Jenner, demonstrating vaccination by injection of products of the organism as a mechanism to induce immunity to small pox and subsequently to other infectious diseases, were well known, it was not until the late nineteenth century that the basic foundations of the science of immunology were laid. In 1882 the Russian scientist Metchnikoff first proved that "phagocytosis" of microorganisms by leucocytes was the major host defense against infection. Soon thereafter, von Behring in 1888 first demonstrated that nonlethal doses of the toxin from diphtheria induced immunity to organism (Steward 1984). Further research on toxin-induced immunity led von Behring and Kitasato to demonstrate in 1890 that immunity to infection following injections of bacterial toxins was the result of the appearance of a "factor" in the serum which neutralized the toxin (Humphrey and White 1964). This antitoxin activity could be transferred to normal animals by injecting serum from immune animals. It thus appeared that the body was able to respond to infectious organisms or injurious substances by producing serum components that would combine with these agents and neutralize their toxic effects. These serum components were called "antibodies" and the agents provoking their production were termed "antigens."

These modest beginnings soon led to the observations that antibodies could specifically lyse, precipitate, and agglutinate bacteria. Furthermore, it was shown that antibodies could be produced against toxins other than those from bacteria and against nontoxic substances such as proteins as long as they were "foreign" to the immunizing species. These early discoveries showed the diagnostic promise of antisera, which was quickly exploited in developing diagnostic tests for typhoid (Widal test 1896) and syphilis (Wassermann test 1906) (Catty 1988). Research in the early part of this century, carried out by such eminent scientists as Landsteiner, Heidelberger, Kabat, and others, demonstrated that antibodies are evolved as a principle means of defense against infection, and that their response has an apparently limitless range of specificity, which can be used to define, isolate, and measure a wide variety of immunogenic molecules not just confined to the products of microorganisms (Butler 1991; Catty 1988). Indeed, most complex molecules with molecular weights greater than 5000 that are not intrinsic to the species immunized were found to be immunogenic. However, it was Landsteiner's (1945) pioneering work, beginning in 1917, on the immunochemistry of low molecular weight compounds and his demonstration that such molecules or "haptens" chemically coupled to a larger carrier could elicit antibodies of exquisite specificity, that established many of the ground rules that form the basis of modern immunodiagnostics.

By 1940, it was firmly established that antigen-antibody reactions follow nearly predictable rules of biochemical interaction and thus could be used to determine accurately the quantity of antigen in a sample. Using by then the newly developed electrophoresis technique, Tisselius and Kabat (1939) demonstrated that the antibody activity of an antiserum is associated with the γ-globulin fraction of the serum. In their study, the electrophoretic patterns of an antiovalbumin serum before and after precipitation with the antigen ovalbumin showed a significant removal of the γ-globulin fraction. Globulins that function as antibodies are now termed "immunoglobulins."

Heidelberger (1939) developed the first quantitative precipitation test and purified antibody for the first time. The practicability of this principle received wide attention and led to its exploitation in simple immunoassays that could measure single antigen systems. Oudin (1946) demonstrated that diffusion of antibodies and antigens in the fluid phase across the interface of opposing agar gels could result in quantitative assessment of both reacting elements by the position of the resulting precipitation bands. Diffusing in a gradient of concentration into each other, the antibody-antigen system formed visible complexes where their relative proportions were conducive to precipitation. This simple gel version of precipitin assay allowed technicians with little immunochemical experience to quantify multivalent antigens in clinical and basic science laboratories. This technological breakthrough by Oudin (1946) quickly led to the application of double immunodiffusion in gel, developed simultaneously by Elek (1948) and Ouchterlony (1948), to determine easily the antigenic relationships between molecules and the specificity of antisera assessed. Subsequent developments in this field led to the introduction of immunoelectrophoresis (Grabar and Williams 1953), radial immunodiffusion with its quantitative properties (Mancini et al. 1965), "rocket" immunoelectrophoresis for the rapid estimation of single antigens from mixtures using unispecific antibody in the gel, and two-dimensional (crossed) immunoelectrophoresis, which allows both qualitative and quantitative assessment of multiple components of antigen mixtures by using multispecific antisera (Laurell 1966).

Although the gel precipitation methods mentioned above were useful in the measurement of antigens and in quality control of specificity and titer of precipitating antibodies, they were limited in sensitivity (5 μg/ml for antigen) and could not be used to quantitate low molecular weight antigens that formed only soluble complexes. To overcome these limitations and to improve the sensitivity of detection, Coombs et al. (1945) described the antiglobulin reaction in which antihuman immunoglobulin sera agglutinated human erythrocytes from certain subjects on whose cells antibodies had previously bound. Coombs pioneered the use of red cells as an indicator system in immunoassays, and later developed his system as an important clinical test (Coombs' antiglobulin test).

The beginning of modern immunodiagnostic tests as they are practiced today could be attributed to the use of a coupled label to provide an indirect signal of antigen-antibody interaction by Coons and collaborators (1941). These researchers used fluorescein-labeled antibodies for antigen localization in lymphoid tissue. Immunocytochemistry now utilizes both fluorescent dyes and, increasingly, enzyme/substrate systems as in the immunoperoxidase method (Catty 1988).

A further milestone in sensitive assays came with the discovery of radioisotopic labeling techniques for antigens, antibodies, and small peptide hormones by Farr (1958) and Yalow and Berson (1959, 1960). The pioneering efforts of the latter group formed the basis of the first radioimmunoassay (RIA). Using ^{131}I-labeled insulin, Yalow and Berson demonstrated the presence of a circulating insulin-binding protein in insulin-dependent diabetics, and showed that the tracer could be displaced from the binding protein by the addition of large quantities of unlabeled insulin. These researchers also recognized that the degree of binding of the tracer was quantitatively related to the total amount of insulin present. Soon after, Ekins (1960) reported an RIA method for serum thyroxine.

The advantages offered by RIA over methods such as agglutination-inhibition and immunoprecipitation, which lacked the desired sensitivity and applicability to the quantitation of low molecular weight compounds, led to the rapid adaptation of RIA by researchers and routine clinical laboratories. The use of RIA also gave rise to a number of in vitro diagnostic companies that have gradually come to provide reagents for a large share of the immunoassay determinations in both research and clinical laboratories.

RIA in its original form was devised to measure antigens, particularly in soluble systems. Refinements made by a number of investigators to improve the separation of the bound and unbound antigen soon led to the discovery by Catt and Tregear (1967) that polystyrene test tubes had the property of absorbing proteins irreversibly and that the absorbed antibody retained its avidity for the antigen. The coating of the polystyrene tubes in a reproducible manner and the separation of the bound and unbound antigen by simply washing the tube opened new vistas in the development of immunodiagnostic tests.

During the same decade, using a diisocyanate derivative as the coupling reagent, Singer and Schick (1960) coupled two protein molecules, viz., ferritin to its antibody for immunologic localization. Subsequently, Avrameas and Uriel (1966) and Nakane and Pierce (1966) demonstrated that enzymes could also be coupled to antibody or antigen. The importance of this discovery is reflected in the now widespread application of chromogenic, fluorogenic, or luminescent enzyme/substrate interaction signals for the detection and measurement of soluble antigens, with an attained sensitivity that approaches that of an RIA.

These discoveries soon led to the subsequent independent introduction of non-isotopic enzyme labels by Engvall and Perlmann (1971) and van Weemen and Schuurs (1971) for the immunoassays. The term "ELISA," first coined by Engvall and Perlmann in 1971, was originally used to describe a reagent-excess enzyme immunoassay of specific antibodies or antigens. When applied to antigens, it is equivalent to IRMA (immunoradiometric assay). However, at present, ELISA is often used interchangeably with enzyme immunoassay (or enzymoimmunoassay) and immunoenzymometric assay.

In both RIA and ELISA, the labeled and unlabeled antigens bound to the antibody must first be separated prior to measuring the activity of the label in either fraction. These methods are, therefore, collectively called "separation-required" immunoassays. Because the separation inevitably involves the use of heterogeneous phases in the assay mixture, the term "heterogeneous" immunoassay is used interchangeably for this type of immunoassay.

A further advance in this now rapidly emerging field came when Rubenstein et al. (1972) developed a new immunoassay using an enzyme as a label in which the antigen-antibody reaction and its measurement are performed in solution without the need of prior separation of the free and antibody-bound components. This "homogeneous" or "separation-free" immunoassay technology uses enzyme inhibition by an antibody as the core concept, and has since been adapted to the measurement of a wide variety of small molecules. This technology precludes the need for separation and in effect reduces the performance of an immunoassay to the simplicity of an enzyme activity measurement. This type of assay is also amenable to automation. Ngo and Lenhoff (1985), however, suggest that this class of immunoassay be referred to as separation-free rather than homogeneous, because the term "homogeneous" does not accurately describe the salient features of the assay. Moreover, some separation-free enzyme immunoassays are not homogeneous and involve heterogeneous phases (Gibbons 1985; Litman 1985).

Recent developments in the immunodiagnostic field include the emergence of fluorescence polarization and pulse-fluorescence immunoassays and lumino-immunoassays (Deshpande and Sharma 1993). Although fluorescence labels, especially fluorescein, have been used routinely in several automated homogeneous immunoassays, conventional fluorometry as applied to immunoassays suffers from several drawbacks such as separation of fluorescence emission from excitation, Rayleigh and Raman scattering, background fluorescence from cuvettes, optics, and samples, nonspecific binding of the reagents, and fluorescence quenching (Soini and Hemmila 1979; Hemmila 1985; Diamandis 1988). Currently no system is available commercially that uses such fluorophores for immunoassay procedures having sensitivity in the subpicomolar range. To overcome the drawbacks associated with the use of conventional fluorophores, the current trend is toward the use of lanthanide chelates as labels for immunoassays (Barnard et al. 1989; Diamandis 1988; Hemmila 1985). Some of the newer, time-resolved fluorescence immunoassay systems using these labels can detect about 0.5 amol of α-fetoprotein in a model noncompetitive immunoassay (Diamandis 1991).

Although RIAs still constitute a major proportion of the immunodiagnostic tests conducted, especially in the clinical laboratories, there are several potential drawbacks to the use of isotopic labels in immunoassays. Their toxic nature necessitates the application of strict regulatory control and special disposal requirements. Their measurement also requires the use of specialized, sophisticated, and hence, expensive equipment. Moreover, the necessity for a separation step has also prevented the development of simple automation. Isotopic labels also have a limited shelf life. These disadvantages have encouraged the search for several alternative labels for use in immunoassays. Since their introduction in the early 1970s and because of the relative ease of use, enzymes have become the most commonly used labels for immunoassays. At present, enzyme immunoassays command over 60% of the diagnostic market as compared to 25% for RIA and 15% for fluorescence and other miscellaneously labeled immunoassays (Deshpande and Sharma 1993).

DIAGNOSTIC MARKETS AND ECONOMICS

Since the advent of the first RIAs in the late 1950s and early 1960s, immunoassays have been well received as a diagnostic and research tool in several disciplines of life sciences. The recent trend toward increasing demand for automation of immunoassays and stable reagents clearly shows a definitive shift to nonisotopic immunoassay systems on a large scale. During the 1980s, many nonisotopic immunoassay technologies were presented in the marketplace both as parts of more or less open immunoassay systems with diagnostic kits and as tools for the researchers to develop their own immunoassays. Historically, enzyme immunoassays have been in a prime position for in-house applications, primarily because of the easy access to inexpensive measuring equipment as well as to a vast body of literature on labeling techniques and applications (Pettersson 1993). In contrast, other nonisotopic technologies have been more difficult to adapt for noncommercial immunoassays.

The immunodiagnostic market is composed of numerous micromarkets. Some of these are clearly differentiated. Such is a case in allergy testing where a primary screening test for total IgE is used for atopic allergy and multiple allergen-specific IgE tests are used to isolate the cause. In other instances, the distinctions between some micromarkets can be less obvious. For example, drug testing is not simply one market. Therapeutic drug monitoring immunoassays are used in routine clinical practice to check that sick patients are receiving the appropriate dose of drug, whereas immunoassays for drugs of abuse testing are used in drug dependence treatment centers and psychiatric clinics. They may also be used in legal situations.

The design and assay performance of immunodiagnostic tests also vary with the endusers. Large, centralized clinical laboratories generally demand sophisticated automation and software to run and interpret these tests. In contrast, doctors' offices and pharmacies require single-use tests with simple and reasonably quick protocols. They cannot afford expensive equipment. At the extreme end of the spectrum is the market for home-use diagnostic tests. For example, home-use pregnancy tests have to be very simple to use. Thus, tests have been developed that do not require any operation other than immersion in a stream of urine.

There are also several examples where a single analyte is presented in a number of different formats to suit the needs of different users (Wild 1994). The most striking example of this is human chorionic gonadotropin (hCG), which is primarily used as a pregnancy detection test. However, monitoring its levels is also useful in screening for Down's syndrome, and in monitoring certain types of tumors. The concentrations of hCG involved in these three applications vary by several orders of magnitude. Whereas sensitivity is crucial in pregnancy testing, other applications require good precision at much higher concentrations, and sensitivity is less important.

At present, the singular most important application for commercial diagnostic products is in the clinical and medical fields. There these products are routinely used for monitoring the levels of various pharmaceutical drugs and endogenous hormone and other marker levels, to detect drugs of abuse, to identify tumor markers, and to monitor bacterial, viral, and parasitic infectious diseases. In several clinical situations, rapid diagnosis of the underlying cause is often of prime importance; thus there are as yet no viable alternatives to immunoassays as a rapid and sensitive diagnostic tool. Similarly, in recent years, several over-the-counter immunoassay products, such as glucose levels kits for diabetics, cholesterol screening tests, and pregnancy detection test kits, are becoming increasingly popular with the general public.

The diagnostic sector, therefore, has an important role in health care and annual sales exceed U.S. $15 billion. Although immunoassays were once considered to be a small and specialized field, their simplicity, specificity, and sensitivity are widely recognized, and sales have increased by about 10% per annum. In several countries, immunoassay products for home use are available in pharmacies and even in supermarkets.

As compared to clinical and therapeutic fields where they have been used routinely for the past 30 years or so, the applications of immunoassays in agriculture, food, veterinary, and environmental sciences developed slowly. Nevertheless, the economics, rapidity, sensitivity, and the easy-to-use approach of these diagnostic tests are already making an impact in these fields. At present, in the diagnostic sector, immunoassay products command an annual worldwide market value in excess of U.S. $10 billion (Ekins 1989; Deshpande and Sharma 1993). According to

Diamandis (1990), the following five developments have led to the widespread use of immunodiagnostic techniques in several diverse disciplines:

1. The introduction of homogeneous immunoassays,
2. The development of "two-site" noncompetitive immunoassay formats,
3. The application of novel solid phases for easy and rapid separation of bound and free label,
4. The introduction of monoclonal antibodies, and
5. The replacement of the widely used ^{125}I radionuclide with alternative, nonisotopic labels.

Driven by an impressive level of scientific and technological innovations and aggressive marketing programs, the immunodiagnostic business is continuously changing. There is fierce competition between many manufacturers. Successful products have impressive growth rates, only to be superseded by new technologies with improved performance or greater automation. For example, the leading enzyme multiplied immunoassay technique (EMIT®) technology lost a huge market share to fluorescence polarization in the therapeutic drug monitoring market, but then, in turn, achieved a revival as an open reagent system that could be used across a range of clinical chemistry analyzers. Occasionally, immunodiagnostic tests are replaced by alternative, noninvasive technologies, such as ultrasound, which is now commonly used instead of estriol and human placental lactogen to check fetoplacental well-being (Wild 1994).

The most critical consideration for any immunodiagnostic product should be its performance. Laboratories cannot afford to provide misleading information to clinicians or regulatory agencies. Manufacturers of immunodiagnostic products, therefore, have invested heavily in improving product performance. As the assay performance of the leading products has improved, the main priority for most laboratories has become the level of automation provided by the analyzer that is at the heart of each system.

Commercially available automated systems vary considerably in design and performance characteristics. Reliability and fluid handling are still a major concern with some systems. Poor fluid handling and washing during separation can lead to cross-contamination, and thus, reduce assay sensitivity in immunometric assays. Automated systems that use stored calibration curves also require a good temperature control.

Although, in the recent past, the various automated systems have offered more features and better performance, they have also become more expensive. Today, the development of a new automated immunoassay system amenable to a wide range of diagnostic tests can cost upwards of $100 million. It has, therefore, become increasingly difficult for smaller companies to compete. In order to survive, they are turning to niche markets that require less automation. It is quite conceivable that several of these companies may eventually withdraw from the crowded immunoassay market, thus leaving it increasingly dominated by a few large companies. At present, Abbott Laboratories alone has about 40% of all the immunodiagnostic test market. Although several other companies compete in this market, it is unlikely that any one company could hold more than a 10% market share.

Several other factors also influence trends in the immunodiagnostic test market. In the laboratory sectors, the cost and quality of labor is a major concern. Automation, in turn, reduces labor costs and requires fewer handling skills than manual methods. Most laboratories are also keen to reduce turnaround times to provide clinicians with a better level of service (Wild 1994). Some clinical applications, such as those in emergency centers, require a particularly fast response. In fact, turnaround time is the main reason for the increasing popularity of random access analyzers that can accept individual samples for a range of analyte tests.

In the coming years, the trend toward automation is certain to continue, reducing labor costs and the level of laboratory skill required by operators. Similarly, random access analyzers, which have proven so successful in the clinical chemistry area, will also command a greater market share in other life science markets. However, because of their cost and size, laboratories may run more tests on fewer analyzers.

According to Wild (1994), the remarkably wide range of immunoassay technologies on the market reflects the immaturity of the immunodiagnostic business. In the near future, therefore, it is not unlikely that fewer formats and signal generation systems will be used in the mainstream products. Although patent issues have precluded the widespread use of monoclonal antibodies, it is inevitable that they will be used much more in the future. Solid-phase assays will also continue to be widely used. Generic coatings, such as streptavidin, for solid phases will lead to lower manufacturing and development costs.

Given the fact that health care is one of the world's largest businesses, and that, in spite of many national and international attempts to control the huge expenditure involved, it continues to grow at an alarming pace, the immunodiagnostic sector still appears to be in its infancy. There is a vast, untapped potential for numerous test sites and micromarkets in the future, especially in the developing world. The use of multiple tests in screening and diagnosis will receive more attention. Although new technologies, such as DNA probes, are growing fast, the immunodiagnostics business still has a long way to go to reach its full potential.

Representative examples from different fields for which immunodiagnostic kits are available commercially, or where such potential exists, are summarized in Table 1.1.

TABLE 1.1.   Representative Examples for Which Immunodiagnostic Kits Are Available Commercially or Where Such Potential Exists

HUMAN CLINICAL AND THERAPEUTICS

Antibiotics/Antimicrobial Drugs
- Gentamicin, tobramycin, amikacin, penicillin, cephalosporin, blasticidin S, viomycin, sulfa drugs, kanamycin, netilmicin, streptomycin, vancomycin

Drugs of Abuse
- Opiates, barbiturates, amphetamines, methadone, cocaine, benzodiazepines, propoxyphene, phencyclidine (PCP), cannabinoids (THC), lysergic acid diethylamide (LSD)

Antiepileptic Drugs
- Phenytoin, phenobarbital, carbamazepine, primidone, ethosuximide, valproic acid

Antiasthmatic Drugs
- Theophylline

Cardioactive Drugs
- Digoxin, digitoxin, lidocaine, procainamide, N-acetylprocainamide, quinidine, propranolol, diisopyramide, flecainide

Chemotherapeutic Drugs
- Methotrexate

Hormones
- Testosterone, estradiol, estrogens, progesterone, cortisol, thyroxine, insulin, human placental lactogen (HPL), thyroid-stimulating hormone (TSH), follicle-stimulating hormone (FSH), luteinizing hormone (LH), human chorionic gonadotropin (hCG)

Immunosuppressants
- Cyclosporin A (CsA), cyclosporin G (CsG, OG37-325), FK506 (tacrolimus)

Serum Proteins
- Albumin, α_1-acid glycoprotein (orosomucoid), serum amyloid P component (SAP), serum retinol binding protein, thyroxine binding globulin (TBG), α_1-antitrypsin, α_2-macroglobulin, anti-DNA antibodies, antithrombin III, apolipoproteins AI and AII, apolipoprotein BI, prealbumin (transthyretin), C1 inactivator, C3 protein, ceruloplasmin, fibronectin, haptoglobin, hemopexin, somatotropin, transferrin, immune complexes, immunoglobulin A, immunoglobulin E, immunoglobulin G and its subclasses, immunoglobulin M, immunoglobulin light chains, rheumatoid factor, β_1-microglobulin, C1-esterase inhibitor, C4-protein, C-reactive protein

Tumor Markers
- α-Fetoprotein, carcinoembryonic antigen (CEA), human chorionic gonadotropin (hCG), β-hCG, pregnancy-specific protein (SP1), placenta-specific protein (PPS), placental alkaline phosphatase (Regan type), isoferritins, tissue polypeptide antigen, Tennessee antigen, pancreatic oncofetal antigen (POA, OPA), prostatic acid phosphatase, carbohydrate antigen 19-9 (sialyl Lewis), carbohydrate antigen 50, cancer antigen 125, cancer antigen 15-3, fecal occult blood, β_2-macroglobulin, neuron specific enolase, squamous cell carcinoma antigen

TABLE 1.1.   (*continued*)

Allergens
- Total serum IgE, allergen-specific IgEs, pollen allergens, epithelial allergens, house dust, occupational dusts, molds, foods, chemicals, drugs

Parasitic and Infectious Diseases and/or Causal Organisms
- HIV, hepatitis, influenza, herpes, *Toxoplasma*, rubella, cytomegalovirus (CMV), adenovirus, coxsackieviruses, arbovirus, malaria, schistosomiasis, trypanosomes, *Trichinella*, *Chlamydia trachomatis*, *Neisseria gonorrhoeae*, amoebiasis, typhoid, leprosy, tuberculosis

Autoimmune Diseases
- Rheumatoid factor (RF), polyarthritis, juvenile chronic polyarthritis, ankylozing spondylitis, Reiter's syndrome, antinuclear antibodies (ANA), anti-DNA antibodies, antihistone antibodies, acetylcholine receptor antibodies, antierythrocyte antibodies, antiplatelet antibodies, thyroglobulin antibodies

Bacterial, Mycoplasmal, and Fungal Antigens and Antibodies
- *Salmonella* O antigens, *Vibrio cholerae* O antigens and exotoxins, *Escherichia coli* O and K antigens, *Haemophilus influenzae* polysaccharide, *Treponema pallidum*, *Brucella* and *Yersinia enterocolitica* O antigens, *Francisella tularensis* O antigen, *Candida albicans* and *Aspergillus fumigatus* cell wall and cytoplasmic antigens, *Streptococcus* M protein, *Mycoplasma*, *Rickettsia*, *Chlamydia*, *Clostridium tetanui* exotoxin, *Corynebacterium diphtheriae* exotoxin

Miscellaneous
- Glucose, cholesterol

AGRICULTURE DIAGNOSTICS
Plant Hormones
- Cytokinins, gibberellins, indole-3-acetic acid, abscisic acid

Spoilage Microorganisms
- *Erwinia* spp., *Fusarium* spp., *Humicola languinosa*, *Legionella pneumophila*, *Ophiostoma ulmi*, *Phytophthora megasperma*, *Pseudocercosporella herpotrichoids*, *Pseudomonas syringae*, *Rhizoctonia solani*, *Xanthomonas campestris*

Viral Agents
- Beet necrotic yellow vein virus, cauliflower mosaic virus, *Citrus* Tristeza virus, cucumber mosaic virus, elongated potato virus, isometric plant viruses, pea seed-borne mosaic virus, potyviruses, soybean mosaic viruses, zucchini yellow mosaic virus

FOOD DIAGNOSTICS
Food Safety: Bacterial Toxins
- *Clostridium botulinum* neurotoxins A, B, E, F, and G; *Staphylococcus aureus* enterotoxins A, B, C, D, and E

Food Safety: Mycotoxins
- Aflatoxins B_1, B_2, B_1 diol, M_1 and Q_1; ochratoxin; T-2 toxin; 3'-OH-T-2 toxin; T-2 tetraoltetraacetate; HT-2 toxin; group A trichothecenes; roridin A; zearalenone; rubratoxin B; sterigmatocystin; deoxyverrucarol; deoxynivalenol

TABLE 1.1.   (*continued*)

Food Safety: Pathogenic Microorganisms
*   *Salmonella, Listeria monocytogenes, Escherichia coli, Vibrio* spp., *Yersinia enterocolitica, Campylobacter jejuni*

Food Safety: Miscellaneous
*   Mushroom poisoning, algal and seafood toxins, potato glycoalkaloids

Food Enzyme/Inhibitor Activity
*   α-Amylase, β-amylase, catalase inhibitor, chymotrypsin, debranching enzyme, lipase, malate dehydrogenase, papain, pepsin, polyphenoloxidase, proteolytic enzymes, trypsin, trypsin inhibitor

Interspecies Meat and Adulterant Identification
*   Beef, sheep, pig, goat, horse, meat products, sausages, processed meats

VETERINARY DIAGNOSTICS
Livestock Diseases and/or Causal Organisms
*   *Toxoplasma gondii, Brucella abortus, Stephanuras dentatus, Mycoplasma bovis, Leptospira interrogans, Trichinella spiralis, Mycobacterium paratuberculosis,* bovine rhinotracheitis, maedi-visna virus, swine fever virus, coronavirus, Aujeszky's disease, swine vesicular disease, enzootic bovine leukemia, foot and mouth disease, avian PMV1, *Rotavirus,* sheep lungworm disease

Anabolic Agents
*   17β-Estradiol, estrone, testosterone, 17α-methyltestosterone, progesterone, trenbolone, diethylstilbestrol, hexoestrol, zeronal

Therapeutic Agents
*   Cephalexin, chloramphenicol, colistin, gentamicin, hydromycin B, monensin, sulfonamides, penicillins, cephalosporins

ENVIRONMENTAL DIAGNOSTICS
Pesticides and Their Metabolites
*   Aldrin, alachlor, atrazine, BAY SIR 8514, S-bioallethrin, chlorosulfuron, cyanazine, 2,4-D, DDT, dichlorfop-methyl, dieldrin, diflubenzuron, endosulfon, iprodione, kepone, maleic hydrazide, metalaxyl, oxfendazole, parathion, paraoxon, paraquat, pentachlorophenol, 2,4,5-T, terbutryn, triadimefon, warfarin

Environmental Pollutants
*   Polychlorinated biphenyls (PCBs), polybrominated biphenyls (PBBs), polynuclear aromatic hydrocarbons (PAHs), nitroaromatics, cyclic ketones, BTEX (benzene, toluene, ethyl benzene, and xylene), nitrosamines, haloalkanes, dioxins, dibenzofurans, TNT

SCOPE OF THE BOOK

The primary aim of writing this book is to provide useful practical information for the development of a successful commercial immunodiagnostic product based on enzyme immunoassay technology. The reader is taken through all stages of this process—from conceptualization, product design, and manufacture to introduction of the product to the market. Every diagnostic product commercially available owes its success to a joint team effort, requiring the expertise of scientists from such diverse disciplines as immunochemistry, organic chemistry, protein chemistry, enzymology, statistics, business management, and marketing. Thus, understanding of the basic concepts and proper manipulation of these techniques are key to a successful product development.

Commercially, a wide variety of immunoassay formats and reagent configurations is available to monitor a quantitative antigen-antibody reaction. It is, therefore, impossible to cover every format in detail. Fortunately, irrespective of the assay format and design, the underlying basic principles are similar.

In the initial stages, it is helpful to have some idea about the basic concept and design of the product that one intends to develop. Knowing whether the intended final product is based on a separation-free or separation-required immunoassay principle is certainly helpful. Limiting the basic design of the product and the choice of solid phase to a manageable few from the vast variety available in the market during the early product development stages also saves considerable time, money, and effort.

It is also essential to understand the regulatory guidelines, if any, for the particular compound for which the product is being developed. For example, is it required to measure absolute concentration of that compound in a given system (e.g., blood, serum, tissue, urine, milk, meat product) and with what sensitivity and detection levels, or does the product need to give only a qualitative "yes-or-no" answer? Immunoassays for hormones and several drugs for which the margin of safety between therapeutic dose and the risk of toxicity is very low fall into the first category. Those in the latter category include products that detect the mere presence or absence of drugs of abuse such as cocaine, amphetamines, and steroids, or certain over-the-counter products such as the pregnancy evaluation test kits.

The intended market for which the product is being developed and its economics, and the format (whether simple or complex) it would prefer in an ideal product are also key marketing considerations. A successful commercial product is generally a user-friendly product. It is, therefore, absolutely essential to understand the technical capability of the endusers who will ultimately be required to use the final product. For example, a truck driver who must collect milk from several different dairy farms prior to delivering it to the dairy plants for processing, and who may prefer to test milk from individual farms for the possible presence of violatory levels of β-lactam antibiotics, will not have the necessary skills to use and interpret a sophisticated and complex immunodiagnostic kit as compared to a well-trained technical person working in a dairy laboratory. In such a case, the former would prefer a simple "yes-or-no" dipstick-type test, whereas the latter would need to identify and quantitate the drug in milk samples. This, however, is not a problem in clinical settings where only qualified and trained professionals are allowed to carry out such tests. Nevertheless, factors such as these primarily govern the level of complexity and sophistication required in a product development process. A thorough market study on the feasibility of the product and its careful evaluation and consideration during the initial phase itself will certainly help to avoid unexpected delays and problems during the latter stages of product development. Also, it is easier to make such changes during the initial stages of this process than to have the specifications of a final product changed to meet these criteria and goals. Even a small change in design and assay format during the latter stages of product development will cause considerable problems with the product performance. In some instances, the entire process may need to be reoptimized, thereby causing considerable delays in marketing the product.

In developing a sound and effective immunodiagnostic product, several key issues at each stage of product development must be critically addressed. The critical phases in the development of a commercially viable immunoassay product include:

1. Immunogen selection and development,
2. Antibody production, processing, and purification,
3. Reagent formatting and assay development,
4. Sample preparation,
5. Data processing,
6. Product evaluation and clinical or regulatory validation, and
7. Standardization and commercial manufacture

These stages are described in this book from a practical viewpoint of a product development process based on enzyme immunoassay technology. For the sake of convenience, the information is presented in two different parts. Part 1 covers the basic considerations including the classification, structure, and function of antibodies; antigen-antibody reactions; general information on various conjugation techniques; antibody development, production and processing; the properties and characteristics of the most widely used enzymes in immunoassays; and the choice of various solid-phase systems available in the market. Several detailed monographs describing the theoretical aspects of each of these categories are available; thus, only those features essential in a product development process will be described here.

Part 2 of the book covers the actual product development process. It includes information on the classification and the various formats of enzyme immunoassays available for product development, reagent formatting, and assay development issues, data processing, standardization, scale up, and commercial manufacture of the product. Finally, various regulatory requirements, such as evaluation and clinical or regulatory validation of the product; the importance of good laboratory, and manufacturing (GLP and GMP) practices; and certain international marketing requirements such as the ISO 9000 certification process are also described in Part 2 of this book.

REFERENCES

AVRAMEAS, S., and URIEL, J. 1966. Methode de marquage d'antigenes et d'anticorps avec des enzymes et son application en immunodiffusion. *C. R. Acad. Sci. Paris* 262:2543–2545.

BANGHAM, D. R. 1983. Reference materials and standardization. In *Principles of Competitive Protein-Binding Assays,* eds. W. D. Odell and P. Franchimont, pp. 85–105, John Wiley, New York.

BARNARD, G.; KOHEN, F.; MIKOLA, H.; and LOVGREN, T. 1989. Measurement of estrone-3-glucuronide in urine by rapid, homogeneous time-resolved fluoroimmunoassay. *Clin. Chem.* 35:555–559.

BUTLER, J. E. 1991. Perspectives, configurations and principles. In *Immunochemistry of Solid-Phase Immunoassay,* ed. J. E. Butler, pp. 3–26, CRC Press, Boca Raton, FL.

CAMPBELL, P. J. 1974. International biological standards and reference preparations. *J. Biol. Standardization* 2:249–258.

CATT, K., and TREGEAR, G. W. 1967. Solid phase radioimmunoassay in antibody-coated tubes. *Science* 158:1570–1572.

CATTY, D. 1988. *Antibodies. A Practical Approach.* IRL Press, Oxford.

CLAUSEN, J. 1988. *Immunochemical Techniques for the Identification and Estimation of Macromolecules.* Elsevier, Amsterdam.

COOMBS, R. R. A.; MOURANT, A. E.; and RACE, R. R. 1945. Certain properties of anti-serums prepared against human serum and its various protein fractions. Their use in the detection of sensitization of human red cells with incomplete Rh antibody and on the nature of this antibody. *Br. J. Exp. Pathol.* 26:255–263.

COONS, A. H.; CREECH, H. J.; and JONES, R. H. 1941. Immunological properties of an antibody containing a fluorescent group. *Proc. Soc. Exp. Biol. Med.* 47:200–202.

DESHPANDE, S. S., and SHARMA, B. P. 1993. Immunoassays, nucleic acid probes, and biosensors. Two decades of development, current status and future projections in clinical, environmental and agricultural applications. In *Diagnostics in the Year 2000,* eds. P. Singh, B. P. Sharma, and P. Tyle, pp. 459–525, Van Nostrand Reinhold, New York.

DIAMANDIS, E. P. 1988. Immunoassays with time-resolved fluorescence spectroscopy: principles and applications. *Clin. Biochem.* 21:139–150.

DIAMANDIS, E. P. 1990. Detection techniques for immunoassay and DNA probing applications. *Clin. Biochem.* 23:437–443.

DIAMANDIS, E. P. 1991. Multiple labeling and time-resolvable fluorophores. *Clin. Chem.* 37:1486–1491.

EKINS, R. P. 1960. The estimation of thyroxine in human plasma by an electrophoretic technique. *Clin. Chim. Acta* 5:453–459.

EKINS, R. P. 1989. Multi-analyte immunoassay. *J. Pharm. Biomed. Anal.* 7:155–168.

ELEK, S. D. 1948. Recognition of toxicogenic bacterial strains in vitro. *Br. Med. J.* 1:493–496.

ENGVALL, E., and PERLMANN, P. 1971. Enzyme-linked immunosorbent assay (ELISA). Quantitative assay of immunoglobulin G. *Immunochemistry* 8:871–874.

FARR, R. S. 1958. A quantitative immunochemical measure of the primary interaction between I*BSA and antibody. *J. Infect. Dis.* 103:239–262.

GIBBONS, I. 1985. Nonseparation enzyme immunoassays for macromolecules. In *Enzyme-Mediated Immunoassay,* ed. T. T. Ngo and H. M. Lenhoff, pp. 121–144, Plenum Press, New York.

GOSLING, J. P. 1990. A decade of development in immunoassay methodology. *Clin. Chem.* 36:1408–1427.

GRABER, P., and WILLIAMS, C. A. 1953. Method permitting the simultaneous study of electrophoretic and immunochemical properties of a mixture of proteins. Application to blood serum. *Biochim. Biophys. Acta* 10:193–194.

HAWCROFT, D.; HECTOR, T.; and ROWELL, F. 1987. *Quantitative Bioassay.* John Wiley, New York.

HEIDELBERGER, M. 1939. Chemical aspects of the precipitin and agglutinin reactions. *Chem. Rev.* 24:323–343.

HEMMILA, I. 1985. Fluoroimmunoassays and immunofluorometric assays. *Clin. Chem.* 31:359–370.

HUMPHREY, J. H., and WHITE, R. G. 1964. *Immunology for Students of Medicine.* Blackwell Scientific, Oxford.

KIRKWOOD, T. B. L. 1977. Predicting the stability of biological standards and products. *Biometrics* 33:736–742.

LANDSTEINER, K. 1945. *The Specificity of Serologic Reactions.* Harvard University Press, Cambridge, MA.

LAURELL, C. B. 1966. Quantitative estimation of proteins by electrophoresis in agarose gel containing antibodies. *Anal. Biochem.* 15:45–52.

LITMAN, D. J. 1985. Test strip enzyme immunoassay. In *Enzyme-Mediated Immunoassay,* ed. T. T. Ngo and H. M. Lenhoff, pp. 155–190, Plenum Press, New York.

MANCINI, G.; CARBONARA, A. O.; and HEREMANS, J. F. 1965. Immunochemical quantitation of antigens by single radial immunodiffusion. *Immunochemistry* 2:235–254.

NAKANE, P. K., and PIERCE, G. B. 1966. Enzyme-labeled antibodies preparation and application for the localization of antigens. *J. Histochem. Cytochem.* 14:929–931.

NGO, T. T., and LENHOFF, H. M. 1985. *Enzyme-Mediated Immunoassay.* Plenum Press, New York.

ODELL, W. D. 1983. Principles of in vitro bioassays. In *Principles of Competitive Protein-Binding Assays,* eds. W. D. Odell and P. Franchimont, pp. 267–279, John Wiley, New York.

ODELL, W. D., and FRANCHIMONT, P. 1983. *Principles of Competitive Protein-Binding Assays.* John Wiley, New York.

OUCHTERLONY, O. 1948. Antigen-antibody reactions in gels. *Acta Path. Microbiol.* 26: 507–515.

OUDIN, J. 1946. Method of immunochemical analysis by specific precipitation in gelled medium. *C. R. Acad. Sci. Paris* 222:115–116.

PETTERSSON, K. 1993. Comparison of immunoassay technologies. *Clin. Chem.* 39: 1359–1360.

RUBENSTEIN, K. E.; SCHNEIDER, R. S.; and ULLMAN, E. F. 1972. "Homogeneous" enzyme immunoassay. A new immunochemical technique. *Biochem. Biophys. Res. Commun.* 47:846–851.

SINGER, S. J., and SCHICK, A. 1961. On the formation of covalent linkages between two protein molecules. *J. Biol. Chem.* 236:2477–2481.

SOINI, E., and HEMMILA, I. 1979. Fluoroimmunoassay: present status and key problems. *Clin. Chem.* 25:353–361.

STEWARD, M. W. 1984. *Antibodies. Their Structure and Function.* Chapman & Hall, New York.

TISELIUS, A., and KABAT, E. 1939. An electrophoretic study of immune serums and purified antibody preparations. *J. Exp. Med.* 69:119–131.

VAN WEEMAN, B. K., and SCHUURS, A. H. W. N. 1971. Immunoassay using antigen-enzyme conjugates. *FEBS Lett.* 15:232–236.

WILD, D. 1994. *The Immunoassay Handbook.* Stockton Press, New York.

YALOW, R. S., and BERSON, S.A. 1959. Assay of plasma insulin in human subjects by immunological methods. *Nature* (London) 184:1648–1649.

YALOW, R. S., and BERSON, S. A. 1960. Immunoassay of endogenous plasma insulin in man. *J. Clin. Invest.* 39:1157–1175.

2

Antibodies: Biochemistry, Structure, and Function

INTRODUCTION

An understanding of the basic principles of the biology of the vertebrate immune system and its regulation is a prerequisite for the successful experimental production of antibodies. Such knowledge helps in avoiding the evils of obtaining highly unpredictable results and undesired effects such as immune unresponsiveness or immune tolerance during antibody production. Moreover, not all antibodies react with their antigens with the same affinity and avidity in in vitro systems. Some classes of immunoglobulins are synthesized in greater amounts than others, while some are better suited for immunoassay development and formulations. Therefore, information on the various antibody (immunoglobulin) classes—their structures, biochemical and physicochemical properties, and functions—is essential to avoid several pitfalls encountered in developing immunodiagnostic products. This chapter deals with the fundamental nature of the vertebrate immune system, and the biochemical, structural, and functional aspects of antibodies belonging to different immunoglobulin classes. Because antibodies are not the only binding agents used in diagnostic products, certain auxiliary binding systems are also described.

VERTEBRATE IMMUNE SYSTEM

The vertebrate immune system can be classified into two functional divisions: the innate or nonadaptive and the adaptive immune systems (Figure 2.1). The former acts as the primary line of defense against infectious agents and potential pathogens. The adaptive system is called upon only when this first line of defense is breached.

Innate or Nonadaptive Immune System

A characteristic feature of the innate immune system is that its resistance is not improved by repeated infection, i.e., it protects against infection in a fixed manner. It comprises the external skin, internal defensive cells, and the soluble, circulating humoral factors (Figure 2.1). Because the adaptive system, as the name implies, takes time to respond to infection, the innate immune system collectively must serve as the first line of defense.

The skin, comprising highly keratinized epithelial cells, provides a passive barrier to penetration by pathogens. The defensive cells include the "phagocytes" which bind, eat, and digest a majority of invading organisms, and the "natural killer cells," which recognize and rapidly remove several kinds of parasitized cells and cancer cells. The soluble circulating factors (sometimes also referred to as "humoral" immune factors) in this system include proteins such as lysozyme, complement, C-reactive protein, interferon, and properdin. These soluble factors

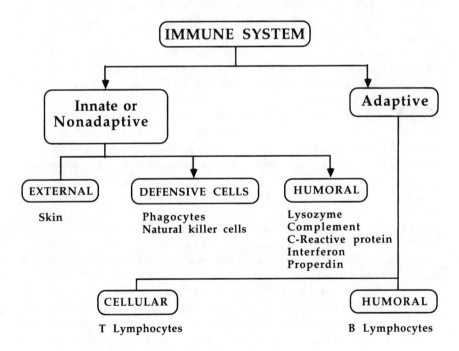

Figure 2.1. Major elements of the vertebrate immune system

bind to bacterial surfaces and initiate elimination reactions. Some of these humoral proteins are always present; others can be rapidly induced upon infection.

Adaptive Immune System

Unlike the nonadaptive system, the adaptive immune system provides resistance that is improved by repeated infections of the same pathogen. It is also characterized by duality, thus providing both cellular and humoral defenses against invading pathogens (Figure 2.1). Although both defenses respond specifically to most foreign substances, generally only one response is favored. The cellular immune response is primarily effective against parasites, intracellular viral infections, cancer cells, and foreign tissue, whereas the humoral response is favored in extracellular phases of bacterial and viral infections.

The cellular immunity in the adaptive system is provided by certain cells (e.g., T lymphocytes) of the lymphoid system, whereas the humoral immunity is provided by the proteins called "antibodies" that circulate through two body fluids, viz., the serum and the lymph. Both cellular and humoral systems thus provide overlapping, yet distinct protection.

All the cells of the vertebrate immune system arise from pluripotent hemopoietic stem cells through two main lines of differentiation: the lymphoid lineage, which produces lymphocytes, and the myeloid lineage, which produce phagocytes and other cells (Figure 2.2). The average human adult has about 10^{12} lymphoid cells and the lymphoid tissue as a whole represents about 2% of the total body weight (Roitt et al. 1989; Hood et al. 1984). Lymphoid cells (approximately 20% B cells involved in humoral immunity and 80% T cells involved in both humoral and cell-mediated immunity) represent about 20% of the total white blood cells or leucocytes present in adult circulation; the majority are phagocytes. Among the various phagocytes, monocytes and polymorphonuclear neutrophils predominate, accounting for 7% and 67%, respectively, of the total circulating leucocyte pool, whereas eosinophils (3%) and basophils (0.5%) are only minor components.

Antibodies are a class of proteins secreted by plasma cells produced by the B lymphocytes of the adaptive immune system. They are primarily produced when the phagocytes are unable to recognize the infectious agent, either because they lack a suitable receptor for it or because the pathogen does not activate complement and thus cannot attach to the phagocyte via the C3b receptor (Roitt 1988; Golub 1981; Male 1986). Antibodies act as flexible adaptor molecules that can attach at one end to the microorganism and at the other end to the phagocyte. The antigen-antibody complexes then are removed from circulation by macrophages through phagocytosis. Because this antigen-antibody reaction is highly specific, antibodies are an important reagent used in immunological research and clinical diagnostics.

Traditionally, the term "antigen" is used to describe any molecule that provokes a specific immune response such as antibody production, cell-mediated

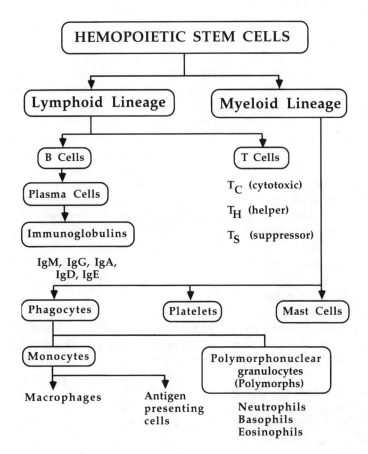

Figure 2.2. Cellular components of the immune system

immunity, or tolerance. However, to distinguish molecules that evoke an immune response and those that are the targets for antibody binding, the modern convention is to use the term "immunogen" for the former and "antigen" for the latter (Catty 1988). To be an immunogen, a molecule must possess certain degree of intrinsic structural complexity (immunogenicity). Macromolecules such as proteins and heteropolysaccharides or their combinations (with or without lipids which are themselves nonimmunogenic) with a molecular weight greater than 1000 and normally above 5000, and containing relatively varied and inflexible structures tend to be natural immunogens as long as they are extrinsic to the immunizing species, because the vertebrate immune system does not produce antibodies against its own constituents irrespective of their composition. However, large but structurally similar molecules such as starch and collagen are generally very weak immunogens. Similarly, flexible molecules even if they are large (e.g., nucleic acids) normally are only very weakly immunogenic. In contrast, smaller molecules (molecular weight normally less than 1000) such as pharmaceutical and therapeutic drugs, steroids, hormones, and peptides can induce an immunogenic response only when they are covalently coupled to larger carrier backbones where they act as haptens on constructed immunogens. This is known as Landsteiner's principle.

A salient feature of the antibodies is that they do not bind to the whole of an infectious agent or antigen. Rather they recognize one of several characteristic surface features ("epitopes" or "antigenic determinants") of the antigen. Each antibody is specific for a particular epitope of the antigen. A particular antigen can have several different epitopes or may have several identical epitopes. Thus, in reality, the antibodies are specific for the epitopes rather than the whole antigen molecule. Because each antigen has its own particular set of epitopes that are usually not shared with other antigens, the collection of antibodies in an antiserum is effectively specific for the antigen. Even when a single pure immunogen with only one antigenic determinant is used, combining antibodies, though specific to the antigen, may be clonally heterogeneous with a range of individual affinities (Hawcroft et al. 1987; Catty 1988).

Thus, although the natural purpose of antibody production is to protect against invading pathogens, antibodies can be produced experimentally by the injection of several substances that are quite unrelated to these organisms. Therefore, antibodies can be produced against any compound that contains at least one antigenic determinant (generally 4–6 amino acid or sugar residues long) and as long as it can be covalently coupled to an immunogenic carrier. This basic principle of the vertebrate immune system has allowed us to develop antibodies against a diverse range of compounds; these are now routinely being used in targeting of drugs, neutralization of toxins, diagnostic assays, and purification processes using affinity chromatography.

Because antibodies are proteins and exhibit distinct structural differences when obtained from different species, they themselves can be used as immunogens. Thus, anti-antibodies can be raised, for example, when a chicken immunoglobulin is injected into a goat. Such an anti-immunoglobulin antibody is generally known as a "secondary antibody." The secondary antibody is usually directed against species-characteristic determinants on the constant regions of the primary or the immunizing antibody to which it binds irrespective of the binding or antigenic specificity of the primary antibody. This fact has been exploited commercially in the diagnostic industry in immunoassay methods that use antibody as a reagent. In such cases, the primary antibody that is being measured as the analyte acts as the antigen, and the secondary antibody raised against it acts as the reagent or the binding antibody. Labeled-secondary antibodies are also required in "sandwich-type" immunodiagnostic products.

Genetic Control

The recognition of genetic loci involved in the control of immune responses (I region) was first studied with inbred mice containing different haplotypes, i.e., combinations of a given allele at different loci (Hood et al. 1984; Hudson and Hay 1989). This region is known as the major histocompatibility complex (MHC) and is found in all mammalian species studied thus far (Jackson et al. 1983; Roitt et al. 1989). Proteins encoded by the MHC genes control several functions of the immune response, including interaction between different lymphoid cells, as well as between lymphocytes and antigen-presenting cells. The MHC region in the mouse is known as H-2 and is located on chromosome 17; that in humans is called the HLA gene cluster and is found on chromosome 6 (Klein 1979; Playfair 1987).

The MHC contains several polymorphic regions, which in turn consist of several subregions. The products of these loci fall into three classes (Bach et al. 1976; Klein 1979). Class I products are strong surface alloantigens expressed on essentially all cells and are responsible in assessing compatibility in blood transfusion and tissue transplantation, class II products are restricted to B cells, macrophages, and some sets of T cells and are responsible for the control of immune responses and regulation of cell interactions, and the class III genes code for components of the complement pathway and control their levels in the serum.

The ability to produce antibodies against a given antigen is determined by the haplotype of the I region of H-2 or the D region of HLA. A given haplotype produces immune responses to a series of antigens but not to others (Hood et al. 1984; Johnstone and Thorpe 1987).

Regulation

Once an immune response is initiated, the components of that response (e.g., B cells that produce antibodies) are capable of immense replication. It is thus evident that immune responses must normally be subject to strict and specific controls, not only for the antigen but also for the type of immune response elicited.

The simplest and best-known mechanism for the regulation of humoral immunity is that by which the circulating antibody itself regulates the production of antibody (feedback inhibition) by two different mechanisms (Hood et al. 1984; Gershon et al. 1981). One is by simply combining with the antigen and thus competing with the antigen receptors of the responding B cells for that determinant. In this mechanism of antigen blocking, B cells with receptors for different epitopes are unaffected, although B cell priming is prevented.

The second mechanism involves receptor cross-linking where low doses of antibody allow cross-linking between a B cell's Fc receptors and its antigen receptors. This inhibits the B cells from entering the phase of antibody synthesis, but does not affect B cell priming.

As far as the regulation of antibody production is concerned, for most immunogens the involvement of T cells is also essential. These T-dependent immunogens have at least two types of determinants (Raff 1970): one for the B cells (the "haptenic" determinant or epitope) and one for the T cells (the "carrier" determinant). Thus, if an animal is primed with a hapten conjugated to a carrier such as ovalbumin, a second injection of the same immunogen at a later date results in the usual secondary response. This is regulated by the T memory cells and is known as the "immunologic memory." However, if at this time, the hapten is coupled to another carrier such as serum albumin, the T cells fail to recognize the carrier and thus no secondary response is seen. The "carrier" effect, however, is transferable. Thus, transfer of serum albumin-primed T cells from another animal of the same species (i.e., with the same MHC constitution) would induce a secondary response in this experimental system. Because the primary-secondary response switch concurs with an IgM-IgG switch, T cells seem to be important for the control of this phenomenon.

This requirement for at least two sites of recognition is responsible for: (a) the necessity of a minimum size for an immunogen, (b) the failure of haptens to elicit immune responses by themselves, and (c) the necessity of the same carrier throughout the production of an antiserum. Moreover, the T cells also exhibit dichotomy. The T helper (T_H) cells promote the production of antibody, while the T suppressor (T_S) cells suppress it (Gershon et al. 1981; Staines et al. 1985).

In contrast to T-dependent antigens, the T-independent antigens share a number of common properties. They all tend to be large polymeric molecules with repeating antigenic determinants and, at high concentrations, they have the ability to activate B cell clones other than those specific for that antigen. This is known as polyclonal B cell activation. T-independent antigens also tend to be resistant to degradation and elicit only a weak primary antibody response. Examples of such antigens include lipopolysaccharides, Ficoll, dextran, levan, poly-D amino acids and polymeric bacterial flagellin.

Other factors involved in this process include macrophages that not only present the immunogen to the T cells but whose MHC products also interact with receptors for these molecules on the T cells, the Ir genes, which control the response of B cells to T-dependent immunogens and thus determine the recognition of carrier determinants, and soluble factors such as lymphokines and monokines that influence B cell activation. Detailed information on the regulation of the vertebrate immune system is available in several excellent monographs (Fabris et al. 1983; Bach et al. 1979; Hood et al. 1984; Roitt et al. 1989).

Immune Tolerance

Immunological tolerance is the acquisition of nonreactivity towards particular antigens (also called "tolerogens" as opposed to "immunogens") and as such is the converse of immunity. As a general rule, tolerance induction by an antigenic epitope is inversely related to its immunogenicity.

As an immature B cell matures into an antibody forming cell, it becomes increasingly resistant to tolerization. At the same time, the form of antigen presentation that will produce tolerance also varies. The type of tolerance induced, therefore, is dependent on the maturity of the B cell, the antigen, and the manner in which the antigen is presented to the immune system (Hood et al. 1984; Howard 1979; Roitt et al. 1989).

Four pathways essentially lead to the tolerization of B cell function (Howard, 1979). They are as follows.

1. *Clonal abortion:* Low concentrations of multivalent antigen may cause the immature clone to abort. The susceptibility of the immature B cells to tolerization, i.e., its tolerizability, is high.

2. *Clonal exhaustion:* Repeated antigenic challenge with a T-independent antigen may remove all mature functional B cell clones. Tolerizability of mature B cells is moderate.

3. *Functional deletion:* Absence of T cell help, concurrent with the presence of T-dependent antigen or with T_S cells, or an excess of T-independent antigen prevents mature B cells from functioning normally.

4. *Antibody-forming cells (AFC) blockade:* An excess of T-independent antigen interferes with the secretion of antibody by the AFCs. Tolerizability of the AFC is low.

The pathways leading to T cell tolerance are superficially similar to those for B cell tolerance (Roitt et al. 1989; Hudson and Hay 1989). However, they do not show marked differences in their tolerizability at different stages of T cell maturation. Similarly, the antigen required to produce tolerance and the circumstances of its presentation are specific to each individual T cell subset. In addition to clonal abortion and functional deletion mechanisms similar to those for B cell tolerance, T_S cells will also actively suppress the actions of other T cell subsets or B cells.

In the immunodiagnostic industry, where immune sera form the heart of the technology, tolerance against a given epitope is not desirable. Hence, a number of conditions that promote tolerance need to be avoided during antibody production process. In this regard, the following points need to be remembered.

- The immunogen dose used for immunization. T-independent immunogens may act as tolerogens if injected at high doses (e.g., 100–1000 times the usual dose; Parks et al. 1979), whereas the T-dependent immunogens induce the so-called low-zone and high-zone tolerance, i.e., the epitope is immunogenic only between these two concentrations (Gershon and Kondo 1971).

- Weakly immunogenic antigens induce tolerance if administered in minute quantity, which is maintained by the action of T_S cells.

- T-dependent immunogens rapidly tolerize the T cells within hours of challenge, adult splenic B cells within 4 days, and those from bone marrow within up to 15 days.
- T-independent immunogens, because of their higher avidity, tend to tolerize the B cells more quickly.
- Antigen persistence. Slowly catabolized antigens are more persistent, and hence, maintain the state of tolerance longer than rapidly catabolized antigens.
- Tolerance is developed for particular antigenic determinants, and not for particular antigens.
- Duration. When tolerance is due to clonal deletion, recovery is related to the time required to regenerate mature lymphocytes from the stem cell population. If it is due to the blockade of AFCs, transfer of the tolerant cells to an environment free of antigen leads to a rapid loss of tolerance.
- Route of administration and the physical form of antigen are also important. Soluble and small immunogens may become tolerogenic when applied intravenously.
- Tolerance may be annulled spontaneously if the tolerogen disappears and the T_H cells are stimulated.
- A cross-reacting immunogen may also break the tolerance if it has carrier determinants not shared with the tolerogen.
- Tolerance against a hapten can be prevented by the use of another carrier that will generate other T_H cells.
- Tolerance is enhanced by immunosuppressive drugs such as cyclosporin A.

ANTIBODY CLASSIFICATION

Antibodies or immunoglobulins are a group of glycoproteins present in the serum and tissue fluids of all mammals. Biologically, antibodies are defined as proteins formed when an animal is immunized with an antigen. They are, therefore, an element of the adaptive immune system. The WHO committee (WHO 1964) defined immunoglobulins as "proteins of animal origin endowed with known antibody specificity, and certain proteins related to them by chemical structure, and hence, antigenic specificity. Related proteins for which antibody activity has not been demonstrated are included, e.g., myeloma proteins, Bence-Jones proteins and naturally occurring subunits of immunoglobulins. Immunoglobulins do not include the components of the complement system." Based on this definition, only three criteria are needed to include a component in the group of immunoglobulins, viz., (1) it is a protein, (2) it comes from an animal source, and (3) it has conformational structures in common with the other members.

Using fractional precipitation with neutral salts combined with ion-exchange or size exclusion chromatography, five distinct classes of immunoglobulins can be isolated from most animal species. These are referred to as immunoglobulins (Ig)

G, A, M, D, and E, and are generally abbreviated as IgG, IgA, IgM, IgD, and IgE, respectively. These differ from each other physicochemically in size, charge, amino acid composition, and carbohydrate content, as well as immunologically.

Electrophoretically, the immunoglobulins show a unique range of heterogeneity which extends from the γ to the α fractions of normal serum (Figure 2.3). Antibodies belonging to the IgG class exhibit the most charge heterogeneity; those belonging to the other four classes show a more restricted mobility in the slow β and the fast γ regions. The role of these fractions in the immune response was very elegantly demonstrated by Tiselius and Kabat in 1939 by a marked depletion in these fractions following absorption with antigen.

The basic structure of all immunoglobulins is composed of four polypeptide chains, two identical heavy (H) chains, and two identical light (L) chains that are held together by disulfide bonds. The class and subclass are determined by the

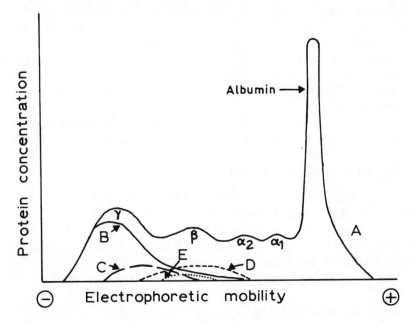

Figure 2.3. The electrophoretic mobility of the immunoglobulin classes
A: Whole serum   B: IgG   C: IgA
D: IgM   E: IgD
Serum proteins are separated according to their charge in an electric field and classified as $\alpha 1$, $\alpha 2$, β, and γ, depending on their mobility. Among the immunoglobulins, IgG exhibits the most charge heterogeneity, the other classes having a more restricted mobility in the slow β and fast γ regions. The IgE class has a similar mobility to that of IgD, but cannot be represented quantitatively because of its low level in serum. These fractions suffer marked depletion following absorption with antigen suggesting a role for them in the immune response.
From: Steward (1984)

heavy chain type. For example, the four human IgG subclasses, viz., IgG1, IgG2, IgG3, and IgG4, have heavy chains called γ1, γ2, γ3, and γ4 that differ only slightly in composition. Nevertheless, the differences between the various subclasses within an immunoglobulin class are less than those between different classes (Clausen 1988; Steward 1984; Catty 1988). Subclasses have been also identified for the human IgA (IgA1 and IgA2) but for none of the remaining three classes. The occurrence and the physicochemical properties of various immunoglobulin classes are briefly described below.

Immunoglobulin G (IgG)

IgG is the predominant immunoglobulin in normal human serum accounting for 70–75% of the total immunoglobulin pool. It is a monomeric protein with a sedimentation coefficient of 6S-7S and a molecular weight range of 146,000–160,000. The four subclasses of human IgG, IgG1, IgG2, IgG3, and IgG4, occur in the approximate proportions of 66%, 23%, 7%, and 4%, respectively. IgG is the only immunoglobulin capable of crossing the placenta in humans, and is thus largely responsible for the protection of the newborn during the early months of life (Hood et al. 1984). The IgG class is distributed evenly between the intra- and extravascular pools, is the major antibody of secondary immune responses, and the exclusive anti-toxin class.

Because of its relative abundance and excellent specificity toward antigens, IgG is the principle antibody used in immunological research and diagnostic products.

Immunoglobulin M (IgM)

IgM accounts for about 10% of the immunoglobulin pool in normal human serum. A pentamer of 970,000 molecular weight, it is the predominant 'primary' antibody frequently directed against antigenically complex infectious organisms. It is also the major immunoglobulin expressed on B lymphocyte surfaces, and is the most efficient complement-fixing immunoglobulin. IgM is largely confined to the intravascular pool, and is infrequently used in diagnostic products.

Immunoglobulin A (IgA)

IgA represents approximately 15–20% of the human serum immunoglobulin pool. Although 80% of it occurs as the basic four-chain monomer, in most other mammals, the serum IgA occurs mostly as a dimer. It is the primary component of seromucous secretions such as saliva, tracheobronchial secretions, colostrum, milk, and genitourinary secretions (Roitt et al. 1989). This secretory IgA (sIgA) exists mainly as an 11S dimer. It is protected from proteolysis by the secretory component associated with it. The main function of sIgA is not to destroy the antigen as much as to prevent its passage into the circulatory system.

Immunoglobulin D (IgD)

IgD is present in large quantities as a receptor on the membrane of most circulatory mature B lymphocytes, where it is presumed to play an important role in antigen-triggered lymphocyte differentiation. It is only a minor component, accounting for less than 1% of the total plasma immunoglobulin pool in human serum.

Immunoglobulin E (IgE)

IgE is a trace serum protein found on the surface of membrane basophils and mast cells. Its primary role is to defend against parasitic invasion. However, it is more commonly associated with hypersensitivity diseases such as asthma and hay fever.

Because of their relatively low concentrations in serum, IgD and IgE are seldom used in diagnostic products. Similarly, because of its tendency to polymerize, the use of IgA is also limited.

Various physicochemical properties of immunoglobulin classes are summarized in Table 2.1. The diversity of structure of the different classes suggests that, in addition to their primary role of antigen binding, they may perform different functions. Nonetheless, in spite of this diversity, they all share a common basic structure.

TABLE 2.1.   Physicochemical Properties of Immunoglobulin Classes[a]

Immuno-globulin	Mol. wt. $\times 10^{-3}$	Mol. wt. $\times 10^{-3}$ (H chain)	No. of H chain domains	Carbo-hydrates %	Sedi-menta-tion constant	Mean serum concen-tration (mg/ml)
IgG						
IgG1	146	51	4	2–3	7S	9
IgG2	146	51	4	2–3	7S	3
IgG3	170	60	4	2–3	7S	1
IgG4	146	51	4	2–3	7S	0.5
IgM	970	65	5	12	19S	1.5
IgA						
IgA1	160	56	4	7–11	7S	3
IgA2	160	52	4	7–11	7S	0.5
sIgA	385	52–56	4	7–11	11S	0.05
IgD	184	69.7	4	9–14	7S	0.03
IgE	184	72.5	5	12	8S	50×10^{-6}

[a]Compiled from Clausen (1988), Hood et al. (1984), Steward (1984) and Roitt et al. (1989).

ANTIBODY STRUCTURE

The basic structure of an immunoglobulin was first elucidated by Porter (1959). Based on his extensive studies on rabbit 7S IgG antibodies using proteolytic digestion with papain in the presence of cysteine followed by carboxymethylcellulose ion-exchange chromatography, he isolated three major fragments I, II, and III of which fragment III could be crystallized. This fragment was subsequently designated as Fc for "fragment crystallizable." The fragments I and II were identical to each other, and unlike Fc, were able to bind the antigen. This accounted for the known valency of 2 for IgG antibodies, and these two fragments were later termed Fab for "fragment antigen binding." Nisonoff et al. (1960) later showed that on treatment of 7S antibody with pepsin, a bivalent 5S fragment is produced, which when reduced yields two monovalent 3.5S fragments. Further studies by Edelman and Poulite (1961) showed that the constituent chains of immunoglobulin can be isolated by reduction of the interchain disulfide bonds with sulfhydryl reagents in urea solution, or by reduction followed by alkylation of the free sulfhydryl groups and subsequent dissociation of noncovalent bonds by gel filtration in acid.

These studies led Porter (1962) to propose a four-chain model for immunoglobulin molecules that is based on two distinct types of polypeptide chains. A typical example of the basic antibody structure as represented by the IgG molecule is shown in Figure 2.4. The smaller, light (L) chain of molecular weight 25,000 is common to all classes of immunoglobulins, whereas the larger, heavy (H) chain of molecular weight 50,000–77,000 is structurally distinct for each class or subclass. These four chains are held together by a combination of noncovalent interactions, and in most antibodies, by covalent interchain disulfide bonds. The pairings of identical light and heavy chains gives a molecule of bilaterally symmetrical structure (Hood et al. 1984; Steward 1984; Roitt et al. 1989).

The five major heavy chain classes that define the corresponding structurally different immunoglobulin classes of IgA, IgG, IgD, IgE, and IgM are designated as α, γ, δ, ε, and μ, respectively. The amino acid sequences in the constant region (see below) of the heavy chains of these classes differ by more than 60% of their residues (Hood et al. 1984). The genes that code for heavy chains are collectively known as the "heavy chain (H) gene family."

In addition, the light chains of most vertebrates exist in two distinct forms. Based on their constant region amino acid sequence homology and antigenicity, they are designated as kappa (κ-type) and lambda (λ-type), and are coded by the κ-gene family and the λ-gene family, respectively. Either of the light chain type may combine with any of the heavy chain types. However, in any one molecule, both light chains are of the same type, and the hybrid molecules do not occur naturally (Roitt et al. 1989). Several characteristics of the subunit structures of the five immunoglobulin classes are summarized in Table 2.2.

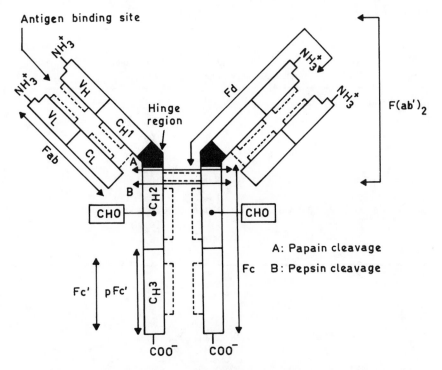

Figure 2.4. The basic four-chain structure of an immunoglobulin showing two identical light (L) polypeptide chains and two identical heavy (H) polypeptide chains linked together by disulfide bonds (dotted lines). The antigen binding site is at the N-terminal end of the molecule. The constant (C) and variable (V) domains are as indicated on both the H and L chains. CHO represents site of attachment for the carbohydrate moiety on the CH2 domain of the H chain. The hinge region is a vaguely defined segment of the H chain between CH1 and CH2 domains. A and B represent the sites at which the immunoglobulin molecule is cleaved by papain and pepsin, respectively, resulting in the generation of various fragments.

This basic four-chain structure is common to IgG, IgD, and IgE, which occur only as monomers of the four-chain units. IgA occurs both in monomeric and polymorphic (either a dimer or trimer of 8 or 12 chains, respectively) forms, whereas IgM is a pentamer in which five four-chain subunits are linked together (Figure 2.5). Heterogeneity in polypeptide chain structures of IgG subclasses in mouse, guinea pig, rabbit, and goat species is shown in Figure 2.6. IgGs from these animal species are widely used in commercial diagnostic products. This heterogeneity in antibody molecules is due to isotypic (different heavy and light chain classes and subclasses), allotypic (variation mostly in the constant region), or idiotypic (variation only in the variable region) variations and is genetically controlled (Roitt et al. 1989).

TABLE 2.2. Subunit Structures of Immunoglobulin Classes

Class	Heavy chain	Subclasses	Light chain	Molecular formula	Antigen binding sites
IgG	γ	$\gamma1, \gamma2, \gamma3, \gamma4$	κ or λ	$(\gamma_2\kappa_2)$	2
				$(\gamma_2\lambda_2)$	2
IgA	α	$\alpha1, \alpha2$	κ or λ	$(\alpha_2\kappa_2)$ or $(\alpha_2\lambda_2)$	2
				$(\alpha_2\kappa_2)_2$ or $(\alpha_2\lambda_2)_2$	4
				$(\alpha_2\kappa_2)_3$ or $(\alpha_2\lambda_2)_3$	6
IgM	μ	None	κ or λ	$(\mu_2\kappa_2)_5$	10
				$(\mu_2\lambda_2)_5$	10
IgD	δ	None	κ or λ	$(\delta_2\kappa_2)$	2
				$(\delta_2\lambda_2)$	2
IgE	ε	None	κ or λ	$(\varepsilon_2\kappa_2)$	2
				$(\varepsilon_2\lambda_2)$	2

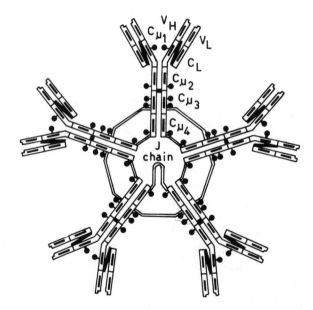

Figure 2.5. Pentameric polypeptide chain structure of human IgM. IgM heavy chains (μ) have five domains (one variable and four constant) with disulfide bonds (solid bars) cross-linking adjacent Cμ3 and Cμ4 domains of different units. The possible location of the J chain is also indicated. Carbohydrate side chains and their locations are shown as solid circles.

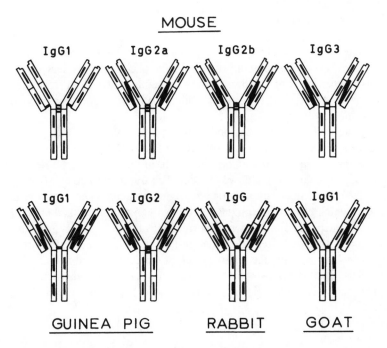

Figure 2.6. Polypeptide chain structure of IgG subclasses in mouse, guinea pig, rabbit and goat. The different IgG molecules in these species vary in their molecular structures and the number of isotypes they carry in their genomes. The Cγ1 domain in the rabbit has two intrachain disulfide bonds.

The amino terminal ends of the antibody are characterized by sequence variability (V) in both the H and L chains, and therefore, are referred to as V_H and V_L regions, respectively (Figure 2.4). The remainder of the molecule has a relatively constant (C) structure. The constant portion of the light chain is designated as C_L, whereas that in the heavy chain (C_H) is further subdivided into three structurally distinct globular regions or domains as C_H1, C_H2, and C_H3. Based on the heavy chain type, a specific nomenclature is used to describe these three domains of different immunoglobulin classes, e.g., Cγ1, Cγ2, and Cγ3 for the domains of IgG. These globular regions in both the variable and constant regions of the H and L chains are stabilized by intrachain disulfide bonds that enclose a peptide loop of approximately 60–70 amino acid residues. Thus, each of the 12 domains in the basic four-chain unit contains one intrachain disulfide bond. The amino acid sequences of these domains exhibit a striking degree of homology (Hood et al. 1984).

The sites at which the antibody binds the antigen are located at the N-terminal ends of the variable domains of the H and L chains (Figure 2.4). The "arms" of these Y-shaped molecules exhibit a great degree of flexibility and thus allow them to operate independently. The "hinge" region responsible for this flexibility is a vaguely defined segment of the H chains between the C_H1 and C_H2 domains and is held together by interchain disulfide bonds. Thus, each monomeric antibody unit contains two antigen binding sites and thus is said to be bivalent. The carbohydrate moieties are attached through the side chains of asparagine residues to the C_H2 domains (Figure 2.4). They are essential in imparting the necessary conformational stability to the molecule.

The structural aspects of antibodies using proteolytic enzymes such as papain and pepsin were extensively studied by Porter (1959) and Nisonoff et al. (1960). Papain cleaves the antibody molecule in the hinge region between the C_H1 and C_H2 domains to yield two identical Fab fragments and one Fc fragment (Figure 2.4). In contrast, proteolytic cleavage with pepsin generates the $F(ab')_2$ fragment, which broadly encompasses the two Fab regions linked by the hinge region, and the pFc' fragment, which corresponds to the C_H3 domain of the molecule. Papain also generates a degraded fragment of C_H3 region after prolonged digestion, which is called the Fc' fragment. Fragments isolated from such proteolytic cleavage of these two and other enzymes have allowed extensive structural and functional comparisons of the various domains of the antibody molecules.

ANTIBODY FUNCTIONS

The primary function of the antibody is to bind the antigen. However, with the exception of their direct neutralizing effects in such cases (e.g., on bacterial toxin or viral penetration of cells), they perform several "effector" functions. As described above, the immunoglobulin molecule comprises a number of domains that tend to be associated with specific biological functions. A summary of these functions is presented in Table 2.3.

Complement activation appears to be the most important function of IgM, IgG1, and IgG3 immunoglobulins. The complement system is a complex group of serum proteins that mediate inflammatory reactions. Once bound to the antigen, these immunoglobulins may activate the complement enzyme cascade. IgG2 appears to be less effective in this regard, while IgG4 does not fix complement. Other important effector functions of immunoglobulins include binding to Fc receptors on various cell types, the facilitation of placental transfer of IgGs and catabolic regulation (Steward 1984; Roitt et al. 1989).

Since their discovery in the context of microbial infections almost a century ago, antibodies have greatly revolutionized our ability to detect and quantitate a diverse array of biologically important molecules. Major commercial applications of antibodies in clinical and diagnostic industries are summarized in Table 2.4.

TABLE 2.3. Functions of Immunoglobulins and the Domains Involved[a]

Function	Immunoglobulin (class/subclass)	Domain
Complement fixation	IgG1 (++), IgG2 (+),	
	IgG3 (+++)	C_H1, C_H2
	IgM (+++)	C_H4
Placental transfer	IgG1 (+), IgG2 (±), IgG3 (+),	
	IgG4 (+)	C_H2 and/or C_H3 (Fc)
Binding to Fc receptors on		
Human macrophages	IgG	C_H3
Mouse macrophages	IgG	C_H2 and C_H3 (Fc)
Monocytes	IgG1, IgG3	C_H3
Neutrophils	IgG1, IgG3, IgG4, IgA	C_H2 and C_H3 (Fc)
Lymphocytes	IgG, IgM	C_H2 and C_H3 (Fc)
Syncytiotrophoblasts	IgG	C_H2 and C_H3 (Fc)
Platelets	IgG	C_H2 and C_H3 (Fc)
Basophils	IgE	C_H2 and C_H3 (Fc)
Mast cells	IgE	C_H2 and C_H3 (Fc)
Reactivity with		
Staphylococcal protein	IgG1, IgG2, IgG4	C_H2 and C_H3 (Fc)
Target for rheumatoid		
factors	IgG, IgM, IgA	C_H2 and C_H3 (Fc)
Catabolism regulation	All classes	C_H2

[a]Compiled from Roitt et al. (1989), Hood et al. (1984), and Steward (1984).

AUXILIARY BINDING SYSTEMS

As analytical reagents, antibodies are perhaps the most useful binding agents produced biologically. However, because a high capacity of molecular recognition is the basis of the structure and operation, several other binding agents are also present in all biological systems. Unlike antibodies, these agents do not have to be induced. Many of these compounds can be exploited commercially for assays in connection with, or similar to, enzyme immunoassays (Tijssen 1985). However, such binding agents are available only for a relatively few ligands, and it is usually not possible to produce them in large quantities. Those exploited commercially in diagnostic products are briefly described below.

Receptors

Receptors are proteins located on surface membranes of cells that specifically bind to particular proteins, peptides, or hormones in the fluid phase. They are

TABLE 2.4.   Major Commercial Applications of Antibodies in Clinical and Diagnostic Industries[a]

IDENTIFICATION AND STUDY OF MOLECULES

Detection of individual antigens in complex mixtures in solution

Testing the purity of separated molecules

Determining sites of production, expression, deposition, and activity of molecules in or on cells and tissues

Determining structure/function correlates of molecules such as enzymes, antibodies, hormones, cytokines, and cell receptors

Definition of differentiation and oncofetal antigens on cells

Definition of cell populations including tumors

Definition of allelic antigens of single genetic loci, e.g., major histocompatibility complex

Determining antigenic relationships between molecules of different species, between hormones, etc.

Lymphocyte (HLA) typing

Erythrocyte (blood group) typing

Probing for recombinant DNA (antigen) expression in clones of *Escherichia coli*, yeast, and other microorganisms

Antigen diagnosis of parasitic, fungal, bacterial, and viral infections by detection of organisms or their products in blood, sputum, urine, feces, swabs, CSF, etc.

Serotyping and subtyping of bacteria and viruses

Definition of antigen variants and gene conversion antigens of parasites

Detection of infectious agents or their antigens in vectors and reservoir hosts

DETECTION AND QUANTITATION OF MOLECULES

Blood components as disease markers: altered levels of composition of immunoglobulins, complement, immune complexes, acute phase proteins, cardiac myosin, α-fetoprotein in neural tube defects, presence of carcinoembryonic antigen, etc.

Hormones: endocrinopathies, estrus cycle, pregnancy, etc.

Specific antibodies: infection, immunity status, autoimmunity, antibody response to allergens, etc.

Mediators of cellular activity: interferons, interleukins, growth factors, chemotactic factors, etc.

Drugs and toxins

PURIFICATION OF MOLECULES AND CELLS

Antibody affinity purification methods for antigens

Precipitation of an RNA/antigen complex from cytosol in construction of enriched gene libraries for the synthesis of antigen by recombinant DNA technology

Antibody labeling of cells for their separation of fluorescence activated cell sorting (FACS)

Binding to discrete cell membrane components for their subsequent isolation

TABLE 2.4. (*continued*)

REAL AND POTENTIAL THERAPEUTIC APPLICATIONS
Immunosuppression: antilymphocyte globulin, antibodies to discrete lymphocyte
  subpopulations, anti-idiotype antibodies, antibodies to other cell receptors
Antibodies to tumor cell antigens, tumor imaging, and cytotoxicity
Anti-idiotype antibodies as "internal image" immunogens
Antibodies to bacterial toxins, bacterial, viral, parasitic, and fungal somatic antigens
Antihuman chorionic gonadotropin hormone and antitrophoblast antibodies in fertility
  control

[a]Adapted from Catty (1988).

identified by their high affinity to specific radiolabeled ligands and by the pharmacology of the specific cellular responses that they elicit (Limbird 1986; O'Brien 1986). Binding assays rely heavily on the use of transmitters or drugs labeled with radioactive isotopes and analytical methods for separating the bound from free radioactivity.

The characteristic features of receptors include structural specificity, high affinity, usually limited capacity, ready saturability, relatively rapid kinetics, and an appropriate degree of reversibility. They are usually "species nonspecific." Receptors exist in low concentration in target cells (0.01% of the total cytoplasmic proteins) and generally occur in two forms: high affinity, low capacity receptors and low affinity, high capacity receptors. The specificity and high affinity ($> 10^9$ M^{-1}) of the receptor, nevertheless, compensates for the low concentration of their respective ligands in the peripheral circulation that is normally sufficient to initiate optimum cell response. Because they are proteins, proteolytic enzymes destroy receptor activity.

At the cellular level, every receptor has at least two primary functions (Gilman et al. 1990; Bonner 1989). The first is to allow the receptor protein to recognize and bind with high affinity its chemical signal (e.g., transmitter or hormone) as well as agonists and antagonists of the receptor. This recognition function forms the basis of receptor-mediated assays. The second function is to trigger a cellular response varying from changes in ion concentration to changes in the levels of a second messenger. For example, the binding of hormones to a specific receptor activates the cellular machinery for that hormonal response (Chard 1978; Schofield 1989). Receptors for several hormones are located in cell cytoplasm, while those for protein hormones lie on cell surface membranes.

Receptors are highly specific and can distinguish molecules on the basis of surface differences such as the presence or absence of a hydroxyl group (Clausen 1988). Specific binding of a ligand to its receptor is effectively displaced only by drugs or ligands known to affect that particular receptor. Their use in assays thus confer similar specificity to in vitro measurements. Therefore, the use of isolated cell receptors as binding agents in diagnostic products, unlike antibodies, which are capable of detecting biologically inactive fragments of the antigen, should, in principle, also provide a direct measure of the functional site of the molecule as well as its biological activity. Cell receptors also have very high affinity constants and, therefore, the potential for yielding a very sensitive assay (e.g., 1–10 pg for corticotrophin) (Chard 1978; Hawcroft et al. 1987).

Receptor assays were first described for estrogens using the uterine cytosol receptor (Korenman 1968). Subsequently, similar assays were described for corticotrophin (Lefkowitz et al. 1970) and gonadotrophins (Catt et al. 1972) by employing cell surface receptors extracted from the appropriate target organs, and for cyclic AMP (Gilman 1970) using a binding protein from skeletal muscle. The sources of the receptors of various polypeptide hormones that are commercially used in some of the diagnostic products are summarized in Table 2.5.

In spite of their high specificity, receptor assays suffer from several drawbacks, and hence, are not exploited as widely as the antibody-based assays in the diagnostic industry. Some of the major drawbacks include:

1. Receptors are not always available readily prepared and their preparation may require access to fresh tissues and extracts. They also need to be purified using complex procedures, thus resulting in considerable batch-to-batch variation.
2. Receptors are generally less stable than antibodies.

TABLE 2.5.   Tissue Receptors for Polypeptide Hormones Used in Diagnostic Industry

Hormone	Tissue source
hCG/LH	Testis, ovary
FSH	Testis, ovary
TSH	Thyroid
Prolactin	Breast, liver, ovary (corpus luteum)
Growth hormones	Cultured human lymphocytes, liver
ACTH	Adrenal
Insulin	Liver, fat, lymphocytes, placenta
LH-RH	Pituitary
TRH	Pituitary
Oxytocin	Uterus, breast
Vasopressin	Kidney
Angiotensin	Kidney, adrenal
Calcitonin	Kidney, bone
Parathormone	Kidney
Glucagon	Liver, β-cells

3. Some receptors can measure an entire class of drugs nonspecifically, e.g., microbial receptors that recognize β-lactam or sulfa family drugs. This is disadvantageous in situations where the individual drug may need to be identified. Similarly, the receptors may show differential affinities to drugs belonging to the same family.

4. The functional specificity of receptors may also embrace molecules that are chemically very different, e.g., the long-acting thyroid stimulator (LATS) can cross-react in a radioreceptor assay for TSH (Pasternak 1975).

5. A substantial proportion of cell surface receptors may not be involved in the biological response of the cell, and hence, their functional specificity may be uncertain (Birnbaumer and Pohl 1973).

6. Receptor assays are limited to a relatively small number of compounds for which recognizable receptors exist. They cannot be used for nonhormonal materials with no target specificity.

Because of these drawbacks, the receptor assays are seriously limited in their practicality. Thus, in spite of their high specificity, assays using cell receptors are not intrinsically superior to those based on antibody recognition. Nonetheless, there are quite a few examples of commercially successful diagnostic products that utilize receptors as the primary binding agents.

Circulating Binding Proteins

Several proteins are used as binding agents in the serum circulation to transport specific ligands to their target organs. Some examples are the cortisol-binding globulin (CBG) for the corticosteroids and progesterone, sex hormone-binding globulin (SHBG) for estrogens and androgens, and the thyroxine-binding globulin (TBG) for thyroxine and triiodothyronine. Because of the presence of these specific binding proteins in the serum, only a small fraction of these potent hormones is available for biological activity, the remainder being found mostly in the bound form.

Ligand assays using circulating binding proteins were first described for thyroxine (Ekins 1960) and cortisol (Murphy et al. 1963). The source material for these, unlike the receptors, is widely available. A very commonly used source of primary material is serum from women in late pregnancy, which contains high levels of all binding proteins (Chard 1978).

As compared to antibody-based immunoassays, those utilizing circulating binding proteins as binding agents suffer from several disadvantages. Their affinity constants are relatively low (10^7–10^8 l/mol) when compared to those for antibodies, and hence, they do not yield a very sensitive assay. Moreover, their affinity constants are highly temperature dependent, and the assays thus have to be conducted under carefully controlled temperature conditions. The low affinity constants also mandate that the binder has to be used at relatively high concentrations. The specificity of assays based on circulating binding proteins is also low.

Protein A (SpA)

Protein A (SpA) is isolated from the cell walls of Cowan 1 or other strains of *Staphylococcus aureus* (Tijssen 1985). It consists of a single polypeptide chain with a molecular weight of 42,000, although, depending on the strain from which it is isolated, it may range from 42,000–56,000 (Bjorck et al. 1972; Sjodal 1977; Sjoquist et al. 1972; Tijssen 1985). It contains little or no carbohydrate.

Protein A has an affinity ($K = 10^8$ l/mol) for Fc fragments of several immunoglobulins, particularly those belonging to the IgG class. This reactivity, similar to that of anti-immunoglobulin antibodies, towards immunoglobulins thus make this protein a very powerful tool for the isolation or removal of immunoglobulins and its subclasses from the antiserum, as well as for use as a tracer in enzyme immunoassays.

This elongated protein has four highly homologous domains, each being capable of binding to the Fc region of immunoglobulins (Sjodal 1977). However, generally only two sites react with the soluble IgG. Protein A is heat stable, and retains its activity even after exposure to denaturing agents such as 4 *M* urea, 4 *M* thiocyanate, or 6 *M* guanidine hydrochloride (Sjoholm 1975).

The reactivity of protein A towards immunoglobulins varies with species, the class and subclass of the immunoglobulins, and ionic conditions (Table 2.6). It is often used as a primary coating protein for some solid-phase immunoassays when the immunoglobulins themselves display a very weak binding affinity for the solid phase. The immunoglobulin-protein A interaction does not inhibit the antigen-antibody reaction.

A genetically engineered recombinant form of protein A (molecular weight 32,000), produced in a nonpathogenic form of *Bacillus,* is currently marketed by the Pierce Chemical Co., Rockford, IL. To eliminate nonspecific binding, most of the nonessential regions in protein A were removed by genetic engineering, leaving only the four IgG-binding domains intact. Recombinant protein A is especially useful in purifying antibodies free of exotoxins (Pierce Chemical Co. 1994).

Protein G

This bacterial cell wall protein is isolated from group G *Streptococcus (*Akerstrom and Bjorck 1986; Akerstrom et al. 1985). Similar to protein A, protein G also binds to most mammalian immunoglobulins through their Fc regions (Eliasson et al. 1988). It generally has much greater binding affinity to most mammalian immunoglobulins than does protein A (Table 2.6). Protein G, however, does not bind to human IgM, IgD, and IgA (Bjorck and Kronvall 1984; Fahnestock 1987; Guss et al. 1986).

The differences in the binding affinities of these two proteins for immunoglobulins are related to unique compositions in their Fc-binding domains. Although the amino acid compositions are significantly different, their tertiary structures

TABLE 2.6. Binding Affinities of Protein A (*Staphylococcus aureus*) and Protein G (Group G *Streptococcus*) to Fc Regions of Various Immunoglobulins[a]

Species	Immunoglobulin class/subclass	Binding affinity	
		Protein A	Protein G
Human	IgG	S	S
	IgM	W	NB
	IgD	NB	NB
	IgA	W	NB
	IgG1	S	S
	IgG2	S	S
	IgG3	W	S
	IgG4	S	S
Mouse	IgG	S	S
	IgG1	W	W/S
	IgG2a	S	S
	IgG2b	S	S
	IgG3	S	S
Rat	IgG	W	W/S
	IgG1	W	W/S
	IgG2a	NB	S
	IgG2b	NB	W
	IgG2c	S	S
Bovine	IgG1	W	S
	IgG2	S	S
Sheep	IgG1	W	S
	IgG2	S	S
Goat	IgG1	W	S
	IgG2	S	S
Rabbit	IgG	S	S
Guinea pig	IgG	S	W
Porcine	IgG	S	W
Horse	IgG	W	S
Dog	IgG	S	W
Chicken	IgG	NB	NB
Cat	IgG	S	W
Monkey (rhesus)	IgG	S	S

Abbreviations used: S = strong affinity, W = weak affinity, W/S = weak overall affinity but much greater than that for protein A, NB = no binding.

[a]Compiled from Akerstrom and Bjorck (1986), Bjorck and Kronvall (1984), Eliasson et al. (1988), Sjobring et al. (1988), Guss et al. (1986), Nilson et al. (1987), Fredrikson et al. (1987), and Pierce Chemical Co. (1994).

appear to be quite similar (Fahnestock 1987; Olsson et al. 1987; Fahnestock et al. 1986; Guss et al. 1986). Protein G has applications similar to those of protein A in the diagnostic industry.

Protein A/G

Protein A/G is a genetically engineered protein produced by a gene fusion product from a nonpathogenic *Bacillus* strain. This 45,000–47,000 molecular weight protein is engineered to contain the four Fc-binding domains of protein A and two of protein G (Eliasson et al., 1988).

Protein A/G binds to the Fc regions of all human IgG subclasses, IgA, IgE, IgM, and, to a lesser extent, to IgD (Pierce Chemical Co. 1994). It also binds to all IgG subclasses of mouse immunoglobulins, but has no affinity for mouse IgA, IgM, and serum albumin. It is, therefore, especially useful in the purification and detection of mouse monoclonal IgG antibodies without interference from other immunoglobulins and proteins.

Lectins

Lectins are specific carbohydrate-recognizing glycoproteins, and are responsible for the hemagglutinating activity of several seed extracts. Commercially important lectins include concanavalin A (or Con A) isolated from *Canavalia ensiformis,* ricin from *Ricinus communis,* and the soybean lectin. Of these, Con A is important in enzyme immunoassays due to its reactivity towards peroxidase enzyme. The intrinsic affinities of lectins are often significantly lower than those of antibodies, and hence, their use in diagnostic products is quite limited. Lectins, however, are an important diagnostic tool in several clinical and immunohistochemical reactions to identify lectin receptors on cells. They have been used for membrane characterization or tagging through their specific affinities for certain carbohydrates (Deshpande and Sathe 1991), thus providing an alternative to using antimembrane antibodies. Because lectins are di- or tetravalent, they can also be labeled with a glycosylated marker such as the peroxidase enzyme prior to their binding to membrane constituents. Their localization of the membrane may then be traced immunochemically or by electron microscopy.

REFERENCES

AKERSTROM, B., and BJORCK, L. 1986. A physicochemical study of protein G molecule with unique immunoglobulin G-binding properties. *J. Biol. Chem.* 261:10240–10247.

AKERSTROM, B.; BRODIN, T.; REIS, K.; and BJORCK, I. 1985. Protein G: A powerful tool for binding and detection of monoclonal and polyclonal antibodies. *J. Immunol.* 135:2589–2592.

BACH, F.; BONAIRDA, B.; and VITETTA, E. 1979. *T and B Lymphocytes. Recognition and Function.* Academic Press, New York.

BACH, F. H.; BACH, M. L.; and SONDEL, P. M. 1976. Differential function of major histocompatibility complex antigens in T-lymphocyte activation. *Nature* (London) 259: 273–281.

BIRNBAUMER, L., and POHL, S. L. 1973. Relation of glucagon-specific binding sites to glucagon-dependent stimulation of adenylyl cyclase activity in plasma membranes of rat liver. *J. Biol. Chem.* 248:2056–2061.

BJORCK, L., and KRONVALL, G. 1984. Purification and some properties of streptococcal protein G. A novel IgG-binding reagent. *J. Immunol.* 133:969–974.

BJORCK, L.; PETERSSON, B. A.; and SJOQUIST, J. 1972. Some physicochemical properties of protein A from *Staphylococcus aureus. Eur. J. Biochem.* 29:579–584.

BONNER, T. I. 1989. The molecular basis of muscarinic receptor diversity. *Trends Neurol. Sci.* 12:148–151.

CATT, K. J.; DUFAU, M. L.; and TSURUHARA, T. 1972. Radioligand-receptor assay of luteinizing hormone and chorionic gonadotropin. *J. Clin. Endocrinol. Metab.* 34:123–132.

CATTY, D. 1988. *Antibodies. A Practical Approach.* IRL Press, Oxford.

CHARD, T. 1978. An Introduction to Radioimmunoassay and Related Techniques. North Holland, Amsterdam.

CLAUSEN, J. 1988. *Immunochemical Techniques for the Identification and Estimation of Macromolecules.* Elsevier, Amsterdam.

DESHPANDE, S. S., and SATHE, S. K. 1991. Toxicants in plants. In *Mycotoxins* and Phytoalexins, eds. R. P. Sharma and D. K. Salunkhe, pp. 671–730, CRC Press, Boca Raton, FL.

EDELMAN, G. M., and POULITE, M. D. 1961. Structural units of the γ-globulins. *J. Exp. Med.* 113:861–884.

EKINS, R. P. 1960. The estimation of thyroxine in human plasma by an electrophoretic technique. *Clin. Chim. Acta* 5:453–459.

ELIASSON, M.; OLSSON, A.; PALMCRANTZ, E.; WIBERG, K.; INGANAS, M.; GUSS, B.; LINDBERG, M.; and UHLEN, M. 1988. Chimeric IgG-binding receptors engineered from staphylococcal protein A and streptococcal protein G. *J. Biol. Chem.* 263: 4323–4327.

FABRIS, N.; GARACI, E.; HADDEN, J.; and MITCHISON, N.A. 1983. *Immunoregulation.* Plenum Press, New York.

FAHNESTOCK, S. 1987. Cloned streptococcal protein G genes. *Tibtech* 5:79–83.

FAHNESTOCK, S.; ALEXANDER, P.; NAGLE, J.; and FILPULA, D. 1986. Gene for an immunoglobulin-binding protein from a Group G Streptococcus. *J. Bacteriol.* 167: 870–880.

FREDRIKSON G.; NILSSON, S.; OLSSON, H.; BJORCK, L.; AKERSTROM, B.; and Belfrage, P. 1987. Use of protein G for preparation and characterization of rabbit antibodies against rat adipose tissue hormone-sensitive lipase. *J. Immunol. Meth.* 97:65–70.

GERSHON, R. K.; EARDLEY, D. D.; DURUM, S.; GREEN, D. R.; SHEN, F. W.; YAMAUCHI, K.; CANTOR, H.; and MURPHY, D. B. 1981. Contrasuppression. A novel immunoregulatory activity. *J. Exp. Med.* 153:1533–1546.

GERSHON, R. K., and KONDO, K. 1971. Degeneracy of the immune response to sheep red cells. *Immunology* 23:321–324.

GILMAN, A. G. 1970. Protein-binding assay for adenosine 3':5'-cycle monophosphate. *Proc. Natl. Acad. Sci.* (USA) 67:305–312.

GILMAN, A. G.; RALL, T. W.; NIES, A. S.; and TAYLOR, P. 1990. Goodman and Gilman's *The Pharmacological Basis of Therapeutics,* 8th ed., Pergamon Press, New York.

GOLUB, E. S. 1981. *The Cellular Basis of the Immune Response.* Sinauer Associates, Sunderland, MA.

GUSS, B.; ELIASSON, M.; OLSSON, A.; UHLEN, M.; FREJ, A.; JORNVALL, H.; FLOCK, J.; and LINDBERG, M. 1986. Structure of the IgG-binding regions of streptococcal protein G. *EMBO J.* 5:1567–1575.

HAWCROFT, D.; HECTOR, T.; and ROWELL, F. 1987. *Quantitative Bioassay.* John Wiley, New York.

HOOD, L. E.; WEISSMAN, I. L.; WOOD, W. B.; and WILSON, J. H. 1984. *Immunology.* 2d ed., Benjamin-Cummings, Menlo Park, CA.

HOWARD, J. G. 1979. Immunological tolerance. In *Defence and Recognition: Cellular Aspects,* vol. 22, ed. E. S. Lennox, University Park Press, Baltimore, MD.

HUDSON, L., and HAY, F. C. 1989. *Practical Immunology.* 3d ed., Blackwell Scientific, Oxford.

JACKSON, S.; CHUSED, T. M.; WILKINSON, J. M.; LEISERSON, W. M.; and KINDT, T. J. 1983. Differentiation antigens identify subpopulations of rabbit T and B lymphocytes. *J. Exp. Med.* 157:34–46.

JOHNSTONE, A., and THORPE, R. 1987. *Immunochemistry in Practice.* 2d ed., Blackwell Scientific, Oxford.

KLEIN, J. 1979. The major histocompatibility complex of the mouse. *Science* 203:516–521.

KORENMAN, S. G. 1968. Radioligand binding assay of specific estrogens using a soluble uterine macromolecule. *J. Clin. Endocrinol. Metab.* 28:127–130.

LEFKOWITZ, R. J.; ROTH, J.; PRICE, W.; and PASTAN, J. 1970. ACTH receptors in the adrenal. Specific binding of ACTH-^{125}I and its relation to adenyl cyclase. *Proc. Natl. Acad. Sci.* (USA) 65:745–752.

LIMBIRD, L. E. 1986. *Cell Surface Receptors: A Short Course on Theory and Methods.* Martinus Nijhoff, Boston, MA.

MALE, D. 1986. *Immunology. An Illustrated Outline.* Gower Medical Publishing, London.

MURPHY, B. E. P.; ENGELBERG, W.; and PATTEE, C. J. 1963. Simple method for the determination of plasma corticoids. *J. Clin. Endocrinol. Metab.* 23:293–300.

NILSON, B.; AKERSTROM, B.; and LOGDBERG, L. 1987. Cross-reacting monoclonal anti-alpha 1-microglobulin antibodies produced by multi-species immunization and using Protein G for the screening assay. *J. Immunol. Meth.* 99:39–45.

NISONOFF, A.; WISSLER, F. C.; and LIPMAN, L. N. 1960. Properties of a major component of a peptic digest of rabbit antibody. *Science* 132:1770–1772.

O'BRIEN, R. A. 1986. *Receptor Binding in Drug Research.* Marcel Dekker, New York.

OLSSON, A.; ELIASSON, M.; GUSS, B.; NILSSON, B.; HELLMAN, U.; LINDBERG, M.; and UHLEN, M. 1987. Structure and evolution of the repetitive gene encoding streptococcal protein G. *Eur. J. Biochem.* 168:319–324.

PARKS, D. R.; BRYAN, V. M.; OI, V. T.; and HERZENBERG, L. A. 1979. Antigen-specific identification and cloning of hybridomas with a fluorescence-activated cell sorter. *Proc. Natl. Acad. Sci.* (USA) 76:1962-1966.

PASTERNAK, C. A. 1975. *Radioimmunoassay in Clinical Biochemistry.* Heyden, London.

PIERCE CHEMICAL CO. 1994. *Life Science and Analytical Research Products Catalog and Handbook.* Rockford, IL.

PLAYFAIR, J. H. L. 1987. *Immunology at a Glance.* Blackwell Scientific, Oxford.

PORTER, R. R. 1959. Hydrolysis of rabbit γ-globulin and antibodies with crystalline papain. *Biochem. J.* 73:119–126.

PORTER, R. R. 1962. *Basic Problems of Neoplastic Disease.* Columbia University Press, New York.

RAFF, M. C. 1970. Role of thymus-derived lymphocytes in the secondary humoral immune response in mice. *Nature* (London) 226:1257–1258.

ROITT, I. M. 1988. *Essential Immunology.* 6th ed., Blackwell Scientific, Oxford.

ROITT, I. M.; BROSTOFF, J.; and MALE, D. 1989. *Immunology.* Gower Medical Publishing, London.

SCHOFIELD, P. R. 1989. The $GABA_A$ receptor: Molecular biology reveals a complex picture. *Trends Pharmacol. Sci.* 10:476–478.

SJOBRING, U.; FALKENBERG, C.; NIELSEN, E.; AKERSTROM, B.; and BJORCK, L. 1988. Isolation and characterization of a 14-KDa albumin-binding fragment of streptococcal protein G. *J. Immunol.* 140:1595–1599.

SJODAL, J. 1977. Structural studies on the four repetitive Fc-binding regions in protein A from Staphylococcus aureus. Eur. J. Biochem. 78:471–490.

SJOHOLM, I. 1975. Protein A from Staphylococcus aureus: *Spectropolarimetric* and spectrophotometric studies. *Eur. J. Biochem.* 51:55–61.

SJOQUIST, J.; MELOUN, B.; and HJELM, H. 1972. Protein A isolated from *Staphylococcus aureus* after digestion with lysostaphin. *Eur. J. Biochem.* 29:572–578.

STAINES, N.; BROSTOFF, J.; and JAMES, K. 1985. *Introducing Immunology.* Gower Medical Publishing, London.

STEWARD, M. W. 1984. *Antibodies. Their Structure and Function.* Chapman & Hall, New York.

TIJSSEN, P. 1985. *Practice and Theory of Enzyme Immunoassays.* Elsevier, Amsterdam.

TISELIUS, A., and KABAT, E. 1939. An electrophoretic study of immune serums and purified antibody preparations. *J. Exp. Med.* 69:119–131.

WHO. 1964. Congress about nomenclature of immunoglobulins. *Bull. W.H.O.* 30:447.

3
Antigen-Antibody Reactions

INTRODUCTION

Enzyme immunoassays are based on antigen-antibody reactions involving enzyme-labeled antigen (or antibody) with antibody (or antigen). An understanding of the fundamentals of antigen-antibody reactions is, therefore, essential for the success of a product development process based on this technology. In the diagnostic industry, antigens represent an extremely diverse spectrum of compounds ranging from simple haptens to macromolecules such as proteins, microorganisms and viruses. In this section, the fundamental aspects of the antigen-antibody reactions are covered only from a process development viewpoint. Wherever possible, references are cited that deal with detailed theoretical, mathematical, and biochemical treatment of this subject.

Classification

Based on the valency of the antigen (e.g., monovalent hapten or polyvalent protein antigens) and the immunoglobulin subclass of the antibody, the antigen-antibody reactions can be classified as primary and secondary reactions (Butler 1980; Steward 1984; Jefferis and Deverill 1991). All reactions between the antigen and its antibody begin with a primary reaction, which is a specific recognition and combination of an antigenic determinant with the binding site of its corresponding antibody. It is the first step in a series of reactions and biochemical processes that may or may not proceed to further reactions. Primary reactions occur rapidly (of the order of milliseconds), and are microscopically invisible. In contrast, the secondary reactions are possible only with multivalent antigens. They also require a longer time to develop. A characteristic feature of the secondary reactions is that their progress can be monitored either visually or with the aid of a microscope.

The tertiary manifestations of antigen-antibody reactions are those that occur as biological reactions that follow primary and secondary reactions described above. These tertiary manifestations include many of the biological effects of complement activation, such as opsonization, phagocytosis, and chemotaxis (Roitt et al. 1989).

Primary antigen-antibody reactions can be followed by using the Farr assay (Farr 1958) based on differential solubility of the antigen, antibody, and the antigen-antibody complex, RIA or EIA. The requirements for these tests include (a) a purified antigen or antibody preparation; (b) a technique to quantitate the antigen or antibody with the use of a radioisotope, enzyme, or fluorescent label; and (c) a method to separate the antigen-antibody reaction complex from free antigen and antibody in solution (Nakamura 1992). These methods generally are more sensitive than ones used to monitor the secondary reactions. The latter can be detected by a number of techniques such as both fluid and gel precipitin reaction tests, direct agglutination, passive hemagglutination, hemolysis, complement fixation, and toxin neutralization tests.

Based on their univalent or polyvalent nature, the antigen-antibody reactions sometimes are classified as Type I and Type II reactions, which are in principle similar to the primary and secondary reactions. Type I reactions are described in connection with the heterogeneity of a population of antibodies for a single antigenic determinant. In contrast, the Type II reactions refer to heterogeneity in an antiserum that results from specificity for two or more epitopes on the same antigen (Butler 1980).

A fundamental understanding of the forces involved and the nature of the antigen-antibody reactions including specificity, affinity, and sensitivity is essential prior to dealing with the kinetic aspects of these reactions and their usefulness in assay development and formulation. Some of these aspects are described below.

ANTIGEN-ANTIBODY BINDING

Several intermolecular forces are involved in antigen-antibody reactions. The binding of antigen to its antibody takes place by the formation of multiple noncovalent bonds between the antigen and the amino acid residues at the binding site on the antibody (Figure 3.1). These attractive forces, viz., hydrogen bonds, electrostatic or coulombic, van der Waals, and hydrophobic, involved in these bonds are individually weak by comparison with covalent bonds. However, a multiplicity of these bonds leads to a considerable binding, thereby leading to the stabilization of antigen-antibody complex. Intermolecular forces involved in these reactions are briefly described below.

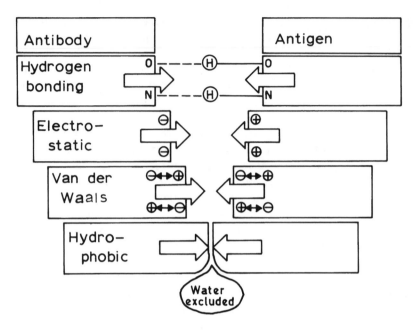

Figure 3.1. The intermolecular forces involved in antibody-antigen reactions. The distance of separation between the interacting groups which produces optimum binding varies for the different types of bonds.
From: Steward (1984)

Hydrogen Bonding

Hydrogen bonding results from the formation of hydrogen bridges between appropriate atoms. It involves the interaction of a hydrogen atom with the unshared electron pair of another electronegative atom (Figure 3.1). Amino, carboxyl, and hydroxyl groups are the major hydrogen donors in antigen-antibody reactions. Hydrogen bonds in these reactions are operative over a very short distance of 1.5–5 Å, primarily because of the competition between the hydrogen donor and acceptor proteins and the surrounding aqueous medium. Hence, hydrogen bonding seldom plays an important role in the primary bonding, although it contributes significantly to the secondary reactions (Jefferis and Deverill 1991).

Electrostatic (Ionic) or Coulombic Interactions

These forces are due to the attraction of oppositely charged groups located on the side chains of two proteins. Such bonding between oppositely charged groups in the antigenic determinant and the antibody binding site is effective over distances of up to 100 Å, and is responsible for primary bonding reactions (Davies and Padlan 1990).

Van der Waals Forces

Van der Waals forces are generated by the interaction of the electron clouds of two polar groups that induce oscillating dipoles. Similar attractive forces, called London dispersion forces, operate between nonpolar, hydrophobic groups. Long range van der Waals forces may operate over distances of up to 1000 Å, and are involved in primary bonding reactions.

Hydrophobic Interactions

Hydrophobic interactions rely upon the association of nonpolar, hydrophobic groups or side chains of amino acids (e.g., side chains of leucine, isoleucine, valine, and phenylalanine that do not form hydrogen bonds with water). Their exceptional stability is due to structural alteration of the aqueous environment when these groups come together with the exclusion of water. Hydrophobic interactions strengthen the existing primary and secondary bonds. The replacement of antigen-water-antibody interactions with direct antigen-antibody bonding by hydrophobic forces greatly increases the binding energy, thereby stabilizing the antigen-antibody complex.

Although both electrostatic and van der Waals forces are responsible for the primary bonding that constitute the specific early part (equivalent to few milliseconds) of the antigen-antibody reaction, their total energy contributes only a small fraction of the total antigen-antibody bond energy (Jefferis and Deverill 1991). In contrast, the secondary bonding, involving the formation of hydrogen bonds and hydrophobic interactions, continues to occur over a longer period, and therefore, contributes substantially to the stabilization of the antigen-antibody complex.

Steric Factor or Steric Repulsive Forces

The intermolecular attractive forces described above are critically dependent on the distance (d) between the interacting groups. For example, the force is proportional to $1/d^2$ for the coulombic or electrostatic interactions and to $1/d^7$ for van der Waals forces. Thus, for these forces to become significant, the interacting groups must be in close proximity to each other.

As compared to these attractive forces, steric repulsion ($1/d^{12}$) is much more sensitive (Steward 1984). This repulsive force between nonbonded atoms arises from the interpenetration of their electron clouds, and thus determines the "good fit" or the "poor fit" between an antigen and its antibody (Figure 3.2). A better complementarity ("good fit") of the electron cloud shapes of the antigenic determinant and the binding site of the antibody will lower the repulsive force, and thus increase the opportunities for the formation of intermolecular attractive forces.

In contrast, a nonhomologous antigenic determinant, because of its overlapping electron cloud with that of the binding site, will generate high repulsive forces. This in turn will minimize any small forces of attraction, thereby resulting in a "poor fit" (Figure 3.2). As will be described below, this phenomenon of "good fit-poor fit" between the antigen and its antibody further defines the concept of antibody affinity.

Because the intermolecular forces involved in the formation and stabilization of the antigen-antibody complex are so varied and heterogeneous, it is difficult to predict the optimum conditions for the formation or dissociation of such a complex. However, some generalizations can be made. Among the forces involved, the electrostatic interactions between the antigen and the antibody are particularly dependent on the pH of the medium. They are often severely disrupted outside the pH 6–8 range (Hughes-Jones et al. 1964). For example, the binding of *p*-aminoazobenzoate by its antibody significantly decreases when the pH of the medium is lowered from 7 to 4, or when its ionic strength is increased from 0.1 to 1 (Eisen 1980). Temperature also has a profound effect on the antigen-antibody reactions.

Chaotropic ions such as SCN^-, I^-, Br^-, and Cl^- are capable of destroying the three-dimensional structure of the protein, and thus may affect the antigen-antibody binding reactions, especially at high concentrations (Dandliker et al. 1967; Edgington 1971). Organic solvents also exert negative effects on the binding phenomenon, whereas organic acids of low surface tension, e.g., acetic and

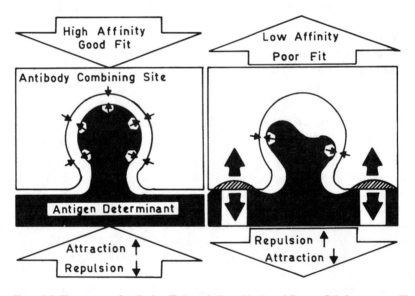

Figure 3.2. The concept of antibody affinity as indicated by "good fit-poor fit" phenomenon. The affinity of interaction between antibody and antigen results from a balance between the attractive and repulsive forces. A high affinity antibody implies a "good fit" and conversely, a low affinity antibody implies a "poor fit."
From: Steward (1984)

propionic acids, disrupt van der Waals bonds and dissociate the complexes quite efficiently (Tijssen 1985).

In-depth reviews on the mechanisms of antigen-antibody binding described above and the chemical and physical nature of the forces involved are available (Davies and Padlan 1990; Getzoff et al. 1988).

ANTIBODY AFFINITY AND AVIDITY

The term "antibody affinity" refers to the strength of the interaction between an antigenic determinant and the homologous antibody binding site (Steward 1984). Its meaning, however, is often misinterpreted in the diagnostic industry. In thermodynamic terms, affinity is expressed as the summation of energy (strength) of close range intermolecular attractive and repulsive forces between the antigenic determinant and its binding site (Figure 3.2).

The antigen (Ag) : antibody (Ab) reactions are reversible, and can be described by the law of mass action under equilibrium conditions as follows:

$$Ag + Ab \underset{K_d}{\overset{k_a}{\rightleftharpoons}} Ag : Ab \tag{1}$$

Rearranging,

$$K_a = k_a / k_d = [Ag : Ab] / [Ag] \cdot [Ab] \tag{2}$$

where k_a and k_d are the respective rate constants for association and dissociation of the bound complex, [Ab] is the concentration of the unbound or free antibody, [Ag] the concentration of unbound antigen, and [Ag : Ab] that of the antigen-antibody complex. K_a, also referred to as the affinity or association constant of equilibrium, is defined in reciprocal molar concentration, M^{-1} or liters per mole (l/mol). It represents the volume into which a mole of the antibody can be diluted to yield 50% binding of the antigen. The larger the K_a, the greater the affinity of the antibody for the antigen.

Thus, a high affinity antibody forms a strong bond with its antigenic determinant. The resulting Ag : Ab complex in such a case has a low tendency to dissociate, i.e., the binding reaction or its equilibrium is shifted towards the right side in the above equation. Therefore, the higher the affinity of the antibody, the greater will be the antigen amount bound to the antibody at equilibrium. In other words, for a constant amount of antibody in the reaction, less antigen is required for a high affinity antibody to bind 50% of the antigen than is required for 50% binding by a low affinity antibody.

In contrast, for a low affinity antibody, the equilibrium of the reaction is shifted to the right, i.e., the complex needs less energy to dissociate into free antigen and free antibody. The total free energy ($\Delta G°$) of the system can be expressed as

$$\Delta G° = -RT\ln K, \tag{3}$$

where R is the gas constant and T the absolute temperature in Kelvins. Thus, a high affinity antibody will yield a greater negative $\Delta G°$ value than a low affinity antibody.

Experimentally, the affinity constant K_a can be determined most precisely by using a monovalent antigen or even a single antigenic determinant—a hapten. As each basic four-chain unit of the antibody has two antigen binding sites, antibodies are potentially multivalent in their reactions. Although one can use the monovalent Fab fragment to obtain a true estimate of K_a, in real world situations, several complications can arise. For example, polyclonal antibody preparations are heterogeneous in affinity towards a single antigenic determinant. The derived K_a in such cases then represents only an average value. Similarly, antiserum of multiple specificity, i.e., specific to many determinants on an antigen, cannot be assessed for affinity.

To avoid such problems, the practical manifestation of the ability of an antibody to bind its antigen is evaluated in terms of "avidity," although in literature these two terms are often used synonymously (Steward 1984; Catty 1988; Jefferis and Deverill 1991). To differentiate it from the "affinity" of the bond between a single antigenic determinant and its individual combining site, "avidity" is defined in terms of the strength with which a multivalent antibody binds a multivalent antigen. Thus, the avidity of an antibody for its antigen is dependent on the affinities of the individual combining sites for the determinants on the antigen, but is greater than the sum of these affinities if both the antigen and the antibody are multivalent. Thus, if we assume an arbitrary value of 10^4 l/mol for the K_a of a Fab fragment, there may be as much as a 10^3-fold increase in the binding energy of an IgG when both valencies or antigen binding sites are utilized, and a 10^7-fold increase when an IgM with 10 binding sites binds an antigen in a multivalent fashion. It should, however, be noted that a monovalent antigen combines with a multivalent antibody with no greater affinity than it does with a monovalent Fab fragment (Roitt et al. 1989).

Thus, strictly speaking, the terms "affinity" and "avidity" are not synonymous. To distinguish the differences in their meanings, they are sometimes referred to as "intrinsic affinity" and "functional affinity," respectively (Steward 1984; Roitt et al. 1989).

The affinity constants can be measured using a variety of techniques, including equilibrium dialysis; neutral salt or antiglobulin precipitation (modified Farr assays); by monitoring changes in fluorescence properties such as quenching, enhancement, or polarization; by electrophoretic separation of the complex; radioimmunoassays; and by size separation. These methods are all applicable to the low molecular weight antigens. Equilibrium dialysis and fluorescence polarization are not suitable for measurement of the interaction between a high molecular weight antigen and its antibody because of the large size of the antigen.

Typical values for the affinity constant K_a of most antibody preparations range from $10^5 - 10^{12}$ l/mol. Low affinity binding proteins and antibodies typically have K_a values on the order of 10^5 to 10^7 l/mol, whereas binders suitable for immunoassays and other competitive protein-binding assays must have affinity constants between 10^8 and 10^{12} l/mol. A higher affinity constant enables one to design assays with sensitivity down to 10^{-9} to 10^{-12} M, provided that the label itself is detectable at such low concentrations.

ANTIBODY SPECIFICITY AND CROSS-REACTIVITY

The antigen-antibody reactions exhibit a high degree of specificity when the binding sites of antibodies directed against determinants on one antigen are not complementary and thus do not recognize (or "cross react") the determinants of another antigen. In contrast, the specificity of an antiserum reflects the many specificities of its constituent antibodies (Catty 1988; Tijssen 1985). An antiserum may bind exclusively to one antigen if the range of its constituent antibody specificities extends to determinants exclusive to that antigen.

Antibodies are capable of expressing remarkable specificity, and are able to distinguish between small differences in the primary structure of the antigen, its optical configuration or spatial (steric) configuration ("configurational specificity"), as well as differences in charge ("charge specificity"). Immunological research during the past decade has shown that a given antibody molecule may be complementary to several unrelated antigens. Using competitive binding techniques, spatially separated positions have been identified within the antigen binding site of the antibody (Roitt et al. 1989; Catty 1988; Steward and Steensgaard 1983; Tijssen 1985). Binding sites that are capable of specifically binding more than one antigenic determinant are termed "polyfunctional" binding sites. These studies have shown that the specificity of a population of antibodies would not necessarily arise because "all" the antibodies have the same exclusive specificity. However, if a large number of different polyfunctional antibodies all had a site that could bind to a particular antigen A, then the net reactivity of those antibodies would be higher for A but low for all other antigens.

The concept of cross-reactivity is complementary to that of specificity. A cross-reactive antigen is defined as one that binds antibodies induced in response to a different molecule by virtue of shared antigenic determinants. The phenomenon of cross-reactivity can be explained in terms of either the binding of struc-

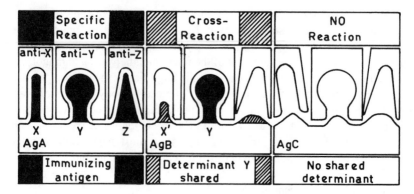

Figure 3.3. The concepts of specificity, cross-reactivity, and nonreactivity. Antiserum (anti-XYZ) to antigen A (AgA) reacts specifically with antigen A (left). Anti-A cross-reacts with antigen B (AgB) through recognizing determinant Y and partial recognition of determinant X' (center). Anti-A shows no reaction with antigen C (AgC), which has no shared determinants (right). Modified from Steward (1984)

turally different determinants of the same antigen by the same antibody, or in terms of the existence of common antigenic determinant(s) on different antigens. In the latter case, the cross-reactivity is not based on true differences in affinity, and hence, is commonly referred to as "shared reactivity."

The concepts of specificity, cross-reactivity, and shared reactivity are illustrated in Figure 3.3.

QUANTITATION OF ANTIGEN-ANTIBODY REACTIONS

The law of mass action [equations (1) and (2) described earlier] provides a useful framework on which to base a theoretical appreciation of the thermodynamic principles underlying immunoassay techniques. However, several assumptions need to be made in understanding such models (Davies 1994; Butler 1980; Thompson, 1984). Some of these include the following.

1. The antibody is homogeneous, and has a single binding site that recognizes only one epitope of the antigen with the same affinity,
2. The antigen should also be homogeneous, consisting of only one chemical species, and possessing only one epitope for binding,
3. The antigen-antibody binding should be uniform and without any positive or negative allosteric effects, i.e., the binding of one antibody-binding site should not influence the binding of the other site,
4. The reaction must be at equilibrium,

5. The separation of the bound from free antigen must be complete, and

6. No nonspecific binding should occur in the assay system.

Thus, in equation (2) described earlier, if Ab_T represents the total concentration of antibody in the system, then the concentration of unbound or free antibody (Ab) becomes

$$Ab = Ab_T - [Ag : Ab] \qquad (4)$$

Further, if the terms B and F are used to denote the bound [Ag : Ab] and free [Ag] antigen, respectively, then equation (2) can be rewritten as

$$K_a = B / (Ab_T - B) * F \qquad (5)$$

and rearranging:

$$B / F = K_a \cdot (Ab_T - B) \qquad (6)$$

or,

$$B / F = K_a \cdot Ab_T - K_a \cdot B \qquad (7)$$

Equation (7) indicates a linear relationship between the ratio of bound/free antigen and the concentration of bound antigen. The graphical representation of this is known as a Scatchard plot (Scatchard 1949). If any two of the three unknowns are known in this equation, the third one can be calculated. Two useful parameters may be derived from the Scatchard plot: the affinity constant K_a from the slope of the line, and the total concentration of antibody-binding sites (Ab_T) from the intercept on the x-axis, because as B/F approaches zero, B equals the total antibody-binding sites.

An example of a typical Scatchard plot using a polyclonal antiserum is shown in Figure 3.4. Because polyclonal antiserum usually contains a mixture of different populations of antibody to a given antigen, the slope of the Scatchard plot is curved. It reflects the relative concentrations of different populations of antibodies with different K_a's in the heterogeneous mix of the antiserum. Three straight lines can be extrapolated from the curve shown in Figure 3.4, indicating division of the antibody population into three subpopulations of differing affinities.

It should be noted that not all the point estimates of B/F and B have equal weighting. Measurement errors have a disproportionate effect at both high and low B/F ratios (Davies 1994). Therefore, caution must be exerted in constructing Scatchard plots to ensure that the concentrations of antigen are spread out to ascertain maximum accuracy. Similarly, the separation of the bound from free anti-

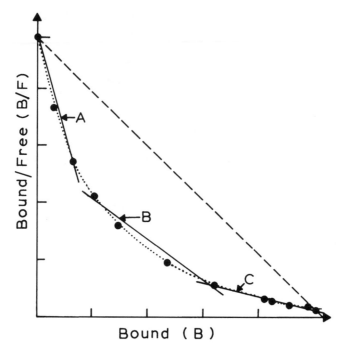

Figure 3.4. Graphical presentation of a typical Scatchard plot of amount of antigen bound (B) versus bound-to-free (B/F) ratio of antigen. The straight interrupted line illustrates the ideal slope if reactants combine homogeneously. The dotted line connecting the experimental data points (•) to give a typical curved line is generally obtained with natural populations of polyclonal antibodies. The short, solid straight lines (A, B, and C) indicate division of the antibody population into three subpopulations of differing affinities (K_a high, intermediate, and low, respectively). The slope of the curve is $-K_a$, the *x*-intercept is the total antibody-binding sites (Ab_T).

gen can rarely be 100% complete. Some residual free antigen concentration, therefore, is usually associated with the bound fraction. The nonspecific binding must be determined with a considerable degree of accuracy. Errors in its determination can have a disproportionate effect on low B/F ratios, i.e., the region where the intercept with the bound fraction provides an estimate of the total antibody concentration.

Because estimating the true K_a of a polyclonal antiserum is difficult from a Scatchard plot, the average K_a can provide the only useful measure of the antibody affinity. The average K_a can be obtained in two ways. The first approximates K_a at half-saturation of antibody-binding sites. Thus, from equation (2) described earlier and when the antibody-binding sites are half-saturated

$$K_a = 1 / [Ag]$$ (8)

From a plot of B versus log F, a line can be drawn at half the apparent maximal binding (Figure 3.5). The point on the x-axis where this corresponds to the intersection with the binding curve provides a reasonable average estimate of $1/K_a$.

The degree of heterogeneity (or homogeneity) of a polyclonal antiserum can also be quantitated with the use of a Sips equation and its corresponding graphical solution (Figure 3.6) (Nisonoff and Pressman 1958). Thus, rearranging equation (2),

$$K_a \cdot [Ag] = [Ag : Ab] / [Ab] \tag{9}$$

and substituting equation (4) for the concentration of free antibody,

$$K_a \cdot [Ag] = [Ag : Ab]/\{[AbT]—[Ag : Ab]\} = B/[Ab_T - B] \tag{10}$$

Because at half-saturation, $[Ag : Ab] = [Ab] = \{[AbT]–[Ag : Ab] \}$, then

$$K_a \cdot [Ag] = 1, \text{ or } K_a = 1/[Ag] \tag{11}$$

When the binding protein or antibody is homogeneous, i.e., there is only one population of uniform binding sites for an antigen, the shape of the Scatchard plot

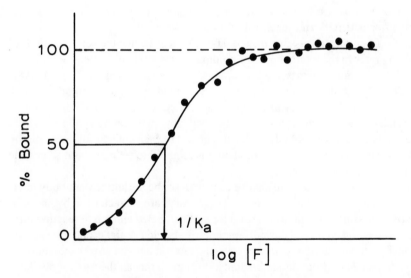

Figure 3.5. Graphical representation of a saturation plot of percent bound versus log free [F] antigen. At half or 50% saturation of the total antibody- binding sites, the average affinity constant K_a is equivalent to the reciprocal of intercept on x-axis.

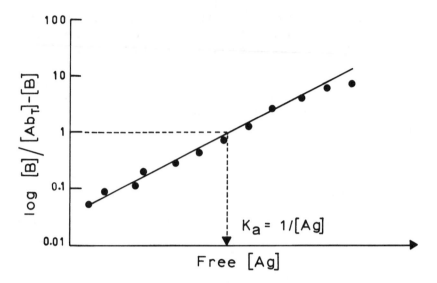

Figure 3.6. Graphical representation of a Sips plot. At half-saturation of antibody- binding sites
(Ab_T), the average affinity constant K_a on the x-intercept equals reciprocal of the free antigen
fraction.

is linear. This usually occurs when the analysis of a binding reaction is performed
with monoclonal antibodies. These antibodies are homogeneous in that they reflect uniform affinity and specificity for the antigen.

In the foregoing discussion, it should be remembered that the concentrations of
antibody refer to the total number of binding sites. For small haptens or antigens,
both IgG binding sites may bind to antigen, i.e., its valency is 2. For some large
protein antigens, steric restrictions may prevent binding of both antibody-binding
sites (Davies 1994, Thompson 1984). Thus, difficulties arise in the interpretation
of Scatchard plots when the antigen is multivalent with repeating epitopes. In such
cases, the numbers of antibody molecules bound to each antigen follow a Poisson
distribution and a more complex modeling is required to estimate both K_a and the
total antibody-binding sites.

The law of mass action can also be used to describe the antigen-antibody interactions in competitive-binding assays. These assays are characterized by the addition of increasing concentrations (or doses) of unlabeled antigen to reaction mixtures containing known, constant amounts of labeled antigen and its specific
binding protein or antibody. In this case, the labeled antigen (Ag^*) and antibody
are added together in equimolar amounts. Presuming that all the Ag^* is bound, the
reaction becomes

$$Ag^* + Ab \rightleftharpoons Ag^* : Ab \tag{12}$$

With the addition of increasing concentrations of unlabeled antigen Ag, two things happen:

1. The unlabeled antigen Ag competes with the labeled antigen Ag* for antibody-binding sites, and
2. There is an excess of the total antigen (Ag and Ag*) in solution.

The concentration of antibody-binding sites is, therefore, limiting with respect to total antigen, thus modifying equation (12) as follows:

$$Ag + Ag^* : Ab \rightleftharpoons Ag : Ab + Ag^* : Ab + Ag^* + Ag \qquad (13)$$

As the concentration of Ag increases, less Ag* is antibody bound as Ag* : Ab and more Ag* is free. The concentration or percentage of Ag* in the bound form can be calculated from the amount of Ag and Ag* present in the assay system.

When the percentage of labeled antigen bound is plotted as a function of the concentration of the unlabeled antigen, it yields a "dose-response" curve (Figure 3.7). The curvature of the dose-response (Figure 3.7A) is attributable to the logarithmic increase in the percentage of Ag that is bound when the concentration (dose) of Ag in the assay increases arithmetically. Therefore, the decrease in bound Ag* is also logarithmic. Conversion of the concentration of Ag to a log scale (Figure 3.7B) makes the relationship more linear.

Such dose-response curves can also be used to predict the changes in B/F ratio with increasing concentration of antigen. This relationship is quadratic (i.e., $ax^2 + bx + c = 0$) in nature, and allows the calculation of B/F for any given quantity of total antigen $[Ag_T]$, provided both K_a and $[Ab_T]$ are accurately known (Yalow and Berson 1971; Feldman and Rodbard 1971; Pesce et al. 1981; Davies 1994). Conversely, and of more interest, $[Ag_T]$ can be solved for a given B/F. Thus, the quantity of an unknown amount of antigen can be determined directly from the ratio of B/F. The latter requires a method of separating the bound antigen from the free, and the ability to measure the relative proportion of each.

The above relationship can be more conveniently expressed as the percentage bound in relation to the total antigen present, as this more directly relates to what is measured in practice. Thus,

$$\% \ B = [B/F]/\{1 + [B/F]\} \times 100 \qquad (14)$$

The above equation can be solved for any theoretical or experimental values of antibody, labeled and unlabeled antigen, and K_a. If any three of these four variables are fixed, then the effect of any one variable on the percent bound fraction can be determined.

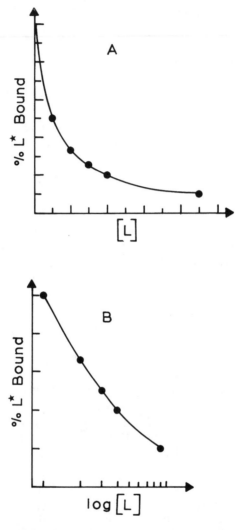

Figure 3.7. An example of a linear dose-response curve (A) and one in which the response is plotted as function of log of ligand or antigen concentration (B). L and L* represent unlabeled and labeled ligand or antigen, respectively.

In Figure 3.8, the above equation has been solved for certain values of these four parameters. The antigen is assumed to have a molecular weight of 100,000 daltons. Thus, 100 ng/ml of the antigen is equivalent to 10^{-9} M. Curve A represents calculated data in the absence of labeled antigen Ag*. This means the label is so intense that only a very small amount is required for detection. It also has no

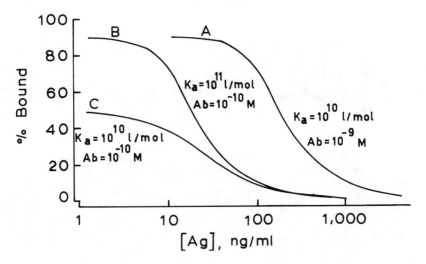

Figure 3.8. Theoretical binding curves for a competitive immunoassay.
From: Pesce et al. (1981)

effect on the total antigen concentration. Curves A (10^{-9} *M)* and C (10^{-10} *M*) in Figure 3.8 represent higher and lower concentrations of the antibody having the same affinity constant K_a of 10^{10} l/mol. Reducing the concentration of antibody shifts the practical range of the assay to a lower concentration of antigen. However, because the slope of curve C is much less than that of A, this shift to higher sensitivity is achieved at the expense of discrimination.

In contrast, if the antibody concentration is kept similar (curves B and C, Figure 3.8), increasing the K_a by 10-fold (curve B) increases the sensitivity of the assay. Because the K_a for a given population of antibody is fixed (generally > 10^8 l/mol for sensitive immunoassays), it cannot be relied upon to improve assay performance. Nonetheless, from Figure 3.8, it is obvious that sensitivity can be improved by reducing the concentration of antibody in the immunoassay within a certain limit.

Two models have been described to define the sensitivity of an immunoassay. Yalow and Berson (1971) have defined the sensitivity of the radioimmunoassay in terms of the slope of arbitrarily selected representations of the dose-response curve. In this model, the maximal slope and assay sensitivity are attained with an initial percent bound of 33% when the antigen or analyte concentration approaches zero. The sensitivity used should have a large K_a at a concentration equal to $0.5/K_a$.

In contrast, the Ekins model (1981a,b) defines the sensitivity of an assay in terms of the lower limit of detection or its essential equivalent, i.e., the precision of measurement of a zero dose. In this model, maximal sensitivity can be obtained

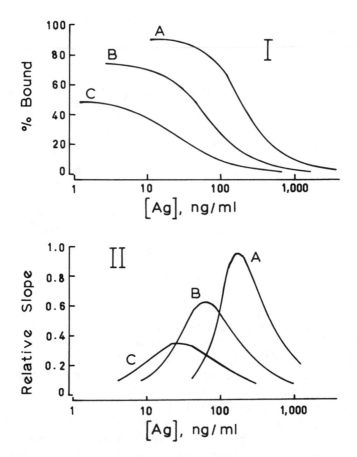

Figure 3.9. Influence of antibody concentration (I) on competitive immunoassay and the relative slopes of the curves A–C (II) in the absence of the labeled antigen, Ag*. The affinity constant K_a of the antibody was 10^{10} l/mol. The antibody concentrations were: A = 1×10^{-9} M, B = 3×10^{-10} M, and C = 1×10^{-10} M.
Modified from Pesce et al. (1981)

with an initial percent binding of 50%. The lower limit of detection expressed as the least detectable dose is inversely proportional to the square root of the specific activity of the label, the affinity constant K_a, and the reaction volume.

The contrasting definitions of the Yalow-Berson and Ekins models lead to different conclusions regarding the optimal concentration of assay reagents required to maximize sensitivity. In most assay systems, the affinity constant K_a of the antibody and the experimental errors in measuring percent bound at very low antigen concentrations are the primary limiting factors in achieving maximal sensitivity.

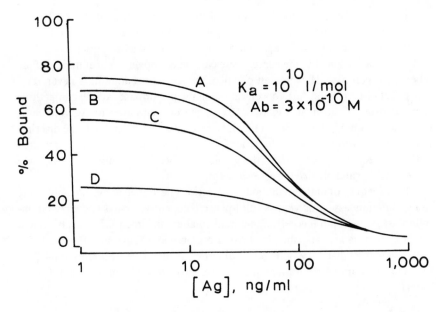

Figure 3.10. Influence of labeled (Ag*) on the competitive immunoassay. Curves A–D represent assay systems with Ag* concentrations of 0, 10, 30, and 100 ng/ml, respectively. From: Pesce et al. (1981

The concept of assay sensitivity is further illustrated in Figures 3.9 and 3.10. In the absence of Ag*, the assay curves A–C (Figure 3.9I) flatten and extend into the lower antigen concentration ranges as the concentration of antibody decreases. The slopes of these curves (Figure 3.9II) further indicate that the ability of the assay to discriminate between two concentrations of antigen increases with an increase in the slope. Theoretically, using less and less antibody, therefore, should produce greater and greater sensitivity. However, in practice, no immunoassay approaches such high precision.

From the curves shown in Figure 3.9, it is quite obvious that when the relative slope becomes less than 0.4 (curve C), the assay becomes less sensitive. Thus, curve C is only slightly more discriminating in the low antigen concentration and much less discriminating in the intermediate and high antigen concentrations than curve B where the antibody concentration is increased 3-fold. In reality, the lowest antibody concentration becomes the most sensitive only when great technical improvements, such as a very high signal-to-noise ratio of the label, are made to the assay system.

Thus, technical improvements that reduce the error associated with measuring the percent bound fraction will improve both the sensitivity and precision of the assay, whereas lowering the antibody concentration improves the sensitivity at the expense of the range and discrimination of the assay (Pesce et al. 1981). Generally, there is an optimal antibody concentration that preserves most of both parameters, and it depends on the characteristics of the actual assay system.

In Figures 3.8 and 3.9, the curves were calculated for zero concentration of the labeled antigen Ag*. However, in practice, the Ag* concentration is never zero. The effect of increasing Ag* concentration on the performance of the competitive immunoassay is shown in Figure 3.10. As the concentration of Ag* increases (from curve A to D), the slopes of the dose-response curves, and thus the discrimination of the assay, decrease. In general, assay sensitivity is not greatly affected until Ag* becomes $> 3/K_a$. Thus, the greater the specific activity of the Ag*, the more sensitive and precise the assay system.

The principles of antigen-antibody interactions described above are applicable to radioimmunoassays as well as solid-phase competitive assays and enzyme immunoassays. In-depth theoretical and mathematical treatment of this subject and its data analysis are described by several researchers (Yalow and Berson 1971; Feldman and Rodbard 1971; Pesce et al. 1981; Butler 1980; Steward 1984; Van Oss and Absolom 1984; Steward and Steensgaard 1983; Thompson 1984; Davies 1994).

REFERENCES

BUTLER, J. E. 1980. Antibody-antigen and antibody-hapten reactions. In *Enzyme Immunoassay,* ed. E. T. Maggio, pp. 5–52, CRC Press, Boca Raton, FL.

CATTY, D. 1988. *Antibodies. A Practical Approach.* IRL Press, Oxford.

DANDLIKER, W. B.; ALONSO, R.; DE SAUSSURE, V. A.; KIERSZENBAUN, F.; LEVISON, S.; and SCHAPIRO, H. C. 1967. The effect of chaotropic ions on the dissociation of antigen-antibody complexes. *Biochemistry* 6:1460–1467.

DAVIES, C. 1994. Principles. In *The Immunoassay Handbook,* ed. D. Wild, pp. 3–47, Stockton Press, New York.

DAVIES, D. R., and PADLAN, E. A. 1990. Antibody-antigen complexes. *Ann. Rev. Biochem.* 59:439–473.

EDGINGTON, T. S. 1971. Dissociation of antibody from erythrocyte surfaces by chaotropic ions. *J. Immunol.* 106:673–680.

EISEN, H. N. 1980. *Immunology. An Introduction to Molecular and Cellular Principles of the Immune Responses.* 2d ed., Harper & Row, Hogerstown, MD.

EKINS, R. P. 1981a. Toward immunoassays of greater sensitivity, specificity and speed. An overview. In *Monoclonal Antibodies and Developments in Immunoassays,* eds. A. Albertini and R. P. Ekins, pp. 3–4, Elsevier/North-Holland Biomedical Press, Amsterdam.

EKINS, R. P. 1981b. Merits and disadvantages of different labels and methods of immunoassays. In *Immunoassays for the 80s,* eds. A. Voller, A. Bartlett, and D. Bidwell, pp. 5–16, University Park Press, Baltimore.

FARR, R. S. 1958. A quantitative immunochemical measure of the primary interaction between I*BSA and antibody. *J. Infect. Dis.* 103:239–262.

FELDMAN, H., and RODBARD, D. 1971. Mathematical theory of radioimmunoassay. In *Principles of Competitive Protein-Binding Assays,* eds. W. D. O'Dell and W. H. Doughaday, pp. 158–203, J.B. Lippincott Co., Philadelphia.

GETZOFF, E. D.; TAINER, J. A.; and LERNER, R. A. 1988. The chemistry and mechanism of antibody binding to protein antigens. *Adv. Immunol.* 43:1–98.

HUGHES-JONES, N. C.; GARDNER, B.; and TELFORD, R. 1964. Effect of pH and ionic strength on the reaction between anti-D and erythrocytes. *Immunology* 7:72–81.

JEFFERIS, R., and DEVERILL, I. 1991. The antigen antibody reaction. In *Principles and Practice of Immunoassay,* eds. C. P. Price and D. J. Newman, pp. 1–18, Stockton Press, New York.

NAKAMURA, R. M. 1992. Concepts of immunochemical assays. In *Immunochemical Assays and Biosensor Technology for the 1990s,* eds. R. M. Nakamura, Y. Kasahara, and G. A. Rechnitz, pp. 3–22, American Society for Microbiology, Washington, D.C.

NISONOFF, A., and PRESSMAN, D. 1958. Heterogeneity of antibody-binding sites in their relative combining affinities for structurally related haptens. *J. Immunol.* 81:126–135.

PESCE, A. J.; FORD, D. J.; and MAKLER, M. T. 1981. Properties of enzyme immunoassays. In *Enzyme Immunoassay,* eds. E. Ishikawa, T. Kawai, and K. Miyai, pp. 27–39, Igaku-shoin, Tokyo.

ROITT, I. M.; BROSTOFF, J.; and MALE, D. 1989. *Immunology.* Gower Medical Publishing, London.

SCATCHARD, G. 1949. The attractions of proteins for small proteins and molecules. *Ann. N.Y. Acad. Sci.* 51:660–672.

STEWARD, M. W. 1984. *Antibodies. Their Structure and Function.* Chapman & Hall, New York.

STEWARD, M. W., and STEENSGAARD, J. 1983. *Antibody Affinity: Thermodynamic Aspects and Biological Significance.* CRC Press, Boca Raton, FL.

THOMPSON, S. G. 1984. Competitive-binding assays. In *Clinical Chemistry: Theory, Analysis and Correlation,* eds., L. A. Kaplan and A. J. Pesce, pp. 211–231, The C.V. Mosby Co., St. Louis.

TIJSSEN, P. 1985. *Practice and Theory of Enzyme Immunoassays.* Elsevier, Amsterdam.

VAN OSS, C. J., and ABSOLOM, D. R. 1984. Nature and thermodynamics of antigen-antibody interactions. In *Molecular Immunology,* eds. M. Z. Atassi, C. J. Van Oss, and D. R. Absolom, Marcel Dekker, New York.

YALOW, R. S., and BERSON, S. A. 1971. Introduction and general considerations. In *Principles of Competitive Protein-Binding Assays,* eds. W. D. O'Dell and W. H. Doughaday, pp. 1–24, J. B. Lippincott Co., Philadelphia.

4
Conjugation Techniques

INTRODUCTION

Preparation of well-defined conjugates during various stages, from hapten immunogen preparation for antibody production to that of tracer or labeling reagents such as enzyme-antigen or enzyme-antibody conjugates that ultimately determine the sensitivity of the product, is vital to the success of an immunoassay product development process. Conjugation or coupling reactions must preserve the immunogenicity of the hapten, i.e., its antigenic determinant, when linked to carrier protein prior to immunization, as well as the nature of the antibodies (primarily heterogeneity, valency, conformation of the antigen binding sites and their biological properties) and of enzymes in terms of a stable conjugate with high retention of both immunoreactivity and enzymatic activity.

Sometimes, the reactive groups available for conjugation of a low molecular weight antigen may also be involved in its antigenic determinant, as well as distinguishing it from closely related molecules. In such instances, the use of these reactive groups for conjugation can not only result in the production of a weak antiserum for its antigenic determinant, but can also decrease the specificity of the antibodies produced. Therefore, alternate reactive groups may need to be introduced in their structures, away from their antigenic determinants, prior to conjugation with carrier protein. In contrast, a failure to select and optimize an appropriate conjugation procedure may result in excessive inter- and intra-cross-linking of the reacting molecules. Excessive cross-linking of proteins, especially antibodies and enzymes, may result in loss of affinity and thus sensitivity of the assay.

A fundamental understanding of the basic conjugation processes and techniques, and a knowledge of their practical implications, therefore, is essential from the very early stages of immunoassay product development. In this chapter, basic aspects of conjugation chemistry; the types of reactions for both hapten-macromolecule (e.g., hapten-carrier protein or hapten-enzyme) and macromolecule-macromolecule (e.g., antibody-enzyme) conjugation and the cross-linking reagents available; and the advantages and disadvantages of these techniques will be described. Because of the sheer number of chemical procedures available for conjugation of any two reactive groups as well as the possibility of in-house modifications of these methods to suit one's needs, the actual methodology is not described. However, wherever possible, appropriate references are cited for the benefit of the reader.

FUNCTIONAL GROUPS ON PROTEINS

Chemical cross-linking methods routinely employed in the diagnostic industry are largely derived from the fields of peptide chemistry and chemical modification of proteins, primarily because of their importance either as carrier immunogens or as enzyme labels. Hence, it is appropriate to review the basic functional groups of proteins and the concepts of protein modification relevant to enzyme immunoassay technology.

Of the 20 amino acids with side chains of different sizes, shapes, charges and chemical reactivity that make up the protein molecules, reactive functional groups are primarily provided by those amino acids that have ionizable side chains. In this regard, the alkyl side chains of the hydrophobic amino acids, which are primarily located in the interior of the protein molecule, are chemically inert and thus not available for modification. Similarly, the aliphatic hydroxyl groups of serine and threonine are considered as water derivatives, and, therefore, have a low reactivity. Furthermore, the reactivity of the available functional groups, defined in terms of their ability to undergo chemical modification, is largely determined by the sequence location and their interactions with the side chains of neighboring amino acids in the overall three-dimensional structure of the protein molecule.

Generally, only those reactive groups located on the protein surface and thus exposed to the aqueous environment are available for chemical modification and conjugation purposes. In proteins, there are essentially eight hydrophilic side chains that are chemically active (Wong 1993; Tijssen 1985). These side chains and their functional groups are as follows:

1. Amino group of *N*-terminal amino acid and ε-amino groups of lysines,
2. Sulfhydryl group of cysteine,

3. Thioether group of methionine,
4. Carboxyl group of *C*-terminal amino acid and the β- and γ-carboxyl groups of aspartic and glutamic acids, respectively,
5. Phenolic group of tyrosine,
6. Imidazolyl group of histidine,
7. Guanidinyl group of arginine, and
8. Indolyl group of tryptophan.

Of these, the first five groups are chemically the most reactive. They are also normally exposed on the protein surface and form the major targets for protein cross-linking and conjugation.

The ε-amino groups of lysyl residues are usually among the most abundant and most accessible of the potential reactive groups on proteins. They are reasonably good nucleophiles above pH 8.0 (pK_a = 9.18) and, therefore, react easily and cleanly with a variety of reagents to form stable bonds.

Other reactive amines that are found on proteins are the α-amino groups of the N-terminal amino acids. They are less basic than lysyl amines and are reactive at around pH 7.0. Because either *N*-terminal amines or lysines are almost always present in any given protein or peptide, and because they are easily reacted, the most commonly used method of protein modification is through these aliphatic amine groups.

In contrast, the free thiol group of the sulfur-containing amino acid cysteine is more nucleophilic than amines, and is generally the most reactive functional group in proteins. Unlike amines, thiols are reactive at neutral pH, and, therefore, can be coupled to other molecules selectively in the presence of amines. This selectivity makes the thiol group the linker of choice for coupling two proteins together, because methods that only couple amines (e.g., glutaraldehyde, dimethyl adipimidate coupling) can result in formation of homodimers, oligomers, and other unwanted products (Brinkley 1992).

Because free thiol groups are relatively reactive, proteins with these groups often exist in their oxidized form as disulfide-linked oligomers or have internally bridged disulfide groups. In such proteins, reduction of the disulfide bonds with reagents such as β-mercaptoethanol or dithiothreitol (DTT or Cleland's reagent) is required to generate the reactive free thiol.

In addition to cysteine and its oxidative product, cystine, proteins also contain the amino acid methionine. It contains sulfur in a thioether linkage. In the absence of cysteine, methionine can sometimes react with thiol-reactive reagents such as iodoacetamides (Means and Feeney 1971). Selective modification of methionine, however, is difficult to achieve, and, therefore, is seldom used as a method of attaching small molecules to proteins. Moreover, the side chains of methionine residues are generally buried in the interior of the protein, and, hence, not commonly available for modification.

The carboxylic acid groups at the carboxy terminal of proteins and within the side chains of the dicarboxylic amino acids, glutamic and aspartic acid, can also be utilized for cross-linking purposes. However, their low reactivity in water usually makes it difficult to use these groups to selectively modify proteins and other biopolymers. Usually, the carboxylic acid group is first converted to a reactive ester by use of a water-soluble carbodiimide and then reacted with a nucleophilic reagent such as an amine or a hydrazide. The amine reagent should be weakly basic in order to react specifically with the activated carboxylic acid in the presence of the other amines on the protein. This is because protein cross-linking can occur when the pH is raised above 8.0, the range where the protein amines are partially unprotonated and reactive. Therefore, hydrazides, which are weakly basic, are preferred in coupling reactions with a carboxylic acid. This reaction can also be used effectively to modify the carboxy terminal group of small peptides.

The phenolic group of the amino acid tyrosine can react in two ways. The phenolic hydroxyl group can form esters and ether bonds, while the aromatic ring can undergo nitration or coupling reactions with reagents such as diazonium salts at the position adjacent to the hydroxyl group. The reaction of protein tyrosyl residues with haptens containing aromatic amines via the latter approach is widely used in the diagnostic industry, especially for preparing carrier protein-hapten immunogen conjugates.

Compared to the functional groups described above, chemical modification of other amino acid side chains in proteins has not been extensively used. Histidines have been subjected to photooxidation and reaction with iodoacetates, whereas the high pK_a of the guanidinyl functional group of arginine ($pK_a = 12–13$) necessitates more drastic reaction conditions than most proteins can survive (Means and Feeney 1971). Similarly, because the indolyl group of the hydrophobic amino acid tryptophan is often buried in the interior of proteins, its modification requires harsh conditions. It is, therefore, seldom used for coupling purposes.

Several proteins also contain carbohydrates that provide a useful site for chemical modification and cross-linking of proteins.

Most chemical modification reactions involving the functional groups of amino acid side chains are nucleophilic substitution reactions. The nucleophile attacks an electron deficient center and displaces a leaving group. The rate of these reactions is dependent on the ability of the leaving group to come off as well as the nucleophilicity of the attacking group (Loudon and Shulman 1941; Bunnett 1963; Edwards and Pearson 1962). In terms of protein modification, the relative chemical reactivity is basically a function of the nucleophilicity of the amino acid side chains (Wong 1993).

The overall nucleophilicity order as formulated by Edwards and Pearson (1962), i.e., $RS^- > ArS^- > I^- > CN^- > OH^- > N_3^- > Br^- > ArO^- > Cl^- >$ pyridine $> AcO^- >$ H_2O suggests that the sulfhydryl (especially its thiolate form) of cysteine is the most potent nucleophile in the protein, followed by the nitrogen in the amino group.

The reactivities of the functional groups on the side chains in proteins vary considerably depending on their locations and the influence of nearby residues with which they interact (Means and Feeney 1990). The relative reactivities of certain types of side chains can be determined from the extent of their reaction with trace levels of one of several simple reagents (Kaplan et al. 1971; Duggleby and Kaplan 1975; Shewale and Brew 1982; Rieder and Bosshard 1980). The intrinsic reactivity and pK_a of each reacting group can be easily determined by comparing its reaction to that of a simple model compound over a range of pH values.

For identical side chain groups at different sequence positions, the observed differences in pK_a and reactivity are assumed to reflect differences in local environment. Different local environments may either suppress or enhance the reactivities of individual side chain groups. Unusually reactive side chains are relatively easy to distinguish from others on the basis of their reactivity and are, in many instances, also those required for biological activity. Rates of inactivation, which may differ from overall rates of modification, can often be used to characterize the reactivity and, sometimes, the number of active site residues (Ray and Koshland 1961). The number of essential residues can also be used to estimate activities remaining at various stages of partial modification of proteins (Tsou 1962).

As a general rule, modifications that have the least effect on side chain character should also have the least effect on protein structure and properties. For example, modifications of lysyl residues that retain their usual cationic charge have generally been found to have relatively little effect on the biological activities and other properties of many proteins (Means and Feeney 1990). Amidination or reductive alkylation of amino groups, both of which also retain the cationic charge, are generally preferred, as both these reactions take place under milder conditions (Hunter and Ludwig 1962; Means and Feeney 1968).

Various types of chemical modification reactions of the active side chains of amino acids are summarized in Table 4.1. Of these, the most important reactions include alkylation (i.e., transfer of an alkyl group to the nucleophilic atom) and acylation (bonding of an acyl group). Because protonation decreases the nucleophilicity of a reactive species, these reactions are dependent on the pH of the reaction medium (Wong 1993). Generally, at any given pH, the most reactive group is usually the one with the lowest pK_a.

CHEMICAL MODIFICATION
OF PROTEIN FUNCTIONAL GROUPS

Chemical modification of the existing functional groups of proteins is sometimes essential for several practical purposes. These include (a) activation of the inactive carbohydrates to functional groups for further chemical reactions, (b) interconversion into one another either to change the specificity or to increase their reac-

TABLE 4.1.   Chemical Modification Reactions of Functional Groups on Amino Acid Side Chains

---

Cysteine (pK_a = 8.3)
   Alkylation, arylation, acylation, oxidation, iodination, esterification, reactions with mercurials and sulfenyl halides
Lysine (pK_a = 10.8)
   Alkylation, arylation, acylation, diazotization, amidation, reactions with dicarbonyl
Methionine
   Alkylation, arylation, oxidation, reaction with cyanogen bromide
Histidine (pK_a = 6.0)
   Alkylation, arylation, acylation, oxidation, iodination, diazotization
Tyrosine (pK_a = 10.9)
   Alkylation, arylation, acylation, oxidation, iodination, diazotization
Tryptophan
   Alkylation, arylation, oxidation, reaction with sulfenyl halides
Aspartic Acid (pK_a = 3.9) and Glutamic Acid (pK_a = 4.3)
   Acylation, esterification, amidation
Arginine (pK_a = 12.5)
   Reaction with dicarbonyls

---

tivity, and (c) incorporation of spacer arms to reduce the steric hindrances as well as to decrease the influence of local environment (Kabakoff 1980; Tijssen 1985; Wong 1993). Modification reactions commonly used in the field of enzyme immunoassay are briefly described below.

Activation of Carbohydrates

The carbohydrates present in glycoproteins contain vicinal hydroxyl groups that are susceptible to oxidation with periodic acid or its sodium or potassium salts. The carbon-carbon bonds containing vicinal hydroxyls in these sugars are cleaved by periodate oxidation to dialdehydes, which subsequently react with amino groups of proteins via Schiff base formation. The latter can be further stabilized by reduction with sodium borohydride.

Interconversion of Functional Groups

Amino groups of lysines and N-terminal amino acids can be converted to their corresponding dicarboxylic acids using dicarboxylic acid anhydrides. Succinic acid anhydride is most commonly used for this purpose.

Similarly, a free thiol can be linked to the amino group in several ways. Thiolation can be achieved by using *N*-acetylhomocysteine thiolactone (White 1972), *S*-acetylmercaptosuccinic anhydride (Klotz and Heiney 1962), thiol-containing imidoesters such as methyl 3-mercaptopropionimidate hydrochloride (Perham and

Thomas 1971) and methyl 4-mercaptobutyrimidate, which cyclizes to form 2-iminothiolane, "Traut's reagent" (Traut et al. 1973; Jue et al. 1978), and thiol-containing succinimidyl derivatives such as succinimidyl 3-(2-pyridyldithio)propionate (SPDP) (Carlsson et al. 1978) and dithiobis (succinimidyl propionate) (DSP) (Lomant and Fairbanks 1976). Since thiols are potent nucleophiles, most of these effective reagents generally contain protected sulfhydryl groups, which can be activated to generate free thiols only after the amino groups have been modified.

Carboxyalkylation is a preferred method for the modification of sulfhydryl groups, because in carboxylic acid derivatives prepared using succinic anhydride, the thioester bond is highly susceptible to hydrolysis. α-Haloacetates are commonly used for this purpose (Gurd 1967). Sulfhydryl groups can also be converted into amines using ethylenimine and 2-bromoethylamine (Raftery and Cole 1963; Lindley 1956).

The carboxyl groups of aspartic and glutamic acids can be converted to more potent cationic amine nucleophiles by using water-soluble carbodiimides such as 1-ethyl-3-(3-dimethylaminopropyl)-carbodiimide (EDC, EDAC) and a diamine such as ethylenediamine (Kurzer and Douraghi-Zadeh 1967). The reaction is relatively mild and provides an extended arm from the protein to avoid steric hindrance. This reaction is commercially used for the preparation of SuperCarrier® Cationized BSA for use as an immunogen (Pierce Chemical Co. 1994). Its extremely high pI of 11–13 is presumed to facilitate uptake by antigen presenting cells of lymphoid origin.

Using similar approaches, the relatively unreactive hydroxyl group of serine and threonine can be converted to the highly reactive thiol. Activation of the hydroxyl group is first achieved by tosyl chloride (toluenesulfonyl chloride), followed by transesterification and the subsequent hydrolysis of the thioester to generate a free thiol (Ebert et al. 1977).

Several excellent monographs and reviews that provide a detailed coverage on the modification of functional groups in proteins are available (Means and Feeney 1971; Lundblad and Noyes 1984; Eyzaguirre 1987; Glazer et al. 1975; Wong 1993).

CROSS-LINKING REAGENTS FOR PROTEIN CONJUGATION

Cross-linking is essentially a chemical modification process in which specific functional groups within a molecule or between two different molecules are linked together using bifunctional cross-linkers with or without a spacer arm which contain two group-specific reagents. Chemical cross-linkers are widely used in the determination of near-neighbor relationships in proteins, their three-dimensional structures, enzyme-substrate orientation, solid phase immobilization, hapten-carrier protein conjugation, and molecular associations in cell membranes (Pierce Chemical Co. 1994). They are particularly useful in the diagnostic industry for the preparation of antibody-enzyme conjugates, immunotoxins, and other labeled protein reagents. Most chemical modifications involve nucleophilic reactions with the amino acid side chains of proteins; thus, the cross-linker must contain characteristics for the nucleophilic attack.

Bifunctional cross-linking reagents can be classified on the basis of the following (Pierce Chemical Co. 1994):

1. Functional groups and chemical specificity,
2. Length of cross-bridge,
3. Whether the cross-linking groups are similar (homobifunctional) or different (heterobifunctional),
4. Whether the groups react chemically or photochemically,
5. Whether the reagent is cleavable, and
6. Whether the reagent can be radiolabeled or tagged with another label.

In immunoassay product development, the choice and design of a bifunctional cross-linker are determined by its specific application as well as the product specifications. Recently, Wong (1993) has suggested the following criteria as a reference guide for the selection of bifunctional cross-linkers for the modification of functional groups in proteins. These criteria essentially follow the principles used in the classification of bifunctional cross-linkers mentioned above.

1. Reaction specificity towards a particular group, e.g., amino, sulfhydryl, carboxyl, guanidinyl, imidazolyl, and other amino acid side chains. If there is available a special functional group on the protein to which another molecule will be linked, the cross-linking reagent must be specific to that group.
2. Hydrophobicity and hydrophilicity of the reagent. A protein in a hydrophobic environment may require a hydrophobic reagent. For example, membrane permeability of a reagent may be necessary for labeling intramembranous proteins.
3. Cleavability of the reagent. It may be desirable in some cases to separate cross-linked proteins for identification of the components. In such instances, the use of cleavable reagents will enable the cross-linking to be reversed.
4. Size and geometry of the reagent. The length of bridge between the reactive groups of cross-linkers may be used to measure the two cross-linked groups for intramolecular topology studies. The cross-linkers can also serve as spacers between two conjugated proteins.
5. Photosensitivity of the reagent. A photoactivatable reagent is essential for photoaffinity labeling studies. This is particularly important when one of the species is unknown, for example, the protein receptor of a hormone. Photoaffinity labels have been used to detect and identify membrane acceptors.
6. Presence of tracer (or reporter) group. Sometimes, it is necessary to follow a cross-linking reaction or detect molecular conformation after conjugation. In such cases,

a tracer or receptor group attached to the cross-linking reagent will facilitate the process. Some examples of such tracers include radiolabels, spin labels, and fluorescent probes.

Bifunctional cross-linkers commonly used for protein modification are described below.

Homobifunctional Cross-Linkers

These compounds contain at least two identical reactive groups that react with the same amino side chain. These reagents induce cross-linking both intramolecularly between two groups within a protein and intermolecularly between two molecules. Although it increases the stability of proteins against thermal and mechanical denaturation, intramolecular cross-linking is undesirable in formulating immunoassay reagents.

Based on their selectivity toward a particular amino acid side chain, homobifunctional cross-linking reagents can be further classified as follows. It should, however, be noted that there is no absolutely group-specific cross-linking reagent and that their specificity and selectivity are often determined by the reactants themselves and the conditions of the reaction.

1. Amino Group Modification.   Homobifunctional cross-linking reagents designed to react with amino groups of protein belong to one of the following classes.

BIS-IMIDOESTERS (BISIMIDATES).   These cross-linkers readily react with amino groups eliminating an alcohol to form amidines (Peters and Richards 1977). Primary amines attack imidates nucleophilically to produce an intermediate that breaks down to amidine at high pH or to a new imidate at low pH. The new imidate can then react with another amino group.

Imidoesters are generally readily soluble in aqueous solutions. They react with amino groups with a high degree of specificity under mild conditions (pH 7–10) to form amidine derivatives (Hunter and Ludwig 1972; Wong 1993). However, they are hydrolyzed rapidly with a pH-dependent half-life ranging from several minutes to half an hour. Their rate of hydrolysis is substantially higher at higher pH values (Wells et al. 1980). To circumvent this problem, incremental additions of the reagents are recommended (Ji 1979).

Commonly used bisimidate cross-linkers are summarized in Table 4.2. Because of the ease of synthesis, several cross-linkers with different spacer arms are available (Peters and Richards 1977; Davies and Stark 1970; Hunter and Ludwig 1972; Wong 1993). Of the compounds listed in Table 4.2, 3 through 12 contain chain lengths ranging from 5 to 14 Å, 13 through 17 provide hydrophilicity to the bridge, 18 through 24 are cleavable compounds containing disulfide bonds, 25

TABLE 4.2. Homobifunctional Cross-Linkers for Amino Group Modification[a]

Number	Cross-Linker (Abbreviation)
	BIS-IMIDOESTERS (BISIMIDATES)
1.	Methyl acetimidate · HCl (MA)
2.	Ethyl acetimidate · HCl (EA)
3.	Diethyl malonimidate · 2 HCl (DEM)
4.	Dimethyl malonimidate · 2 HCl (DMM)
5.	Dimethyl succinimidate · 2 HCl (DMSC)
6.	Dimethyl glutarimidate · 2 HCl (DMG)
7.	Dimethyl adipimidate · 2 HCl (DMA)
8.	Dimethyl pimelimidate · 2 HCl (DMP)
9.	Dimethyl suberimidate · 2 HCl (DMS)
10.	Dimethyl azelaimidate · 2 HCl
11.	Dimethyl sebacimidate · 2 HCl
12.	Dimethyl dodecimidate · 2 HCl
13.	Dimethyl 3,3'-oxydipropionimidate · 2 HCl (DODP)
14.	Dimethyl 3,3'-(methylenedioxy)-dipropionimidate · 2 HCl (DMDP)
15.	Dimethyl 3,3'-(dimethylenedioxy)-dipropionimidate · 2 HCl (DDDP)
16.	Dimethyl 3,3'-(tetramethylenedioxy)-dipropionimidate · 2 HCl (DTDP)
17.	Dimethyl 3,3'-(diethyletherdioxy)-dipropionimidate · 2 HCl
18.	Diisothionyl 3,3'-dithiobispropionimidate · 2 HCl
19.	Dimethyl 3,3'-dithiobispropionimidate · 2 HCl (DTBP)
20.	Dimethyl 4,4'-dithiobisbutyrimidate · 2 HCl (DTBB)
21.	Dimethyl 5,5'-dithiobisvalerimidate · 2 HCl (DTBV)
22.	Dimethyl 7,7'-dithiobisenenthimidate · 2 HCl (DTBE)
23.	Dimethyl 3,3'-(dithiodimethylenediosy)-bisimidate · 2 HCl
24.	Dimethyl 3,3'-(dithiodimethylenediamido)-bispropionimidate · 2 HCl
25.	*N,N*'-bis (2-carboximidomethyl)-tartarimidate dimethyl ester · 2 HCl (CMTD); Tartryldi (methyl-2-aminoacetimidate) · 2 HCl (TDAA)
26.	*N,N*'-bis (2-carboximidoethyl)-tartarimidate dimethyl ester · 2 HCl (CETD); Tartryldi (methyl-3-aminopropionimidate) · 2 HCl
27.	3,4,5,6-Tetrahydroxy suberimidate · 2 HCl (THS)
28.	Dimethyl 3,3'-(*N*-2,4-dinitrophenyl)-bispropionimidate · 2 HCl
29.	Dimethyl 3,3'-(*N*-2,4-dinitro-5-carboxyphenyl)-bispropionimidate · 2 HCl
30.	Dimethyl 3,3'-[*N*-(5-(*N,N*'-dimethylamino)naphthyl)-sulfonyl] bispropionimidate · 2 HCl
	BIS-N-SUCCINIMIDYL DERIVATIVES
31.	Disuccinimidylsuberate (DSS); *N*-hydroxysuccinimidylsuberate (NHS-SA)
32.	Bis (sulfosuccinimidyl)-suberate (BSSS)
33.	Succinate bis-(*N*-hydroxy-succinimide ester), spin-labeled bis-(*N*-hydroxy-succinimide ester)
34.	3-Oxy-2,2-bis-[6-((*N*-succinimidyloxy)carbonyl)hexanyl]-4,4-dimethyloxazolidine ($n = 6$)

TABLE 4.2.   (*continued*)

Number	Cross-Linker (Abbreviation)
35.	3-Oxy-2,2-bis-[8-((*N*-succinimidyloxy)carbonyl)octanyl]-4,4-dimethyloxazolidine (spin-labeled derivatives available) ($n = 8$)
36.	3-Oxy-2,2-bis-[6-(((3-((*N*-succinimidyloxy)carbonyl)propyl)amino)-carbonyl)hexanyl]-4,4-dimethyloxazolidine ($n = 3$)
37.	3-Oxy-2,2-bis-[6-(((5-((*N*-succinimidyloxy)carbonyl)pentanyl)amino)-carbonyl)hexanyl]-4,4-dimethyloxazolidine ($n = 5$)
38.	Bis (sulfo-*N*-succinimidyl)-doxyl-2-spiro-5'azelate (BSSDA)
39.	Bis (sulfo-*N*-succinimidyl)-doxyl-2-spiro-4'pimelate (BSSDA)
40.	Ethyleneglycol bis-(succinimidylsuccinate) (EGS)
41.	Bis-[2-(succinimidooxycarbonyl)ethyl]-sulfone (BSES)
42.	Bis-[2-(succinimidooxycarbonyloxy)ethyl]-sulfone (BSOCOES)
43.	3,3'-Dithiobis-(succinimidylpropionate) (DTSP, DSP) (Lomant's reagent)
44.	2,2'-Dithiobis-(succinimidylpropionate) (2,2'-DSP)
45.	3,3'-Dithiobis-(sulfosuccinimidylpropionate) (DTSSP)
46.	Disuccinimidyl-(*N*,*N*'-diacetylhomocystine)
47.	Disuccinimidyl tartarate (DST)
48.	*N*,*N*'-Bis (3-succinimidyloxycarbonylpropyl)tartaramide (BSOPT)

ARYL HALIDES

49.	1,5-Difluoro-2,4-dinitrobenzene (FFDNB, DFBN, DFDNB)
50.	1,5-Dichloro-2,4-dinitrobenzene
51.	1,5-Dibromo-2,4-dinitrobenzene
52.	Bis (3,5-dibromosalicyl) fumarate
53.	Bis (3,5-dibromosalicyl) succinate
54.	4,4'-Difluoro-3,3'-dinitrodiphenylsulfone; bis (3-nitro-4-fluorophenyl) sulfone

ACYLATING AGENTS
A. Diisocyanates and Diisothiocyanates

55.	1,6-Hexamethylene diisocyanate (HMDI)
56.	1,3-Dicyanatobenzene
57.	1,4-Dicyanatobenzene
58.	Toluene-2,4-diisocyanate
59.	Toluene-2-isocyanate-4-isothiocyanate; 2-cyanato-4-isothiocyanatotoluene
60.	Xylene diisocyanate
61.	Benzidine diisocyanate (BDI)
62.	2,2'-Dimethoxybenzidine diisocyanate
63.	Diphenylmethane-4,4'-diisocyanate
64.	3-Methoxydiphenylmethane-4,4'-diisocyanate
65.	Dicyclohexylmethane-4,4'-diisocyanate
66.	Hexahydrobiphenyl-4,4'-diisocyanate
67.	2,2'-Dicarboxy-4,4'-azophenyldiisocyanate

TABLE 4.2. (*continued*)

Number	Cross-Linker (Abbreviation)
68.	2,2'-Dicarboxy-4,4'-azophenyldithioisocyanate
69.	2,2'-Dicarboxy-6,6'-azophenyldiisocyanate
70.	Bis-*p*-(2-carboxy-4'-azophenylisocyanate)
71.	Bis-*p*-(2-carboxy-4'-azophenylisothiocyanate)
72.	*p*-Phenylene diisothiocyanate; 1,4-phenylene diisothiocyanate
73.	Diphenyl-4,4'-diisothiocyanato-2,2'-disulfonic acid
74.	4,4'-diisothiocyanato-2,2'-disulfonic acid stilbene (DIDS)
75.	4,4'-diisothiocyanato dihydrostilbene-2,2'-disulfonic acid

B. Sulfonyl Halides

76.	Phenol-2,4-disulfonyl chloride
77.	α-Naphthol-2,4-disulfonyl chloride
78.	Naphthalene-1,5-disulfonyl chloride

C. Bis-Nitrophenol Esters

Bis-(*p*-nitrophenyl ester) of carboxylic acids

79.	Bis-(*p*-nitrophenyl) adipate (*n* = 4)
80.	Bis-(*p*-nitrophenyl) pimelate (*n* = 5)
81.	Bis-(*p*-nitrophenyl) suberate (*n* = 8)
82.	Carbonyl bis (L-methionine *p*-nitrophenyl ester) (CBMNPE)

D. Acylazides

83.	Tartryl diazide (TDA)
84.	Tartryl di(glycylazide) (TDGA) (*n* = 1)
85.	Tartryl di(β-alanylazide) (TDAA) (*n*= 2)
86.	Tartryl di(γ-aminobutyrylazide) (TDBA) (*n* = 3)
87.	Tartryl di(δ-aminovalerylazide) (TDVA) (*n* = 4)
88.	Tartryl di(ε-aminocaproylazide) (TDCA) (*n* = 5)
89.	*p*-Bis-(ureido)azidooligoprolylazobenzene (PAPA)
90.	Trimethyl-tris-β-alanylazide (TTA)

DIALDEHYDES

91.	Glyoxal
92.	Malondialdehyde (MDA)
93.	Succinialdehyde
94.	Adipaldehyde
95.	α-Hydroxyadipaldehyde
96.	Glutaraldehyde
97.	3-Methyl glutaraldehyde
98.	2-Methoxy-2,4-dimethyl glutaraldehyde
99.	*o*-Phthalaldehyde
	P^1,P^2-Bis (5'-pyridoxal)-polyphosphate derivatives
100.	P^1,P^2-Bis (5'-pyridoxal)-diphosphate (Bis-PLP)

TABLE 4.2.   (*continued*)

Number	Cross-Linker (Abbreviation)
101.	P^1,P^2-Bis (5'-pyridoxal)-triphosphate
102.	Bis-(5'-pyridoxalpyrophospho)methane
103.	Bis-(5'-pyridoxalpyrophospho)-1,6-fructose
104.	Bis-(5'-pyridoxalpyrophospho)-2,3-glycerate
105.	Formaldehyde
	DIKETONES
106.	2,5-Hexanedione
107.	3,4-Dimethyl-2,5-hexanedione
	MISCELLANEOUS
108.	*p*-Benzoquinone
109.	2-Iminothiolane
110.	Erythreitolbiscarbonate (EBC)
111.	Mucobromic acid
112.	Mucochloric acid
113.	Ethylchloroformate
114.	*p*-Nitrophenylchloroformate

[a]Compiled from Peters and Richards (1977), Ji (1979), Han et al. (1984), Means and Feeney (1990), Brinkley (1992), Wong (1993), and Pierce Chemical Co. (1994).

through 27 contain vicinal diols cleavable by periodate, and 28 through 30 are bridge-labeled reagents, with 28 and 29 providing the colored dinitrophenyl group and 30 being a fluorescent probe.

BIS-SUCCINIMIDYL DERIVATIVES.   These cross-linkers are synthesized by condensing *N*-hydroxysuccinimide (NHS) with the corresponding dicarboxylic acids in the presence of dicyclohexylcarbodiimide (DCC) (Anderson et al. 1964). They react preferentially with amino groups eliminating NHS as the leaving group (Cuatrecasas and Parikh 1972). The reaction is relatively rapid (10 to 20 minutes at pH 6–9), although rapid hydrolysis at alkaline pH effectively competes against the reaction (Vanin and Ji 1981; Lomant and Fairbanks 1976).

Commonly used bis-succinimidyl derivatives (compounds 31 through 48) are summarized in Table 4.2. The parent compound 31 is only sparingly soluble in water. Solubility can be increased by introduction of the sulfonate group on the succinimidyl ring (compounds 32, 38, 39, and 45), and by incorporation of hydrogen bonding atoms such as esters of oxygen (compound 40) or hydroxyl groups (compounds 47 and 48). Compounds 34 and 39 contain spin labels; while

40 is cleavable by hydroxylamine, 41 and 42 by base, 43 through 46 by free mercaptans, and 47 and 48 are cleavable by periodate.

ARYL HALIDES.   Bifunctional aryl halides (compounds 49–54, Table 4.2) react preferentially with amino and tyrosine phenolic groups. They are also reactive towards thiol and imidazolyl groups. The electron withdrawing carboxyl and nitro groups on the benzene ring activate the halogen for nucleophilic substitutions (Wong 1993). Rapid reaction rates are possible at high pH values, with fluoro-derivatives being most reactive, followed by chloro- and then bromo-derivatives. Aryl halides, however, are insoluble in water, and hence, first need to be solubilized in organic solvents such as acetone, dioxane, alcohol, or other water-miscible organic solvents prior to their addition to protein solutions.

Among the aryl halides listed in Table 4.2, only the salicyl (compounds 52 and 53) and the sulfone derivatives (54) are cleavable by base or by Ni-catalytic reduction (Wold 1972; Chatterjee et al. 1986; Wong 1993).

ACYLATING AGENTS.   Homobifunctional acylating agents directed against amino side chains of proteins can be broadly classified as follows:

1. Diisocyanates and diisothiocyanates (compounds 55–75, Table 4.2),
2. Sulfonyl halides (compounds 76–78),
3. Bis-nitrophenyl esters (compounds 79–82), and
4. Acylazides (compounds 83–90)

The compounds belonging to the first category are aromatic derivatives that form stable urea and thiourea derivatives, respectively, upon reaction with primary amines. Their reactions with sulfhydral, imidazolyl, and phenolic groups give relatively unstable bonds that undergo spontaneous hydrolysis (Wong 1993). The isocyanates are not stable in aqueous solutions, whereas the aromatic isocyanates are somewhat more reactive than the aliphatic ones. Among the compounds listed in Table 4.2 under this category, 67 and 71 contain azo bond cleavable in the presence of dithionite, the stilbene derivative (compound 74) is fluorescent, while its reduced analog, dihydrostilbene (75), is non-fluorescent.

The sulfonyl halides (compounds 76–78), upon reaction with amino groups yield sulfonamide derivatives with chloride as the leaving group. These reagents are quite insoluble in water, and hydrolyze rapidly in aqueous solutions.

Bis-nitrophenyl esters are produced upon activation of compounds 79–82, which react with amino groups very rapidly. Their specificity, however, is not very high (Busse and Carpenter 1974). The reaction involves nucleophilic attack at the ester carbonyl carbon displacing nitrophenol (Wong 1993). These compounds are rapidly hydrolyzed in aqueous solutions. Compound 82 is a derivative of methionine that is cleavable by cyanogen bromide.

The fourth category of acylating agents includes acylazides (compounds 83–90, Table 4.2) that readily react with amino groups to produce amide bonds. The tartryl diazides of different lengths (compounds 83–88) contain vicinal hydroxyl groups and thus are cleavable by mild periodate reaction (Lutter et al. 1974).

DIALDEHYDES.   Simple dialdehydes (compounds 91–105, Table 4.2), particularly glutaraldehyde, are extensively used in protein cross-linking reactions. However, their reactions are known to produce undesirable polymers. o-Phthalaldehyde (compound 99) is a fluorogenic aromatic dialdehyde, while compounds 100–104 contain pyridoxal phosphate, the pyrophosphate bond of which may be hydrolyzed by acid or base. Formaldehyde (compound 105) is not a dialdehyde. It is, however, capable of reacting bifunctionally (Ji 1983). In concentrated aqueous solution, it exists as a series of low molecular weight polymers (formalin) that revert to monomeric form in dilute solutions. Cross-linking reactions using formaldehyde involve the attack of an amino group to form a quaternary ammonium salt, which loses a molecule of water to produce an immonium cation. This strongly electrophilic cation then reacts with a number of nucleophiles in the protein via a methylene-bridged cross-link (Ji 1983).

DIKETONES.   Ketones (compounds 106 and 107, Table 4.2) are capable of forming Schiff bases with amines. They react with the amino group of lysine and cyclize to form a pyrrole, which subsequently leads to cross-linking of proteins (Sager 1989; Graham et al. 1984).

MISCELLANEOUS REAGENTS.   Cross-linkers belonging to this group have been shown to react with primary amines. p-Benzoquinone (compound 108) has been used in protein-protein and enzyme-antibody coupling (Ternynck and Avrameas 1977; Avrameas et al. 1978). 2-Iminothiolane (Traut's reagent, compound 109) is a thiolating agent that introduces free thiols (Traut et al. 1973), whereas erythreitolbiscarbonate (compound 110) reacts with amino groups to yield a bis-carbonate derivative (Coggins et al. 1976). After decarboxylation to form vicinal diols, the cross-linked product may be cleaved by periodate.

Compounds 111 and 112 contain a cyclic lactone and react with the amino group via Schiff base formation (Robinson 1964). Ethylchloroformate (113) and p-nitrophenylchloroformate (114) react similarly to cross-link proteins (Avrameas and Ternynck 1967).

2. Sulfhydryl Group Modification.   Homobifunctional cross-linkers directed against sulfhydryls can be classified as follows:

1. Mercurial cross-linkers
2. Disulfide forming cross-linkers

3. Bismaleimides, and
4. Alkylating agents

Of these, only the mercurial derivatives and disulfide and disulfide forming compounds are thiol specific. The N-maleimido and alkylating reagents can also react with other neutrophiles in proteins.

MERCURIAL CROSS-LINKERS.   Mercuric ion (compound 1, Table 4.3) reacts reversibly with sulfhydryl groups (Arnon and Shapira 1969). The first thiol group reacts very quickly followed by a slower reaction with the second. Cross-linking is favored when the ratio of Hg^{2+}/-SH is 0.5 or less (Wong 1993). The 3,6-bis-(mercurymethyl) dioxane derivatives (compounds 2–4) react similarly to but faster than the mercuric ion. Compound 5 also cross-links two fast-reacting thiol groups (Vas and Csanady 1987).

DISULFIDE FORMING CROSS-LINKERS.   Compound 6 (Table 4.3) is a spin label that uses disulfide-thiol interchange as a cross-linking reaction. The reactions of polymethylene bis-methane thiosulfonate reagents (compounds 7–11) with thiols involve disulfide formation and elimination of methanesulfinate, which is subsequently oxidized to methane sulfonic acid (Bloxham and Sharma 1979; Bloxham and Cooper 1982). Such cross-linked peptides, although cleavable by thiols, are stable to cyanogen bromide cleavage and tryptic digestion. Carbescein (compound 12) is a fluorescent derivative of fluorescein containing two free sulfhydryls that have been shown to add across disulfide bonds of reduced antibody (Mahoney and Azzi 1987).

BISMALEIMIDES.   N-substituted bismaleimides (compounds 13-31, Table 4.3) are perhaps the most commonly used sulfhydryl reagents (Wong 1993; Means and Feeney 1971; Hashida et al. 1984). They are synthesized from their corresponding amine with maleic anhydride and either acetic anhydride (Cava et al. 1961) or dicyclohexyl carbodiimide (Trommer and Hendrick 1973). Compounds listed in Table 4.3 under this category react rapidly with thiol groups at pH 7 to 8 to form sulfides through Michael addition (Wong 1993), and much more slowly with amino and imidazolyl groups. These reagents are generally insoluble in water and need to be added to the reaction medium after dissolving in a water-miscible organic solvent.

Among those listed in Table 4.3, the azo bond of compound 24 can be cleaved by dithionite and the ester bond of compound 27 can be hydrolyzed by base (Sato and Nakao 1981). The acetal, ketal and ortho ester bonds of compounds 28–31 are susceptible to hydrolysis at acidic pHs (Srinivasachar and Neville 1989). Naphthalene dimaleimide (23) and the stilbene compounds (25 and 26) are fluorogenic.

TABLE 4.3.   Homobifunctional Cross-Linkers for Sulfhydryl Group Modification[a]

Number	Cross-Linker (Abbreviation)

MERCURIAL REAGENTS

1.   Mercuric ion
     3,6-Bis-(mercurymethyl) dioxane derivatives
2.   3,6-Bis-(acetoxymercurymethyl) dioxane
3.   3,6-Bis-(chloromercurymethyl) dioxane
4.   3,6-Bis-(nitromercurymethyl) dioxane
5.   1,4-Bis-(bromomercuri) butane

DISULFIDE FORMING REAGENTS

6.   3-Oxy-2,2-bis-[{((2-((3-carboxy-4-nitrophenyl)dithio)ethyl)amino)
     carbonyl}hexanyl]-4,4-dimethyloxazolidine
     Polymethylenebis-(methanethiosulfonate) derivatives
7.   5,5'-Pentamethylenebis-(methanethiosulfonate) ($n = 5$)
8.   6,6'-Hexamethylenebis-(methanethiosulfonate) ($n = 6$)
9.   8,8'-Octamethylenebis-(methanethiosulfonate) ($n = 8$)
10. 10,10'-Decamethylenebis-(methanethiosulfonate) ($n = 10$)
11. 12,12'-Dodecamethylenebis-(methanethiosulfonate) ($n = 12$)
12. Carbescein

BISMALEIMIDES

13. *N,N'*-Methylenebismaleimide
14. *N,N'*-Trimethylenebismaleimide
15. *N,N'*-Hexamethylenebismaleimide; bis-(*N*-maleimido)-1,6-hexane (BMH)
16. *N,N'*-Octamethylenebismaleimide; bis-(*N*-maleimido)-1,8-octane (BMO)
17. *N,N'*-Dodecamethylenebismaleimide; bis-(*N*-maleimido)-1,12-dodecane
     (BMD)
18. Bis-(*N*-maleimidomethyl)ether
19. *N,N'*-(1,3-Phenylene)-bismaleimide
20. *N,N'*-(1,2-Phenylene)-bismaleimide
21. *N,N'*-(1,4-Phenylene)-bismaleimide
22. Bis-(*N*-maleimido)-4,4'-bibenzyl (BMB)
23. Naphthalene-1,5-dimaleimide (NDM)
24. Azophenyldimaleimide
25. 4,4'-Dimaleimidostilbene
26. 4,4'-Dimaleimidylstilbene-2,2'-disulfonic acid (DMSDS)
27. Maleimidomethyl-3-maleimidopropionate (MMP)
28. 2,2-Bis-(maleimidoethoxy)-propane
29. 2,2-Bis-(maleimidomethoxy)-propane
30. 1,1'-[{3,9-Diethyl-2,4,8,10-tetraoxaspiro[5.5]undecane-3,9-diyl}bis
     (oxymethylene)]-bis-1*H*-pyrrole-2,5-dione
31. 1,1'-[{3,9-Diethyl-2,4,8,10-tetraoxaspiro[5.5]undecane-3,9-diyl}bis (oxy-2,1-
     ethane-diyl)]-bis-1*H*-pyrrole-2,5-dione

TABLE 4.3. (*continued*)

Number	Cross-Linker (Abbreviation)

ALKYLATING AGENTS

A. Bishaloacetyl Derivatives

32. 1,3-Dibromoacetone
    N,N'-Bis(iodoacetyl)-polymethylenediamine derivatives
33. *N,N'*—Bis(iodoacetyl)-ethylenediamine; *N,N'*-ethylene-bis(iodoacetamide) (*n* = 2)
34. *N,N'*-Bis(iodoacetyl)-hexamethylenediamine; *N,N'*-hexamethylene-bis(iodoacetamide) (*n* = 6)
35. *N,N'*-Bis(iodoacetyl)-undecamethylenediamine; *N,N'*-undecamethylene-bis(iodoacetamide) (*n* = 11)
36. *N,N'*-Di(bromoacetyl)-phenylhydrazine
37. 1,2-Di(bromoacetyl)-amino-3-phenylhydrazine
38. γ-(2,4-dinitrophenyl)-—bromoacetyl-L-diaminobutyric acid bromoacetylhydrazide (DIBAB)
39. Bis-{α-bromoacetyl-—(2,4-dinitrophenyl)-lysylpropyl}-ethylenediamine
40. 2,2'-Dicarboxy-4,4'-diiodoacetamidoazobenzene
41. 2,2'-Dicarboxy-4,4'-dibromoacetamidoazobenzene
42. *N,N'*-Bis(α-iodoacetyl)-2,2'-dithiobisethylamine (DIDBE)
43. 4,5'-Di[{(iodoacetyl)-amino}-methyl]-fluorescein
44. *p*-Bis-(ureido)-(1-iodoacetamido-2-ethylamino)-oligoprolylazobenzene

B. Dialkyl Halides

45. α-α—Dibromo-*p*-xylene sulfonic acid
46. α-α—Diiodo-*p*-xylene sulfonic acid
47. Di(2-chloroethyl)sulfide
48. Di(2-chloroethyl)sulfone
49. Di(2-chloroethyl)methylamine
50. Tri(2-chloroethyl)amine (TCEA)
51. *N,N'*-Bis-(β-bromoethyl)benzylamine
52. Di(2-chloroethyl)-*p*-methoxyphenylamine
    Nitrosourea derivatives
53. 1,3-Bis-(2-chloroethyl)-1-nitrosourea
54. 1,3-Bis-(*trans*-4-hydroxyhexyl)-1-nitrosourea

C. *s*-Triazines

55. 2,4-Dichloro-6-methoxy-*s*-triazine
56. 2,4,6-Trichloro-*s*-triazine; cyanuric chloride
57. 2,4-Dichloro-6-(3'-methyl-4-aminoanilino)-*s*-triazine
58. 2,4-Dichloro-6-amino-*s*-triazine
59. 2,4-Dichloro-6-(sulfonic acid)-*s*-triazin-2-yl-6-sulfonic acid
60. 2,4-Dichloro-6-(5'-sulfonic acid-naphthaleneamino)-*s*-triazine; 5-{(4,6-dichloro-*s*-triazin-2-yl)amino}-naphthalene-1-sulfonic acid

TABLE 4.3.   (*continued*)

Number	Cross-Linker (Abbreviation)
61.	5-{(4,6-Dichloro-s-triazin-2-yl)amino}-fluorescein (5-DTAF)
62.	5-{(4,6-Dichloro-s-triazin-2-yl)-amino}-fluorescein diacetate
	D. Aziridines
63.	2,4,6-Tri(ethyleneimino)-*s*-triazine
64.	*N*,*N*'-Ethyleneiminoyl-1,6-diaminohexane
65.	Tri-{1-(2-methylaziridenyl)}-phosphine oxide
	E. Bis-Epoxides
66.	1,2:3,4-Diepoxybutane
67.	1,2:5,6-Diepoxyhexane
68.	Bis-(2,3-epoxypropyl)-ether
69.	1,2-Butadioldiglycidoxyether
70.	3,4-Isopropylidene-1,2:5,6-dianhydromannitol
	MISCELLANEOUS
71.	Divinyl sulfone

[a]Compiled from Peters and Richards (1977), Ji (1979), Han et al. (1984), Means and Feeney (1990), Brinkley (1992), Wong (1993), and Pierce Chemical Co. (1994).

ALKYLATING CROSS-LINKERS.   Similar to bismaleimides, alkylating agents are not specific to thiol groups. They may also target modification of amino groups, particularly at high pH. Alkylating agents used for sulfhydryl modifications can be categorized as:

1. Bis-haloacetyl derivatives,
2. Bis-alkyl halides,
3. *s*-Triazines,
4. Aziridines,
5. Bis-epoxides (bisoxiranes), and
6. Miscellaneous reagents.

The bis-haloacetyl derivatives (compounds 32–44, Table 4.3) react primarily with sulfhydryl, imidazolyl, and amino groups. Reactions involving compounds 40 and 41 are reversible, compound 42 is cleavable through thiol-disulfide exchange with mercaptans, while compound 43 is a fluorescent derivative (Wold 1972; Ludena et al. 1982; Fasold et al. 1964).

Among the bis-alkyl halides, benzyl halides (compounds 45 and 46) are activated by the benzene ring through resonance. They react similarly as haloacetyl ketones (Hiremath and Day 1964). Among those belonging to this class, only the sulfone derivative (compound 48) is cleavable. TCEA (50) is a trifunctional cross-linking reagent, whereas the two nitrosourea derivatives (compounds 53 and 54), although used as tumor therapeutic agents, have not yet been tried as in vitro bifunctional cross-linkers (Ali-Osman et al. 1989; Wong 1993).

Cross-linkers belonging to the *s*-triazine class (compounds 55–62, Table 4.3) are very reactive towards nucleophiles, including hydroxyl groups of carbohydrates (Han et al. 1984). Because of the high reactivity of their chlorine atom, they are rapidly hydrolyzed in aqueous solutions, resulting in poor cross-linking. Compounds 61 and 62 are fluorogenic derivatives of fluorescein.

Aziridines (compounds 63–65) contain a strained three-membered heterocyclic nitrogen ring, and are highly reactive towards nucleophiles by ring opening (Wong 1993). Compounds 63 and 65 are trifunctional reagents.

Similar to aziridines, bis-epoxides (compounds 66–70) also undergo ring opening reactions with nucleophiles, including hydroxyl groups (Uckert et al. 1984; Skold 1983). Divinyl sulfone (compound 71) is an alkylating agent with nucleophilic addition to the double bond.

3. Carboxyl Group Modification.   As compared to amino and sulfhydryl groups, relatively few homobifunctional cross-linkers are available commercially for the modification of carboxyl side chains of proteins (compounds 1 and 2, Table 4.4). The zero-length cross-linkers belonging to the carbodiimide class can activate the carboxyl groups of proteins in the presence of primary amines, hydrazides, or cleavable cystamines to form *O*-acylisoureas, which are cross-linked by diamines. In these reactions, no cross-bridge ("zero-length") is formed, and the amide bond created is the same as a peptide bond. The reversal of such cross-linking using carbodiimides, therefore, is impossible without the destruction of the protein.

At neutral to slightly acidic pH, the carboxylate ion can effectively compete with other nucleophiles (e.g., amino, thiol, and phenolate ions) in proteins, as the latter are generally present in less reactive protonated forms. Water-soluble carbodiimides, which are also capable of coupling heterobifunctional reagents, are widely used in reagent formulations in the diagnostic industry.

4. Phenolate and Imidazolyl Group Modification.   The phenolate and imidazolyl side chains of tyrosyl and histidyl residues in protein readily react with diazonium salts by electrophilic substitution reactions (Gold et al. 1966). Proteins that are deficient in aromatic amino acids do not react with these reagents. bis-Diazonium reagents (compounds 3–14, Table 4.4) can be easily synthesized by treatment of aryl diamines with sodium nitrite in acidic conditions. Compounds 3–12 are routinely used to couple a diverse array of antigens or haptens to carrier proteins, while 11 and 12 have been used to prepare antigens and insoluble enzymes (Wong 1993; Koch and Haustein 1983; Anderer and Schlumberger 1969, Tijssen 1985; Kabakoff 1980). The disulfide-containing diazonium compounds 9 and 10 are cleavable by mercaptans.

TABLE 4.4.   Carboxyl, Phenolate, Imidazolyl, and Guanidinyl Specific and
Photoactivatable Homobifunctional Cross-Linkers in Protein Modification[a]

Number	Cross-Linker (Abbreviation)
	CARBOXYL GROUP-SPECIFIC CROSS-LINKERS
1.	1,1-Bis-(diazoacetyl)-2-phenylethane
2.	Bisdiazohexane
	PHENOLATE AND IMIDAZOLYL SPECIFIC CROSS-LINKERS
3.	*p*-Phenylenediamine
4.	Bisbenzidine
5.	3,3'-Dimethoxybenzidine; *o*-Dianisidine
6.	Benzidine-2,2'-disulfonic acid
7.	Benzidine-3,3'-disulfonic acid
8.	4,4'-Diaminodiphenylamine
9.	4,4'-Diaminodiphenyldisulfide
10.	2,2'-Dinitro-4,4'-diaminodiphenyldisulfide
11.	Poly-(4,4'-diaminodiphenylamine-3,3'-dicarboxylic acid)
12.	Poly-(*p*-amino-D,L-phenylalanyl-L-leucine)
13.	Potassium nitrosyl disulfonate
14.	Tetranitromethane
	GUANIDINYL GROUP SPECIFIC CROSS-LINKERS
15.	*p*-Phenylene diglyoxal
	MISCELLANEOUS REAGENTS
16.	*cis*-Dichlorodiamino platinum (II) (*cis*-DDP)
17.	Adipic acid dihydrazide
18.	*N*,*N*'-Bis-(β-aminoethyl)-tartramide
	PHOTOACTIVATABLE CROSS-LINKERS
19.	4,4'-Diazidobiphenyl (DABP)
20.	1,5-Diazidonaphthalene (DAN)
21.	*N*,*N*'-Bis-(*p*-azido-*o*-nitrophenyl)-1,3-diamino-2-propanol
22.	4,4'-Dithiobisphenylazide

[a]Compiled from Peters and Richards (1977), Ji (1979), Han et al. (1984), Means and Feeney (1990),
Brinkley (1992), Wong (1993), and Pierce Chemical Co. (1994).

5. Guanidinyl Group Modification.   Compounds such as *p*-phenylene digly-
oxal (compound 15, Table 4.4), which contain vicinal diones, are capable of reacting with the guanidinyl group of arginyl side chains in proteins. Such reagents
are also used for cross-linking guanosine bases in nucleic acids (Wagner and Garrett 1978; Hancock and Wagner 1982).

Among the miscellaneous homobifunctional cross-linkers, *cis*-dichlorodiamine
platinum (II) (compound 16, Table 4.4) is used to cross-link methionine in pro-

tein (Roche et al. 1988; Pizzo et al. 1986; Gonias et al. 1984). The cross-linking can be reversed by diethyldithiocarbamate, which is a potent platinum chelator. Adipic acid dihydrazide (compound 17) has been used to cross-link glycoproteins, while compound 18 was used to cross-link guanidinated β-casein, which has the lysine protected to prevent ε-(γ-glutamyl)lysine cross-links (Gorman and Folk 1980).

Photoactivatable Homobifunctional Cross-Linkers

The only photoactivatable homobifunctional cross-linking reagents are derived from aryl azides (compounds 19–22, Table 4.4); alkyl azides are not favored, while acyl, sulfonyl, and phosphoryl azides are generally not used as photoactivatable reagents because of their nucleophilic reactivity in the dark (Lutter et al. 1974; Wong 1993). The aryl azides can be photolyzed at wavelengths of 300–400 nm such that the biological components (e.g., proteins and nucleic acids) are not damaged by photoirradiation.

Heterobifunctional Cross-Linkers

Heterobifunctional cross-linkers contain two or more reactive groups of different specificity. They allow for sequential conjugations with specific groups of proteins, thereby minimizing undesirable polymerization and self-conjugation reactions. The two reactive functionalities in these reagents can be any combination of the conventional group-selective moieties (Wong 1993). For example, one end of the cross-linker may be selective for an amino group while the other end is directed to a sulfhydryl group.

A general rule for the use of heterobifunctional cross-linkers in protein and/or hapten conjugation is that the most labile group of the reagent should be reacted first to ensure effective cross-linking and to avoid undesirable polymerization (Pierce Chemical Co. 1994). Similarly, reagents in which the reactivity can be controlled and that contain one group that is spontaneously nonreactive have distinct advantages. This allows for specific attachment of the labile group first; the second reaction can then be initiated when appropriate.

Based on their group specificity, the commonly used heterobifunctional cross-linkers are summarized in Table 4.5. Similar to homobifunctional reagents, these compounds do not exhibit absolute group specificity. Thus, cross-reactive reactions between different nucleophiles are possible. However, cross-linking of the desired reactants can be controlled both selectively and specifically by a proper selection of the cross-linker as well as the reaction conditions.

1. Amino and Sulfhydryl Group Directed Reagents. Heterobifunctional cross-linkers belonging to this class contain two specific reactive moieties directed toward amino and sulfhydral groups. Most of the amino group directed rel-

TABLE 4.5.   Group-Specific Heterobifunctional Cross-Linkers[a]

Number	Cross-Linker (Abbreviation)
	AMINO AND SULFHYDRYL GROUP DIRECTED
1.	N-Succinimidyl 3-(2-pyridyldithio) propionate (SPDP)
2.	N-Succinimidyl maleimidoacetate (AMAS)
3.	N-Succinimidyl 3-maleimidopropionate (BMPS)
4.	N-Succinimidyl 4-maleimidobutyrate
5.	N-Succinimidyl 6-maleimidocaproate; N-succinimidyl 6-maleimidylhexanoate (SMH)
6.	N-Succinimidyl 4-(N-maleimidomethyl)-cyclohexane-1-carboxylate (SMCC)
7.	N-Sulfosuccinimidyl 4-(N-maleimidomethyl)-cyclohexane-1-carboxylate (Sulfo-SMCC)
8.	N-Succinimidyl 4-(p-maleimidophenyl)-butyrate (SMPB)
9.	N-Sulfosuccinimidyl 4-(p-maleimidophenyl)-butyrate (Sulfo-SMPB)
10.	N-Succinimidyl o-maleimidobenzoate
11.	N-Succinimidyl m-maleimidobenzoate (SMB); m-maleimidobenzoyl-N-Hydroxysuccinimide ester (MBS)
12.	N-Sulfosuccinimidyl m-maleimidobenzoate (Sulfo-SMB); m-Maleimidobenzoyl-N-hydroxysuccinimide ester (Sulfo-MBS)
13.	N-Succinimidyl p-maleimidobenzoate
14.	N-Succinimidyl 4-maleimido-3-methoxybenzoate
15.	N-Succinimidyl 5-maleimido-2-methoxybenzoate
16.	N-Succinimidyl 3-maleimido-4-methoxybenzoate
17.	N-Succinimidyl 3-maleimido-4-(N,N-dimethyl)-aminobenzoate
18.	Maleimidoethoxy-{p-(N-succinimidylpropionato)-phenoxy}-ethane
19.	N-Succinimidyl 4-{(N-iodoacetyl)amino}-benzoate (SIAB)
20.	N-Sulfosuccinimidyl 4-{(N-iodoacetyl)amino}-benzoate (Sulfo-SIAB)
21.	N-Succinimidyliodoacetate
22.	N-Succinimidylbromoacetate
23.	N-Succinimidyl 3-(2-bromo-3-oxobutane-1-sulfonyl)-propionate
24.	N-Succinimidyl 3-(4-bromo-3-oxobutane-1-sulfonyl)-propionate
25.	N-Succinimidyl 2,3-dibromopropionate
26.	N-Succinimidyl 4-[{N,N-bis-(2-chloroethyl)}amino]-phenylbutyrate; chlorambucil-N-hydroxysuccinimide ester
27.	p-Nitrophenyl 3-(2-bromo-3-oxobutane-1-sulfonyl)-propionate
28.	p-Nitrophenyl 3-(4-bromo-3-oxobutane-1-sulfonyl)-propionate
29.	p-Nitrophenyl 6-maleimidocaproate
30.	(2-Nitro-4-sulfonic acid-phenyl)-6-maleimidocaproate
31.	p-Nitrophenyliodoacetate
32.	p-Nitrophenylbromoacetate
33.	2,4-Dinitrophenyl-p-(β-nitrovinyl)-benzoate
34.	N-(3-Fluoro-4,6-dinitrophenyl)-cystamine
35.	Methyl 3-(4-pyridyldithio)-propionimidate · HCl
36.	Ethyl iodoacetimidate · HCl
37.	Ethyl bromoacetimidate · HCl

TABLE 4.5.   (*continued*)

Number	Cross-Linker (Abbreviation)
38.	Ethyl chloroacetimidate · HCl
39.	*N*-(4-Azidocarbonyl-3-hydroxyphenyl)-maleimide; 2-hydroxy-4-(*N*-maleimido)-benzoylazide (HMB)
40.	4-Maleimidobenzoylchloride
41.	2-Chloro-4-maleimidobenzoylchloride
42.	2-Acetoxy-4-maleimidobenzoylchloride
43.	4-Chloroacetylphenylmaleimide
44.	2-Bromoethylmaleimide
45.	*N*-[4-{(2,5-Dihydro-2,5-dioxo-3-furanyl)methyl}-thiophenyl-2,5-dihydro-2,5-dioxo-1*H*-pyrrole]-1-hexanamide
46.	Epichlorohydrin
47.	2-(*p*-Nitrophenyl)-allyl-4-nitro-3-carboxyphenylsulfide
48.	2-(*p*-Nitrophenyl)-allyltrimethylammonium iodide
49.	α-α-Bis-[{(*p*-chlorophenyl)sulfonyl}methyl]-acetophenone
50.	α-α-Bis-[{(*p*-chlorophenyl)sulfonyl}methyl]-*p*-chloroacetophenone
51.	α-α-Bis-[{(*p*-chlorophenyl)sulfonyl}methyl]-4-nitroacetophenone
52.	α-α-Bis-{(*p*-tolysulfonyl)methyl}-4-nitroacetophenone
53.	α-α-Bis-[{(*p*-chlorophenyl)sulfonyl}methyl]-*m*-nitroacetophenone
54.	α-α-Bis-{(*p*-tolylsulfonyl)methyl}-*m*-nitroacetophenone
55.	4-[2,2-Bis-{(*p*-tolylsulfonyl)methyl}acetyl]-benzoic acid
56.	*N*-[4-[2,2-{(*p*-tolylsulfonyl)methyl}acetyl]benzoyl]-4-iodoaniline
57.	α-α-Bis-{(*p*-tolylsulfonyl)methyl}-*p*-aminoacetophenone
58.	*N*-[{5-(Dimethylamino)naphthyl}sulfonyl]-α-α-bis-{(*p*-tolylsulfonyl)methyl}-*p*-aminoacetophenone
59.	*N*-[4-{2,2-Bis-(*p*-tolylsulfonyl)methyl}-acetyl]-benzoyl-1-(*p*-aminobenzyl)-diethylenetriaminepentaacetic acid
CARBOXYL AND ETHER SULFHYDRYL OR AMINO GROUP DIRECTED	
60.	Pyridyl-2,2'-dithiobenzyldiazoacetate (PDD)
61.	1-Diazoacetyl-1-bromo-2-phenylethane
62.	*p*-Nitrophenyl diazoacetate
63.	*p*-Nitrophenyl diazopyruvate
CARBONYL AND SULFHYDRYL GROUP DIRECTED	
64.	1-(Aminooxy)-4-{(3-nitro-2-pyridyl)dithio}-butane
65.	1-(Aminooxy)-4-{(3-nitro-2-pyridyl)dithio}-but-2-ene
MISCELLANEOUS REAGENTS	
66.	2-Methyl-*N*'-benzenesulfonyl-*N*^{4}-bromoacetylquinonediimide
67.	N-Hydroxysuccinimidyl-*p*-formylbenzoate
68.	Methyl-4-(6-formyl-3-azidophenoxy)-butyrimidate . HCl (FAPOB)
69.	Acrolein

[a]Compiled from Peters and Richards (1977), Ji (1979), Han et al. (1984), Means and Feeney (1990), Brinkley (1992), Wong (1993), and Pierce Chemical Co. (1994).

ative functionalities are acylating agents, while those directed against sulfhydryls are alkylating agents (Wong 1993).

Compounds 1–26 listed in Table 4.5 are activated by the hydroxysuccinimide ester moiety that undergoes nucleophilic substitution reaction, liberating *N*-hydroxysuccinimide (Ji 1979; Kitagawa et al. 1981; Srinivasachar and Neville 1989; Wong 1993; Pierce Chemical Co. 1994). Similar compounds include *p*-nitrophenyl esters (27–33) and imidoesters (35–38). Acyl azides (39), acyl chlorides (40–42), and aryl halides (34) are also directed toward the amino group. These compounds are highly active, and may also react with hydroxyl groups (Wong 1993).

Haloketones (43) and alkyl halides (44) are examples of nonspecific cross-linkers. The functional moieties of these two groups of compounds react much faster with thiols (compounds 19–28, 31, 32, 36–38, and 46, Table 4.5). Other thiol directed groups include maleimido moiety (compounds 2–18, 29, 30, 39–45), disulfides (1, 34, and 35) and other alkylating moieties (33, 47–59). Among these compounds, SMCC and its sulfo-analog are probably the most stable compounds, as their maleimido group attached to an aromatic ring is labile at neutral pH (Liberatore et al. 1990).

The nitrovinyl group of compound 33 reacts with the thiol through Michael addition to the double bond more readily than the maleimido derivatives under acidic conditions (Fujii et al. 1985). The dinitrophenyl ester at the other end of this compound also reacts with amines at a much faster rate than *N*-hydroxysuccinimide esters under basic conditions.

Among the compounds listed in Table 4.5, 43 and 44 can also act as homobi-functional cross-linkers for sulfhydryl groups if the thiol groups in a protein are in excess. Cleavable heterobifunctional cross-linkers include the disulfide-containing compounds 1 and 35; sulfone-containing 23, 24, 27, and 28 which are cleavable by dithionite, and compound 45, which is cleavable under mildly acidic conditions (Blatter et al. 1985).

The amino and sulfhydryl directed heterobifunctional reagents are extensively used in the diagnostic industry to cross-link a wide variety of haptens to carrier protein immunogens, and for antibody and antigen labeling (Ishikawa et al. 1983, 1988). Among these, SPDP and SMCC with its sulfo-analog appear to be the most popular reagents (Wong 1993; Pierce Chemical Co. 1994).

2. Carboxyl and Ether Sulfhydryl or Amino Group Directed Reagents.
Compounds 60–63 (Table 4.5) containing diazoacetyl groups are potential cross-linkers for carboxyl groups. They are reactive towards the carboxyl groups at acidic pHs in the dark (Harrison et al. 1989). Being photosensitive, these cross-linkers are also used as photoaffinity labels (Bayley and Knowles 1977).

Compound 60 is thiol specific; its disulfide bond will undergo thiol-disulfide interchange with sulfhydryl groups of proteins. In contrast, the alkylating and acylating cross-linkers (compounds 61–63) will cross-link the carboxylate ion with any other reactive nucleophile. However, with the exception of the acylating cross-linkers that form more stable products with amino groups, others act primarily as thiol specific in the presence of sulfhydryl groups.

3. Carbonyl and Sulfhydryl Group Directed Reagents. Compounds 64 and 65 contain a disulfide bond and a free alkoxylamino group that are specific toward the sulfhydryl and carbonyl groups, respectively (Wong 1993). Protein sulfhydryls can form new disulfide bonds through thiol-disulfide interchange with these compounds while eliminating aminothiopyridine (Webb and Kaneko 1990). The alkoxylamino moiety of these cross-linkers reacts readily with ketones and aldehydes to produce stable alkoximes. This group is also capable of forming alkoximes with dialdehydes generated from glycoproteins upon periodate treatment (Wong 1993).

4. Miscellaneous Cross-Linkers. In addition to categories described above, several reagents are available to cross-link nucleophiles. Compound 66, also known as Cyssor reagent, is an alkylating reagent used to cross-link antibodies under acidic conditions (Liberatore et al. 1989). Compounds 67–69 contain an aldehyde group that will form a Schiff base with amino groups (Maassen and Terhorst 1981; Wong 1993). The succinimidyl ester of 67 and the imidoester of 68 are also reactive toward the amino group, and thus are capable of behaving as homobifunctional reagents. The double bond of acrolein (compound 69) is subject to Michael addition with nucleophiles, and is used for cross-linking proteins (Cater 1963).

Photoactivatable Heterobifunctional Cross-Linkers

Photosensitive labels, which represent one of the largest classes of heterobifunctional cross-linkers, are generally classified according to the active species they produce. Their functional moieties are inert till they are photolyzed; thus, they are first linked to the protein in the dark through a group directed reaction. Upon photoirradiation, generally with a radiation in the 300–400-nm wavelength region so as not to harm the biological component, their photosensitive group is activated, which then reacts indiscriminately with its environment (Pierce Chemical Co. 1994; Wong 1993).

The active species generated by the photosensitive labels include primarily the nitrenes generated from azides and the carbenes from diazo compounds. Among the photoactivatable cross-linkers listed in Table 4.6, only compounds 42–45 and 54 contain carbene precursors. Their ability to undergo a variety of reactions, including the very efficient reaction with water, is a major limitation in the use of these labels. Moreover, their parent diazoacetyl compounds are generally unstable, particularly at low pH, and hence, are reactive toward nucleophiles such as the carboxylate anion (Bayley and Knowles 1977).

TABLE 4.6.   Photoactivatable Heterobifunctional Cross-Linkersa

Number	Cross-Linker (Abbreviation)
	AMINO GROUP ANCHORED PHOTOSENSITIVE REAGENTS
1.	N-Succinimidyl-4-azidobenzoate (NHS-ABA, HSAB)
2.	N-Succinimidyl-4-azido-salicylate (NHS-ASA)
3.	N-Succinimidyl-N'-(4-azidosalicyl)-6-aminocaproate (NHS-ASC)
4.	N-Succinimidyl-5-azido-2-nitrobenzoate (NHS-ANBA); N-5-Azido-2-nitrobenzoyloxysuccinimide (ANB-NOS)
5.	N-Succinimidyl-4-azidobenzoylglycinate (NHS-ABG)
6.	N-Succinimidyl-4-azidobenzoylglycylglycinate (NHS-ABGG)
7.	N-Succinimidyl-4-azidobenzoylglycyltyrosinate (NHS-ABGT)
8.	N-Succinimidyl-(4-azido-2-nitrophenyl)-glycinate
9.	N-Succinimidyl-(4-azido-2-nitrophenyl)-γ-aminobutyrate (NHS-ANAB)
10.	N-Succinimidyl-6(-4-azido-2'-nitrophenylamino)-hexanoate (SANPAH) (Lomant's reagent II)
11.	Sulfosuccinimidyl-6(-4-azido-2'-nitrophenylamino)-hexanoate (Sulfo-SANPAH)
12.	N-Succinimidyl-N'-(4-azidonitrophenyl)-dodecanoate
13.	N-Succinimidyl-2-{(4-azidophenyl)dithio}-acetate (NHS-APDA)
14.	Sulfosuccinimidyl-3-{(4-azidophenyl)dithio}-propionate (Sulfo-SADP)
15.	N-Succinimidyl-3-{(4-azidophenyl)dithio}-propionate (NHS-APDP); N-succinimidyl-(4-azidophenyl)-1,3'-dithiopropionate (SADP)
16.	N-{4-(p-Azidophenylazo)-benzoyl}-3-aminopropyl-N'-oxysuccinimide ester
17.	N-{4-(p-Azido-o-iodophenylazo)-benzoyl}-3-aminopropyl-N'-oxysuccinimide ester
18.	N-{4-p-Azidophenylazo)-benzoyl}-3-aminohexyl-N'-oxysuccinimide ester
19.	N-{4-p-Azidophenylazo)-benzoyl}-3-amino-undecyl-N'-oxysuccinimide ester
20.	3-{(2-Nitro-4-azidophenyl)-2-aminoethyldithio}-N-succinimidylpropionate (NAP-AEDSP): N-succinimidyl-3-{(2-nitro-4-azidophenyl)-2-aminoethyldithio}-propionate (SNAP)
21.	Sulfosuccinimidyl-2-(p-azidosalicylamino)-ethyl-1,3'-dithiopropionate (SASD)
22.	Sulfosuccinimidyl-2-(m-azido-o-nitrobenzamido)-ethyl-1,3'-dithiopropionate (SAND)
23.	N-Succinimidyl-N-{N'-(4-azidobenzoyl)-tyrosyl}-β-alanine
24.	N-Succinimidyl-N-{N'-(3-azidobenzoyl)-tyrosyl}-β-alanine
25.	N-Succinimidyl-N-{N'-(3-azido-5-nitrobenzoyl)-tyrosyl}-β-alanine
	SALICYLATE AZIDES
26.	N-Succinimidyl-N-[2-{(4-azidosalicyloyl)oxy}ethyl]-succinimate (n = 2)
27.	N-Succinimidyl-N-[2-{(4-azidosalicyloyl)oxy}ethyl]-adipamate (n = 4)
28.	N-Succinimidyl-N-[2-{(4-azidosalicyloyl)oxy}ethyl]-suberamate (n = 6)
29.	Methyl-4-azidobenzimidate · HCl (MABI)
30.	Methyl-3-(4-azidophenyl)acetimidate · HCl (MAPA)
31.	Ethyl-N-(5-azido-2-nitrobenzoyl)-aminoacetimidate · HCl (ABNA)

TABLE 4.6.   (*continued*)

Number	Cross-Linker (Abbreviation)
32.	Methyl-4-(6-formyl-3-azidophenoxy)-butyrimidate · HCl (FAPOB)
33.	Methyl-{3-(4-azidophenyl)dithio}-propionimidate · HCl (MADP)
34.	Methyl-{3-(4-azidophenyl)dithio}-butyrimidate · HCl (MADB)
35.	Ethyl-(4-azidophenyl)-1,4-dithiobutyrimidate · HCl (EADB)
36.	4-Azidoiodobenzene
37.	4-Fluoro-3-nitrophenylazide (FNA, FNPA)
38.	2,4-Dinitro-5-fluorophenylazide (DNFA)
39.	*p*-Azidophenylisothiocyanate
40.	1-Azido-5-naphthaleneisothiocyanate; 5-Isothiolcyanato-1-naphthalene azide
41.	Benzophenone-4-isothiocyanate
42.	2-Diazo-3,3,3-trifluoropropionyl chloride (DTPC)
43.	*p*-Nitrophenyl-2-diazo-3,3,3-trifluoropropionate (NDTFP)
44.	*p*-Nitrophenyldiazoacetate
45.	*p*-Nitrophenyl-3-diazopyruvate

SULFHYDRYL GROUP ANCHORED PHOTOACTIVATABLE REAGENTS

46.	4-Azidophenylmaleimide (APM)
47.	*p*-Azidophenacyl bromide (APB)
48.	4-(Bromoaminoethyl)-3-nitrophenylazide (BANPA)
49.	4-Azidophenylsulfenyl chloride
50.	2-Nitro-4-azidophenylsulfenyl chloride (NAPSCI)
51.	*N*-(4-Azidophenylthio)-phthalimide (APTP)
52.	Di-*N*-(2-nitro-4-azidophenyl)-cystamine-*S*,*S*-dioxide (DNCO)
53.	4,4'-Dithiobisphenylazide
54.	Pyridyl-2,2'-dithiobenzyl diazoacetate (PDD)
55.	*N*-(4-Azidobenzoyl-2-glycyl)-*S*-(2-thiopyridyl)-cysteine (AGTC)
56.	*N*-(3-Iodo-4-azidophenylpropionamide-*S*-(2-thiopyridyl)-cysteine
57.	3-(4-Azido-2-nitrobenzoylseleno)-propionic acid (ANBSP); 2-Carboethylseleno-4-azido-2-nitrobenzoate
58.	Benzophenone-4-maleimide
59.	Benzophenone-4-iodoacetamide
60.	*N*-(Maleimidomethyl)-2-(*o*-methoxy-*p*-nitrophenoxy)-carboxamidoethane

GUANIDINYL GROUP ANCHORED PHOTOACTIVATABLE REAGENTS

61.	4-Azidophenylglyoxal (APG)

**CARBOXYL AND CARBOXAMIDE GROUP ANCHORED
PHOTOACTIVATABLE REAGENTS**

62.	*N*-(4-Azido-2-nitrophenyl)-ethylenediamine; *N*-(β-Aminoethyl)-4-azido-2-nitroaniline
63.	*N*-(5-Azido-2-nitrophenyl)-ethylenediamine
64.	*N*-{β-(β'-Aminoethyldithioethyl)}-4-azido-2-nitroaniline
65.	*N*-(4-Azido-2-nitrophenyl-β-aminoethyl)-*N*'-(β-aminoethyl)-tartramide

TABLE 4.6. (*continued*)

Number	Cross-Linker (Abbreviation)
	PHOTOAFFINITY REAGENTS
66.	3'-Arylazido-β-alanine-δ-azido-ATP; 3'-*o*-[3-{*N*-Azido-(2-nitrophenyl)amino}-propionyl]-8-azidoadenosine 5'-triphosphate
67.	5'-(*p*-Fluorosulfonylbenzoyl)-8-azidoadenosine (FSBAzA)

[a]Compiled from Wong (1993) and Pierce Chemical Co. (1994).

Azido derivatives of the aryl, alkyl, and acyl azides constitute the majority of the photoactivatable cross-linking reagents available for conjugation of heterologous compounds (Table 4.6). Of these, alkyl azides have several limitations, the major ones being:

1. Absorption maxima in the UV region at which photoirradiation to activate them may damage proteins, nucleic acids, and other biological components,
2. The alkylnitrene intermediates produced in reactions using these labels readily undergo rearrangement to form inactive imines, and
3. They are highly reactive, and thus are readily capable of undergoing nucleophilic substitution reactions.

Similar reasons also account for the limited use of photosensitive acyl azide cross-linkers, which are primarily used as acylating agents (Lutter et al. 1974). Therefore, only the aryl azides are extensively used as photosensitive acylating agents for conjugation purposes (Grob et al. 1983; Puma et al. 1983; Ji and Ji 1982; Schmidt and Betz 1989; Sorensen et al. 1986; Imai et al. 1990a,b; Wong 1993; Pierce Chemical Co. 1994).

Aryl azides have a low activation energy, and thus can be photolyzed in the long UV region (Peters and Richards 1977; Knowles 1972). The presence of electron-withdrawing substituents such as nitro- and hydroxyl-groups in these cross-linkers further increases the wavelength of absorption into the 300-nm region (Bayley and Knowles 1977). Aryl azides, however, are susceptible to reduction to amino groups, and are not stable in the presence of thiols (Wong 1993). Nevertheless, their half-life, as compared to that for aryl nitrenes (10^{-2} to 10^{-4} second), is much longer, being of the order 5 to 15 min in 10 mM dithiothreitol (pH 8.0) and over 24 hr in 50 mM mercaptoethanol (pH 8.0) (DeGraff et al. 1974; Staros et al. 1978).

Other chemical classes of photoactivatable cross-linkers include benzophenone derivatives (compounds 41, 58, and 59, Table 4.6), which can form covalent adducts upon irradiation with nearby amino acid residues leading to cross-linking, and nitrophenyl ethers (compound 60), which react quantitatively with amines at slightly alkaline conditions (pH 8.0) on irradiation with 366-nm light. The former group of compounds can almost yield a theoretical maximum 100% cross-linking efficiency, primarily because of the ability of the excited triplet state of benzophenones to revert back to parent compound in the absence of photoreaction. In contrast, reactions involving aryl azides are irreversible upon photoirradiation. The nitrophenyl ethers are attached to the protein through the Michael addition reaction of a thiol group at the maleimide ring.

Depending on the group selectivity and specificity, photoactivatable reagents can be classified as amino group specific (compounds 1–45), sulfhydryl directed (compounds 46–60, and those specific to the guanidinyl group of arginine (compound 61), and to the carboxyl and carboxamide side chains of aspartate, glutamate, and their amines (compounds 62–65, Table 4.6).

The amino group specific labels contain such amino group-selective functional moieties as *N*-hydroxysuccinimide esters (compounds 1–28), imidoesters (29–35), aryl halides (36–38), isothiocyanates (39–41), acyl chlorides (42), and *p*-nitrophenyl esters (43–45). Among these, compounds 2, 3, 21, and 23–28 contain phenolic rings, and hence, can be iodinated using reagents such as chloramine T. Compounds 13–15, 20, 21, and 33–35 contain disulfide bonds cleavable by mercaptans, while the azo derivatives (16–19) can be cleaved by dithionite. Compound 32 (FAPOB) is a trifunctional photoaffinity label.

Among the photoaffinity labels directed against sulfhydryls, compounds 52–54, being involved in disulfide-exchange reactions, are truly specific to sulfhydryls. Their cross-linked products are, therefore, cleavable by mercaptans. In contrast, reactions of labels containing maleimido groups (46–58), alkyl halides (47, 48, and 59), and thiol ethers (49–51), although specific to thiolate ions, may also involve other nucleophiles, particularly at high pHs.

The guanidinyl-specific label (compound 61) contains a glyoxal function directed against the arginyl side chain, while photoactivatable reagents 62–65 contain a free amino group that can be coupled to proteins through the carboxyl groups or the γ-carboxamide side chain of glutamine (Wong 1993). The two photoaffinity reagents (compounds 66 and 67) contain ATP/adenosine analogs, and thus can be used for conjugation of ATP-binding proteins (Schafer 1986).

Zero-Length Cross-Linkers

Although similar to the homo- and heterobifunctional cross-linking reagents, zero-length cross-linkers join two intrinsic chemical groups of proteins without the introduction of any extrinsic material (Wong 1993). These cross-linking reactions eliminate atoms from the reactants, thereby shortening the distance between the two linked moieties. This is in contrast with other cross-linking reagents that always introduce a spacer between the two cross-linked groups. Zero-length cross-linkers are routinely used to form disulfide bonds between sulfhydryl groups, amide bonds between carboxyl and primary amino groups, esters involving hydroxyl and carboxyl groups, and thioesters between thiols and carboxyls. The commonly used zero-length cross-linkers, summarized in Table 4.7, are briefly described below.

1. Carboxyl Group Activating Reagents Zero-length cross-linker mediated condensation reactions of carboxyl and amino groups to form amide bonds proceed in two steps. In the first step, the reagent forms a highly reactive adduct with the carboxyl group. During the subsequent reaction, nucleophilic attack at the activated species eliminates the activating moiety, resulting in the formation of a bond that does not involve the incorporation of the cross-linking agent.

The most commonly used zero-length cross-linkers for carboxyl group activation include carbodiimides (Carraway and Koshland 1972; Kurzer and Douraghi-Zadeh 1967), Woodward's reagent K (Woodward et al. 1961; Woodward and Olofson 1961), N-ethylbenzisoxazolium tetrafluoroborate (Kemp and Woodward 1965; Goodfriend et al. 1966), ethyl chloroformate (Patramani et al. 1969; Avrameas and Ternynck 1967), diethylpyrocarbonate (Wolf et al. 1970), and carbonyl diimidazole (Chang and Hammes 1986).

TABLE 4.7. Zero-Length Cross-Linking Reagents[a]

Number	Cross-Linker (Abbreviation)
	CARBOXYL GROUP ACTIVATING REAGENTS
1.	Dihexylcarbodiimide (DCC)
2.	1-Ethyl-3-(3-dimethylaminopropyl)-carbodiimide hydrochloride (EDC)
3.	1-Ethyl-3-(4-azonia-4,4-dimethylpentyl)carbodiimide iodide (EAC); 1-Ethyl-3-(3-dimethylaminopropyl)-carbodiimide methiodide
4.	1-Cyclohexyl-3-[2-morpholinyl-(4)-ethyl]-carbodiimide metho-*p*-toluenesulfonate (CMC); N-Cyclohexyl-N'-[β-(N-methylmorpholine)-ethyl]-carbodiimide *p*-toluene sulfonate
5.	N-Benzyl-N'-3-dimethylaminopropyl-carbodiimide hydrochloride; 1- Benzyl-3-(3-dimethylaminopropyl)-carbodiimide hydrochloride (BDC)
6.	N-Ethyl-5-phenylisoxazolium-3'-sulfonate (Woodward's Reagent K)
7.	N-Ethylbenzisoxazolium tetrafluoroborate
8.	Ethylchloroformate
9.	*p*-Nitrophenylchloroformate
10.	1,1'-Carbonyldiimidazole
11.	N-(Ethoxycarbonyl)-2-ethoxy-1,2-dihydroquinoline (EEDQ)
12.	N-(Isobutoxycarbonyl)-2-isobutoxy-1,2-dihydroquinoline (IIDQ)
	DISULFIDE FORMING REAGENTS
13.	Cupric di(1,10-phenanthroline) (CuP)

[a]Compiled from Ji (1979), Means and Feeney (1990), Wong (1993), and Pierce Chemical Co. (1994).

Although a large number of carbodiimides has been synthesized (Kurzer and Douraghi-Zadeh 1967; Sheehan et al. 1961), the most commonly used are compounds 1–5 listed in Table 4.7. These reactions involve activation of the carboxyl groups by the carbodiimide derivative to an *O*-acylisourea intermediate that can react further in a subsequent step in several ways (Wong 1993). Nucleophilic attack of water hydrolyzes the intermediate to regenerate the free carboxyl group. Reaction with an amino group from a second protein will lead to a cross-link between the two proteins. Intramolecular cross-linking may also occur if the nucleophile is from the same protein. Without productive nucleophilic reaction, the *O*-acylisourea may undergo an intramolecular *O*- to *N*-acyl shift to form a more stable *N*-acylurea.

Carbodiimide reagents differ greatly in their stability in aqueous solutions. For example, EDC (compound 2, Table 4.7) has a half-life of 37 hr at pH 7.0, whereas EAC (compound 3) is about 10-fold less stable (Wong 1993). Phosphate and phosphate-containing reagents as well as hydroxylamine and other amine derivatives also increase the rate of loss of carbodiimides, and in some instances, dramatically (Gilles et al. 1990).

Among other zero-length cross-linkers used for carboxyl group activation, woodward's reagent K (compound 6, Table 4.7) is first converted to a reactive ketoketenimine under alkaline conditions (Woodward et al. 1961). The latter then reacts with the carboxylate anion to form an enol ester intermediate that subsequently reacts with an amino group to form an amide bond. *N*-Ethylbenzisoxazolium tetrafluoroborate (compound 7) also reacts with a similar mechanism.

The homobifunctional chloroformates (compounds 8 and 9) form an active anhydride intermediate on activation of the carboxyl group, which then reacts with an amino nucleophile to form an amide bond. Similarly, carbonyl diimidazole (compound 10) also functions both as a homobifunctional and a zero-length cross-linker.

N-Carbalkoxydihydroquinolines (compounds 11 and 12, Table 4.7) also function as a zero-length cross-linker by activating the carboxyl group through a mixed anhydride reaction (Belleau et al. 1969; Bertrand et al. 1988). In addition, several homobifunctional cross-linkers described earlier are also capable of acting as zero-length cross-linkers, although their mechanism of reaction is not well understood.

2. Disulfide formation.   Free sulfhydryl groups may be cross-linked by oxidation to form disulfide bonds. Such linkage may occur both inter- and intramolecularly. Any oxidizing agent that facilitates the disulfide bond formation is, therefore, regarded as a zero-length cross-linker. These include air, iodide, or hydrogen peroxide. The reaction is also catalyzed by bis-1,10-phenanthroline complex of cupric ion (compound 13, Table 4.7) (Quinlan and Franke 1982).

3. Carbohydrate Activation.   Cross-linking by reductive alkylation between aldehydes derived from carbohydrates and amino groups is an example of a zero-length coupling process, as the Schiff base formation does not incorporate any extrinsic atoms. Reagents such as periodate that oxidize vicinal diols of carbohydrates to dialdehydes may also be regarded as zero-length cross-linkers (Wong 1993).

Most of the bifunctional cross-linkers listed in Tables 4.2–4.7 are available commercially from Pierce Chemical Co., Rockford, IL; Sigma Chemical Co., St. Louis, MO; and Aldrich Chemical Co., Milwaukee, WI. The product cataloge of Pierce Chemical Co. (1994) also describes the ideal reaction conditions for several of these cross-linkers as well as their applications in immunologic research and in the diagnostic industry. Pierce also provides literature with each cross-linker they supply, detailing the reaction conditions as well as practical tips on achieving the desired conjugates.

An excellent monograph on the chemistry of protein conjugation and cross-linking techniques was recently published (Wong 1993). It describes several critical aspects of protein cross-linking and conjugate preparation, including the criteria for the selection and design of reagents as well as the analysis of the cross-linked species. The journal *Bioconjugate Chemistry,* published by the American Chemical Society, also deals with several chemical aspects of conjugate preparation and characterization, in vivo applications of conjugate methodology, molecular biological aspects of antibodies including genetically engineered fragments and other immunochemicals, as well as the relationships between conjugation chemistry and the biological properties of conjugates. In recent years, it has become the primary source of latest information on conjugation chemistry.

The major applications of the bifunctional cross-linking reagents in the diagnostic industry include the preparation of carrier protein-hapten conjugates, protein-protein coupling, solid phase immobilization of the reagents, and reagent formulations for product kits. These cross-linkers, however, have found several other important applications in the areas of cell surface cross-linking, subunit cross-linking for the study of protein interactions and associations, cell membrane structural studies, immunotoxin preparations, and DNA/RNA cross-linking to proteins.

PRACTICAL ASPECTS

Successful coupling reaction procedures require a thorough knowledge of protein reactivity and the available reagents for the desired type of protein modification. It is also important to understand the practical aspects of carrying out reactions between highly reactive small organic molecules and large, complex, conformationally sensitive, water-soluble proteins and enzymes.

An ideal conjugation procedure should yield 100% conjugate of well-defined composition without the inactivation of the enzyme or the antibody, produce a stable link, and be practical in terms of cost and simplicity (Tijssen 1985). However, none of the procedures currently available fully responds to these requirements. In the diagnostic reagent formulations, large differences exist among the conjugation efficiencies and in the relative detectabilities by the conjugates.

In this section, some of the general rules, problems, and pitfalls encountered in cross-linking methodology are briefly discussed.

Cross-Linking Reactions

Protein-protein or protein-hapten coupling reactions can be broadly divided into four categories: one-step, two-step, three-step, and multistep reactions. The basic characteristics of these reactions are briefly described below.

1. One-Step Reactions. These are the simplest and the oldest of all conjugation reactions, in which the bifunctional cross-linker is added to a mixture of the conjugating species. In the preparation of immunoconjugates, one-step reactions, however, are often undesirable, as they yield both homo- and heteropolymers of the reacting species. Moreover, the reactivity of the conjugating species may be different to different cross-linking reagents, thereby resulting in selective homopolymerization reaction with one of the conjugating species.

In one-step conjugation reactions, the rate of addition of the cross-linking reagent greatly influences the yield and efficiency of conjugation (Modesto and Pesce 1971). Slow addition of the cross-linker over a period of time, rather than its addition all at once, often increases the yield of coupled proteins.

2. Two-Step Reactions. In this procedure, one of the reactants to be conjugated is first activated with the cross-linker. The unreacted reagent is then removed prior to addition of the second reacting species. This method has the advantage of using the differential reactivities of functional groups in heterobifunctional reagents as well as the differential selectivity of homobifunctional reagents towards the two conjugating species to be coupled (Wong 1993). Two-step conjugation reactions are, therefore, widely used for almost all heterobifunctional cross-linkers, including the photoaffinity labels.

3. Three-Step Reactions. This procedure involves an extra step for the preparation of proteins to be coupled. An example of this type of reaction is the labeling of IgG with pyrrole——acyl azide and that of albumin with bis-diazotized *p*-phenylene diamine in an acidic condition, respectively. After isolation, the pyrrole ring of the modified IgG is reacted with the diazo-albumin at pH 6.0. This approach minimizes the undesirable side reactions by adjusting the coupling conditions. For example, in the above example, the first diazotized group is reactive, whereas the second group is reactive only at higher pH (Howard and Wild 1957). Three-step reaction procedures are often used for the preparation of many immunotoxins and immunoconjugates.

4. Multistep Reactions.   Multistep reactions, although sometimes used for the preparation of immunotoxins, are not very common in the diagnostic industry. These procedures involve preparation procedures of proteins prior to the actual coupling process.

Solvent Systems

Cross-linking conditions are highly dependent on the type of reagents used and the particular system under investigation. Solvent selection is often an important consideration in these reactions, because many cross-linkers are either not soluble in aqueous buffers in which proteins are dissolved or undergo rapid hydrolysis. In this regard, the following factors need to be considered.

1. Buffers.   Conjugations should be carried out in a well-buffered system at a pH optimal for the reaction. In most cases, the ionic strength of the buffers needs to be adjusted between 20–100 mM. Thiol groups and α-amino groups can be selectively modified at physiological pH of 7.0–7.5; phosphate buffers are ideally suited. In contrast, the more strongly basic lysyl amines require a slightly higher alkaline pH of 8.0–9.5, where phosphate solutions do not buffer well. Hence, for these reactions, carbonate/bicarbonate or borate buffers are preferred.

In some instances, the choice of buffers is dictated by compatibility of the protein as well as the type of cross-linker used. For example, amine-containing buffers such as Tris and glycine should not be used in conjunction with amino group-specific cross-linkers.

2. Cosolvents.   Generally, if the hapten and/or the cross-linker that is to be attached to proteins is readily soluble at millimolar concentrations in water or buffer, no cosolvent is needed. They can be added in small quantities as a concentrated aqueous solution to the buffered reaction medium. Haptens and cross-linkers that are insoluble and highly reactive in aqueous systems and/or undergo rapid hydrolysis, however, need to be dissolved in water-miscible cosolvents. At the same time, the cosolvent should not cause any irreversible denaturation or precipitation of the proteins. Cosolvents that have been successfully utilized in protein modification reactions include methanol, ethanol, 2-propanol, 2-methoxyethanol, dioxane, dimethyl formamide (DMF), dimethyl sulfoxide (DMSO), and acetonitrile.

DMF and DMSO are the most versatile of these cosolvents widely used in conjugation reactions. They are inert to many of the reactive reagents used in preparing conjugates, are miscible with water in all proportions, and are compatible with most aqueous protein solutions up to 30% v/v ratios (Brinkley 1992). Ideally, the concentration of these solvents must be less than 5–10% by volume of the reaction medium to prevent protein precipitation and enzyme inactivation reactions. The cosolvents should be carefully dried and stored over a drying agent to prevent competing hydrolysis of the reactive modification reagent.

Reaction Conditions

As a general rule, cross-linking reactions should be done at 0–10°C, as the rate of reaction of most conjugation reagents is rapid at low temperatures. The use of low temperatures also tends to increase the selectivity of the reaction, and results in fewer undesirable side reactions and more consistent and reproducible results. A convenient procedure is to add the reagent to a gently stirring buffered solution of the protein in an ice bath, and then allow it to warm to room temperature over a 2–4-hr period. Highly reactive reagents such as sulfonyl chlorides should be reacted under more carefully controlled conditions, such as 4°C for 1 hr.

Addition of the reagent should be carried out dropwise and as slowly as possible. Gradual addition of the cross-linker to the protein solution often increases the selectivity of the reaction. The kinetics of conjugation reactions is bimolecular, whereas the hydrolysis rates of most cross-linkers follow a pseudo first order reaction. Moreover, dilution of the reaction medium results in competition between conjugation and loss of reagent by hydrolysis. Some cross-linkers are hydrolyzed rapidly, and thus may need to be used in excess quantity. For proper cross-linking, protein concentrations above 4 mg/ml, and ideally 8–12 mg/ml, are strongly recommended.

Conjugation reaction times may also vary considerably, ranging from a few minutes to several hours. Usually 1–2 hr is sufficient time for most cross-linking reactions to go to completion. Longer reaction times, if convenient, are acceptable, because the degree of labeling is generally limited by the ratio of the reagent to protein, rather than the reaction time. Generally, the more reactive the cross-linker, the shorter the reaction time.

Purification of Conjugates

Once the cross-linking reactions are completed, the desired conjugate needs to be isolated and purified. Generally, the small molecular weight byproducts of the cross-linking compounds can be readily removed by dialysis or gel filtration. The conjugates can then be further purified using any of the classical protein purification techniques to remove any undesirable homo- and heteropolymers that may have formed during the coupling process.

Among the various techniques used to purify conjugates, dialysis is the simplest and is inexpensive. It is also the most time-consuming method of purifying protein conjugates. Also, not all molecules dialyze efficiently; the rate being dependent on their relative affinity for the protein versus the dialysis solution. Dialysis is not suited for the removal of hydrophobic molecules. It works best when the labeling reagent and its unreacted byproducts are hydrophilic. Typically, a dialysis buffer volume of at least 100 times the volume of the conjugate solution should be used. Ideally, during the dialysis process, the buffer should be changed at least five times at 4–6 hr intervals.

In contrast, gel filtration or size exclusion chromatography is much faster and effectively removes most hydrophilic and hydrophobic labeling reagents. The purified labeled protein solution will often contain a mixture of species with variable degrees of substitution. If required, further separation of the lightly and heavily labeled fractions can be done by ion-exchange chromatography.

Certain cross-linking reagents have a very strong affinity for proteins, and hence, cannot be removed completely by gel filtration chromatography alone. These conjugates can be further purified by treatment with microporous, hydrophobic polystyrene beads. The conjugate is simply mixed with the beads to selectively adsorb the small hydrophobic molecules into the micropores, while the larger conjugates are excluded.

Several of the reagents commonly used for protein modification, including NHS esters, isothiocyanates, and sulfonyl chlorides, can react with tyrosines to form esters. These adducts are unstable and slowly hydrolyze at physiological pH, thereby causing loss of label over a period of time. Because any measurable loss of label can interfere with the intended use of many conjugates, it is often desirable to remove such esters that may have formed in the conjugation reaction. In most cases, this can be effectively done by treating the conjugate before purification with hydroxylamine at a final concentration of 0.1 M at pH 8.0 for 1 hr at room temperature. The conjugate can then be purified by either dialysis or gel filtration.

Degree of Substitution

Several methods are available for determining the degree of substitution of modified proteins. The incorporation of several haptens that absorb strongly in the UV or visible range of the spectrum can be determined by spectroscopic techniques. However, such methods often give an erroneous degree of substitution, as there is no way to know precisely how the spectral characteristics change when it is conjugated to the protein. Generally, it is assumed that the extinction coefficient of the protein-bound hapten at its absorption maximum is about the same as that of the free hapten.

If the modification involves amino groups or creation of thiol residues, as is often the case with bifunctional reagents, it is relatively easy to determine the degree of substitution of protein conjugates. Either the remaining amino groups on the protein conjugate or quantitation of thiols by any of the several colorimetric procedures available often gives a more accurate value for the label substitution on the proteins.

Storage of Conjugates

Conjugates should be stored in a similar way as the parent protein. If the protein is stable to freezing, then lyophilization is strongly recommended for long term storage. Preservatives such as sodium azide or thimerosol should be added to conjugates that need to be stored frozen. The former is a potent inhibitor of the widely used enzyme horseradish peroxidase (HRP). Therefore, other preservatives need to be substituted where the conjugate is derived from HRP or when its anticipated use is in the presence of HRP.

For the preparation of carrier protein-hapten conjugates, the major concern is the effect of cross-linking reagents and the reaction conditions on the immunogenicity of the antigen. The coupling reaction must protect the antigenic determinant(s) on the molecule. With small molecular weight antigens, cross-linking reagents that provide a spacer arm of 5–10 Å length are especially useful. They allow for the extension of the antigen molecule farther away from the carrier protein, thus preventing its reactions with the neighboring amino acid side chains. Such an approach often leads to an enhanced immunogenicity of the molecule, as well as the production of a desirable, high affinity antibody.

Yet another important factor in this regard is the determination of the optimum degree of conjugation. Whereas a high degree of conjugation is normally desired to increase the immunogenicity of the hapten while preparing immunogen conjugates, a low to moderate degree of conjugation is often required for the conjugation of the antigen to an antibody or an enzyme to ensure that the biological activity of the protein is retained. In the latter case, excessive labeling often results in decreased solubility of the enzyme conjugate, thereby reducing its overall activity.

The number of available reactive groups on the surface of the protein also determines the conjugation efficiency. For example, the availability of a large number of reactive groups may require the use of a lower cross-linker-to-protein ratio. Corollary, for a limited number of potential targets, a higher cross-linker-to-protein ratio may be required.

In general, the following factors must be remembered to prepare desirable conjugates as reagents in the diagnostics industry (Tijssen 1985).

1. Concentration of the molecules to be conjugated. Conjugation follows the law of mass action. This factor is extremely important in the preparation of carrier protein-hapten conjugates used for immunization purposes. Too high a molar ratio of hapten to carrier protein may lead to a conjugate containing an excessive number of hapten molecules on the carrier protein. Such an immunogen may suppress the antibody formation by immunotolerance.

2. The relative rate of intramolecular over intermolecular cross-linking increases at lower protein concentrations.

3. Molecular concentration ratio of the two molecules to be conjugated, i.e., the formation of different conjugates follows Poisson and not normal distribution.

4. The relative reaction rates of the cross-linking agent with the two molecules must be considered for the possibility of formation of homopolymers versus heteropolymers.

5. Effective concentration of the cross-linking agent: the active fraction of the reagent depends on a large number of parameters such as pH, temperature, the reaction medium, etc.

6. The purity of buffer solutions and reactivity of its components with the cross-linking agents. For examples, amine-containing buffers such as Tris or glycine cannot be used in conjunction with amino group-specific cross-linkers.

7. Ionic strength and pH of the buffer solution must be optimized properly to obtain the highest probability of the desired conjugation.

8. The conjugation procedures must protect the groups involved in the biological activity of the conjugating species.

REFERENCES

ALI-OSMAN, F.; CAUGHLAN, J.; and GRAY, G. S. 1989. Diseased DNA intrastrand cross-linking and cytotoxicity induced in human brain tumor cells by 1,3-bis(2-chloroethyl)-1-nitrosourea after in vitro reaction with glutathione. *Cancer Res.* 49:5954–5958.

ANDERER, F. A., and SCHLUMBERGER, H. D. 1969. Antigenic properties of proteins cross-linked by multidiazonium compounds. *Immunochemistry* 6:1–10.

ANDERSON, G. W.; ZIMMERMAN, J. E.; and CALLAHAN, F. M. 1964. Esters of N-hydroxysuccinimide in peptide synthesis. *J. Am. Chem. Soc.* 86:1839–1842.

ARNON, R., and SHAPIRA, E. 1969. Crystalline papain derivative containing an intramolecular mercury bridge. *J. Biol. Chem.* 244:1033–1038.

AVRAMEAS, S., and TERNYNCK, T. 1967. Biologically active water-insoluble protein polymers. I. Their use for isolation of antigens and antibodies. *J. Biol. Chem.* 242:1651–1659.

AVRAMEAS, S.; TERNYNCK, T.; and GUESDON, J. L. 1978. Coupling of enzymes to antibodies and antigens. *Scand. J. Immunol. (Suppl. 7)* pp. 7–23.

BAYLEY, H., and KNOWLES, J. R. 1977. Photoaffinity labeling. *Methods Enzymol.* 46:69–114.

BELLEAU, B.; DI TULLIO, V.; and GODIN, D. 1969. The mechanism of irreversible adrenergic blockade by N-carbethoxydihydroquinolines. Model studies with typical serine hydrolases. *Biochem. Pharm.* 18:1039–1044.

BERTRAND, R.; CHAUSSEPIED, P.; and KASSAB, R. 1988. Cross-linking of the skeletal myosin subfragment 1 heavy chain to the N-terminal actin segment of residues 40–113. *Biochemistry* 27:5728–5736.

BLATTER, W. A.; ENZI, B. S.; LAMBERT, J. M.; and SENTER, P. D. 1985. New heterobifunctional protein cross-linking reagent that forms an acid-labile link. *Biochemistry* 24:1517–1524.

BLOXHAM, D. O., and COOPER, G. K. 1982. Formation of a polymethylene bis(disulfide) intersubunit cross-link between cys-281 residues in rabbit muscle glyceraldehyde-3-phosphate dehydrogenase using octamethylene bis(methane[^{35}]thiosulfonate). *Biochemistry* 21:1807–1812.

BLOXHAM, D. P., and SHARMA, R. P. 1979. The development of S-S'-polymethylene bis(methanethiosulfonates) as reversible cross-linking reagent for thiol groups and their use to form stable catalytically active cross-linked dimers with glyceraldehyde-3-phosphate dehydrogenase. *Biochem. J* 181:355–366.

BRINKLEY, M. 1992. A brief survey of methods for preparing protein conjugates with dyes, haptens, and cross-linking reagents. *Bioconjugate Chem.* 3:2–13.

BUNNETT, J. F. 1963. Nucleophilic reactivity. *Ann. Rev. Phys. Chem.* 14:271–290.

BUSSE, W. D., and CARPENTER, F. H. 1974. Carbonylbis(L-methionine-*p*-nitrophenyl ester). A new reagent for the reversible intramolecular cross-linking of insulin. *J. Am. Chem. Soc.* 96:5947–5949.

CARLSSON, J.; DREVIN, H.; and AXEN, R. 1978. Protein thiolation and reversible protein-protein conjugation. *N*-Succinimidyl 3-(2-pyridyldithio)-propionate, a new heterobifunctional reagent. *Biochem. J.* 173:723–737.

CARRAWAY, K. L., and KOSHLAND, D. E. 1972. Carbodiimide modification of proteins. *Methods Enzymol.* 25:616–623.

CATER, C. W. 1963. The evaluation of aldehydes and other difunctional compounds as cross-linking agents for collagen. *J. Soc. Leather Trades Chem.* 47:259–272.

CAVA, M. P.; DEANA, A. A.; MUTH, K.; and MITCHELL, M. H. 1961. *N*-Phenylmaleimide. *Org. Synth.* 41:93–95.

CHANG, S. I., and HAMMES, G. G. 1986. Interaction of spin-labeled nicotinamide adenine dinucleotide phosphate with chicken liver fatty synthase. *Biochemistry* 25:4661–4668.

CHATTERJEE, R.; WELTY, E. V.; WALDER, R. Y.; PRUITT, S. L.; ROGERS, P. H.; ARONE, A.; and WALDER, J. A. 1986. Isolation and characterization of a new hemoglobinderivative cross-linked between α chains (lysine 99α$_{1\to}$ lysine 99α$_2$). *J. Biol. Chem.* 261:9929–9937.

COGGINS, J. R.; HOOPER, E. A.; and PERHAM, R. N. 1976. Use of DMS and novel periodate-cleavable bis-imidoesters to study the quaternary structure of the pyruvate dehydrogenase multienzyme complex of *E. coli. Biochemistry* 15:2527–2533.

CUATRECASAS, P., and PARIKH, I. 1972. Adsorbents for affinity chromatography. Use of N-hydroxysuccinimide esters of agarose. *Biochemistry* 11:2291–2299.

DAVIES, G. E., and STARK, G. R. 1970. Use of dimethyl suberimidate, a cross-linking reagent, in studying the subunit structure of oligomeric proteins. *Proc. Natl. Acad. Sci.* (USA) 66:651–656.

DEGRAFF, B. A.; GILLESPIE, D. W.; and SUNDBERG, R. J. 1974. Phenyl nitrene. A flash photolytic investigation of the reaction with secondary amines. *J. Am. Chem. Soc.* 96:7491–7496.

DUGGLEBY, K. G., and KAPLAN, H. 1975. A competitive labeling method for the determination of the chemical properties of solitary functional groups in proteins. *Biochemistry* 14:5168–5175.

EBERT, C.; EBERT, G.; and KNIPP, H. 1977. On the introduction of disulfide cross-links into fibrous proteins and bovine serum albumin. *Adv. Exp. Med. Biol.* 86A:235–245.

EDWARDS, J. O., and PEARSON, R. G. 1962. The factors determining nucleophilic reactivities. *J. Am. Chem. Soc.* 84:16–24.

EYZAGUIRRE, J. 1987. *Chemical Modification of Enzymes: Active Site Studies.* John Wiley & Sons, New York.

FASOLD, H.; GROSCHEL-STEWART, U.; and TURBA, F. 1964. Synthese and Reaktionen eines wasserloslichen, spaltbarenreagens zur Verknupfung frier SH-gruppen in Proteinen. *Biochem. Z.* 339:287–291.

FUJII, N.; HAYASHI, Y.; KATAKURA, S.; AKAJI, K.; YAJIMA, H.; INOUYE, A.; and SEGAWA, T. 1985. Studies on peptides. CXXVIII. Application of new heterobifunctional crosslinking reagents for the preparation of neurokinin (A and B)-BSA (bovine serum albumin) conjugates. *Int. J. Peptide Prot. Res.* 26:121–129.

GILLES, M. A.; HUDSON, A. Q.; and BORDERS, C. L. 1990. Stability of water-soluble carbodiimides in aqueous solution. *Anal. Biochem.* 184:244–248.

GLAZER, A. N.; DELANGE, R. J.; and SIGMAN, D. S. 1975. *Chemical Modification of Proteins.* North-Holland/Elsevier, Amsterdam.

GOLD, P.; JOHNSON, J.; and FREEDMAN, S. O. 1966. The effect of the sequence of erythrocyte sensitization on the bis-diazotized benzidine hemagglutination reaction. *J. Allergy* 37:311–317.

GONIAS, S. L.; OAKLEY, A. C.; WALTHER, P. J.; and PIZZO, S. V. 1984. Effect of diethyldithiocarbamate and nine other nucleophiles on the intersubunit protein crosslinking and inactivation of purified human α_2-macroglobulin by *cis*-diamminedichloroplatinum (II). *Cancer Res.* 44:5764–5770.

GOODFRIEND, T.; FASMAN, G.; KEMP, D.; and LEVINE, L. 1966. Immunochemical studies of angiotensin. *Immunochemistry* 3:223–231.

GORMAN, J. J., and FOLK, J. E. 1980. Transglutaminase amine substrates for photochemical labeling and cleavable cross-linking of proteins. *J. Biol. Chem.* 255:1175–1180.

GRAHAM, D. G.; SZAKAL-QUIN, G.; PRIEST, J. W.; and ANTHONY, D. C. 1984. In vitro evidence that covalent cross-linking of neurofilaments occurs in γ-diketone neuropathy. *Proc. Natl. Acad. Sci. (*USA) 81:4979–4982.

GROB, P. M.; BERLOT, C. H.; and BOTHWILL, M. A. 1983. Affinity labeling and partial purification of nerve growth factor receptors from rat pheochromocytoma and human melanoma cells. *Proc. Natl. Acad. Sci. (*USA) 80:6819–6823.

GURD, F. R. N. 1967. Carboxymethylation. *Methods Enzymol.* 11:532–541.

HAN, K. K.; RICHARD, C.; and DELACOURTE, A. 1984. Chemical cross-links of proteins by using bifunctional reagents. *Int. J. Biochem.* 16:129–145.

HANCOCK, J., and WAGNER, R. 1982. A structural model of 5S RNA from *E. coli* based on intramolecular cross-linking evidence. *Nucleic Acid Res.* 10:1257–1269.

HARRISON, J. K.; LAWTON, R. G.; and GNEGY, M. E. 1989. Development of a novel photoreactive calmodulin derivative. Cross-linking purified adenylate cyclase from bovine brain. *Biochemistry* 28:6023–6027.

HASHIDA, S.; IMAGAWA, M.; INOUE, S.; RUAN, K. H.; and ISHIKAWA, E. 1984. More useful maleimide compounds for the conjugation of Fab' to horseradish peroxidase through thiol groups in the hinge. *J. Appl. Chem.* 6:56–63.

HIREMATH, C. B., and DAY, R. A. 1964. Introduction of covalent cross-linkages into lysozyme by reaction with α-α'-dibromo-*p*-xylenesulfonic acid. *J. Am. Chem. Soc.* 86:5027–5028.

HOWARD, A. N., and WILD, F. 1957. A two-stage method of cross-linking proteins suitable for use in serological techniques. *Br. J. Exp. Pathol.* 38:640–643.

HUNTER, M. J., and LUDWIG, M. L. 1962. The reaction of imidoesters with proteins and related small molecules. *J. Am. Chem. Soc.* 84:3491–3504.

HUNTER, M. J., and LUDWIG, M. L. 1972. Amidination. *Methods Enzymol.* 25:585–596.

IMAI, N.; KOMETANI, T.; CROCKER, P. J.; BOWDEN, J. B.; DEMIR, A.; DWYER, L. D.; MANN, D. M.; VANAMAN, T. C.; and WATT, D. S. 1990a. Photoaffinity heterobifunctional cross-linking reagents based on N-(azidobenzoyl)tyrosines. *Bioconjugate Chem.* 1:138–143.

IMAI, N.; DWYER, L. D.; KOMETANI, T.; JI, T,; VANAMAN, T. C.; AND WATT, D. S. 1990b. Photoaffinity heterobifunctional cross-linking reagents based on azide-substituted salicylates. *Bioconjugate Chem.* 1:144–148.

ISHIKAWA, E.; HASHIDA, S.; KOHNO, T.; and TANAKA, K. 1988. Methods for enzyme-labeling of antigens, antibodies and their fragments. In *Nonisotopic Immunoassays,* ed. T. T. Ngo, pp. 27–55, Plenum Press, New York.

ISHIKAWA, E.; IMAGAWA, M.; HASHIDA, S.; YOSHITAKE, S.; HAMAGUCHI, Y.; and UENO, T. 1983. Enzyme-labeling of antibodies and their fragments for enzyme immunoassay and immunohistochemical staining. *J. Immunoassay* 4:209–327.

JI, T. H. 1979. The application of chemical cross-linking for studies on cell membranes and the identification of surface reporters. *Biochem. Biophys. Acta* 559:39–69.

JI, T. H. 1983. Bifunctional reagents. *Methods Enzymol.* 91:580–609.

JI, T. H., and JI, I. 1982. Macromolecular photoaffinity labeling with radioactive photoactivable heterobifunctional reagent. *Anal. Biochem.* 121:286–289.

JUE, R.; LAMBERT, J. M.; PIERCE, L. R.; and TRAUT, R. R. 1978. Addition of sulfhydral groups of Escherichia coli ribosomes by protein modification with 2-iminothiolane (methyl 4-mercaptobutyrimidate). *Biochemistry* 17:5399–5406.

KABAKOFF, D. S. 1980. Chemical aspects of enzyme immunoassay. In *Enzyme Immunoassay,* ed. E. T. Maggio, pp. 71–104, CRC Press, Boca Raton, FL.

KAPLAN, H.; STEVENSON, K. J.; and HARTLEY, B. S. 1971. Competitive labeling, a method for determining the reactivity of individual groups in proteins. *Biochem. J.* 124:289–299.

KEMP, D. S., and WOODWARD, R. B. 1965. N-Ethylbenzisoxazolium cation. I. Preparation and reactions with nucleophilic species. *Tetrahedron* 21:3019–3035.

KITAGAWA, T.; SHIMOZONO, T.; AIKAWA, T.; YOSHIDA, T.; and NISHIMURA, H. 1981. Preparation and characterization of heterobifunctional cross-linking reagents for protein modifications. *Chem. Pharm. Bull.* 29:1130–1135.

KLOTZ, I. M., and HEINEY, R. E. 1962. Introduction of sulfhydral groups into proteins using acetyl-mercaptosuccinic anhydride. Arch. Biochem. Biophys. 96:605–612.

KNOWLES, J. R. 1972. Photogenerated reagents for biological receptor site labeling. *Acc. Chem. Res.* 5:155–160.

KOCH, N., and HAUSTEIN, D. 1983. Association of surface IgM with two membrane proteins on murine B lymphocytes detected by chemical cross-linking. *Mol. Immunol.* 20:33–37.

KURZER, F., and DOURAGHI-ZADEH, K. 1967. Advances in the chemistry of carbodiimides. *Chem. Rev.* 67:107–152.

LIBERATORE, F. A.; COMEAU, R. D.; and LAWTON, R. G. 1989. Heterobifunctional cross-linking of a monoclonal antibody with 2-methyl-N^1-benzenesulfonyl-N^4-bromoacetyl-quinonediimide. *Biochem. Biophys. Res. Commun.* 158:640–645.

LIBERATORE, F. A.; COMEAU, R. D.; MCKEARIN, J. M.; PEARSON, D. A.; BELONGA, B. Q.; BROCCHINI, S. J.; KATH, J.; PHILLIPS, T.; OSWELL, K.; and LAWTON, R. G. 1990. Site-directed chemical modification and cross-linking of a monoclonal antibody using equilibrium transfer alkylating cross-link reagents. *Bioconjugate Chem.* 1:36–50.

LINDLEY, H. 1956. A new synthetic substrate for trypsin and its application to the determination of the amino acid sequence of proteins. *Nature* (London) 178:647–648.

LOMANT, A. J., and FAIRBANKS, G. 1976. Chemical probes of extended biological structures: Synthesis and properties of the cleavable protein cross-linking reagent [^{35}S]-dithiobis(succinimidyl propionate). *J. Mol. Biol.* 104:243–261.

LOUDON, J. D., and SHULMAN, N. 1941. Mobility of groups in chloronitrodiphenyl sulfones. *J. Chem. Soc.* 157:722–727.

LUDUENA, R. F.; ROACH, M. C.; TREKA, P. P.; and WEINTRAUB, S. 1982. Bi-iodoacetyldithioethylamine: A reversible cross-linking reagent for protein sulfhydral group. *Anal. Biochem.* 117:76–80.

LUNDBLAD, R. L., and NOYES, C. M. 1984. *Chemical Reagents for Protein Modification,* vols. 1 and 2, CRC Press, Boca Raton, FL.

LUTTER, L. C.; ORTANDERL, F.; and FASOLD, H. 1974. Use of a new series of cleavable protein-cross-linkers on the *Escherichia coli* ribosome. *FEBS Lett.* 48:288–292.

MAASSEN, J. A., and TERHORST, C. 1981. Identification of a cell-surface protein involved in the binding site of sindbis virus on human lymphobastoic cell lines using a heterobi-functional cross-linker. *Eur. J. Biochem.* 115:153–158.

MAHONEY, C. W. and AZZI, A. 1987. The synthesis of fluorescent chlorotriazinylamino-fluorescein-concanavalin A and its use as a glycoprotein stain on sodium dodecyl sulfate/polyacrylamide gels. *Biochem. J.* 243:569–574.

MEANS, G. E., and FEENEY, R. E. 1968. Reductive alkylation of amino groups in proteins. *Biochemistry* 7:1366–1371.

MEANS, G. E., and FEENEY, R. E. 1971. *Chemical Modification of Proteins.* Holden-Day, San Francisco, CA.

MEANS, G. E., and FEENEY, R. E. 1990. Chemical modification of proteins: History and applications. *Bioconjugate Chem.* 1:2–12.

MODESTO, R. R., and PESCE, A. J. 1971. The reaction of 4,4'-difluoro-3,3'-dinitrodiphenyl sulfone with γ-globulin and horseradish peroxidase. *Biochim. Biophys. Acta* 229:384–395.

PATRAMANI, I.; KATSIRI, K.; PISTEVOU, E.; KALOGERAKOS, T.; PAWLATOS, M.; and EVAN-GELOPOULOS, A. E. 1969. Glutamic-aspartic transaminase-antitransaminase interaction: A method for antienzyme purification. *Eur. J. Biochem.* 11:28–36.

PERHAM, R. N., and THOMAS, J. O. 1971. Reaction of tobacco mosaic virus with a thiol-containing imido ester and a possible application to X-ray diffraction analysis. *J. Mol. Biol.* 62:415–418.

PETERS, K., and RICHARDS, R. M. 1977. Chemical cross-linking. Reagents and problems in studies of membrane structure. *Ann. Rev. Biochem.* 46:523–551.

PIERCE CHEMICAL CO. 1994. *Life Science and Analytical Research Products Catalog and Handbook.* Rockford, IL.

PIZZO, S. V.; ROCHE, P. A.; FELDMAN, S. R.; and GONIAS, S. L. 1986. Further characterization of platinum-reactive component of the α_2-macroglobulin-receptor recognition site. *Biochem. J.* 238:217–225.

PUMA, P.; BUXSER, S. E.; WATSON, L.; KELLCHER, D. J.; and JOHNSON, G. L. 1983. Purification of the receptor for nerve growth factor from A875 melanoma cells by affinity chromatography. *J. Biol. Chem.* 258:3370–3375.

QUINLAN, R. A., and FRANKE, W. W. 1982. Heteropolymer filaments of vimentin and desmin in vascular smooth muscle tissue and cultured baby hamster kidney cells demonstrated by chemical cross-linking. *Proc. Natl. Acad. Sci.* (USA) 79:3452–3456.

RAFTERY, M. A., and COLE, R. D. 1963. Tryptic cleavage at cysteinyl peptide bonds. *Biochem. Biophys. Res. Commun.* 10:467–472.

RAY, W. J., and KOSHLAND, D. E. 1961. A method for characterizing the type and numbers of groups involved in enzyme action. *J. Biol. Chem.* 236:1973-1979.

RIEDER, R., and BOSSHARD, H. R. 1980. Comparison of the binding sites on cytochrome c for cytochrome c oxidase, cytochrome bc_1 and cytochrome c_1. Differential acetylation of lysyl residues in free and complexed cytochrome c. *J. Biol. Chem.* 255:4732–4739.

ROBINSON, I. D. 1964. Role of cross-linking of gelatin in aqueous solutions. *J. Appl. Polymer Sci.* 8:1903-1918.

ROCHE, P. A.; JENSEN, P. E. H.; and PIZZO, S. V. 1988. Intersubunit cross-linking by *cis*-dichlorodiammineplatinum (II) stabilizes an α_2-macroglobulin "nascent" state: Evidence that thiol ester bond cleavage correlates with receptor recognition site exposure. *Biochemistry* 27:759–764.

SAGER, P. R. 1989. Cytoskeletal effects of acrylamide and 2,5-hexanedione: Selective aggregation of vimentin filaments. *Toxicol. Appl. Pharmacol.* 97:141–155.

SATO, S., and NAKAO, M. 1981. Cross-linking of intact erythrocyte membrane with a newly synthesized cleavable bifunctional reagent. *J. Biochem.* 90:1177–1185.

SCHAFER, H. J. 1986. Divalent azido-ATP analog for photoaffinity cross-linking of F_1 subunits. *Methods Enzymol.* 126:649–660.

SCHMIDT, P. R., and BETZ, H. 1989. Cross-linking of β-bungarotoxin to chick brain membranes. Identification of subunits of a putative valtage-gated K^+ channel. *Biochemistry* 28:8346–8350.

SHEEHAN, J. C.; CRUICKSHANK, P. A.; and BOSHART, G. L. 1961. A convenient synthesis of water-soluble carbodiimides. *J. Org. Chem.* 26:2525–2528.

SHEWALE, J. G., and BREW, K. 1982. Effects of Fe^{3+} binding on the microenvironments of individual amino groups in human serum transferrin as determined by differential kinetic labeling. *J. Biol. Chem.* 257:9406–9415.

SKOLD, S. E. 1983. Chemical crosslinking of elongation factor G to the 23S RNA in 70S ribosomes from *Escherichia coli. Nucleic Acid Res.* 11:4923–4930.

SORENSEN, P.; FARBER, N. M.; and KRYSTAL, G. 1986. Identification of the interleukin-3 receptor using an iodinatable, cleavable, photoreactive cross-linking agent. *J. Biol. Chem.* 261:9094–9097.

SRINIVASACHAR, K., and NEVILLE, D. M. 1989. New protein cross-linking reagents that are cleaved by mild acid. *Biochemistry* 28:2501–2509.

STAROS, J. V.; BAYLEY, H.; STANDRING, D. N.; and KNOWLES, J. R. 1978. Reduction of aryl azides by thiols: Implications for the use of photoaffinity reagents. *Biochem. Biophys. Res. Commun.* 80:568–572.

TERNYNCK, T., and AVRAMEAS, S. 1977. Conjugation of *p*-benzoquinone treated enzymes with antibodies and Fab fragments. *Immunochemistry* 14:767–774.

TIJSSEN, P. 1985. *Practice* and Theory of Enzyme Immunoassays. *Elsevier, Amsterdam.*

TRAUT, R. R.; BOLLEN, A.; SUN, T. T.; HERSHEY, J. W. B.; SUNDBERG, J.; and PIERCE, L. R. 1973. Methyl 4-mercaptobutyrimidate as a cleavable cross-linking reagent and its application to *Escherichia coli* 30S ribosome. *Biochemistry* 12:3266–3273.

TROMMER, W. E., and HENDRICK, M. 1973. Formation of maleimides by a new mild cyclization procedure. *Synthesis* 8:484–485.

TSOU, C. 1962. Kinetic determination of essential side chains in proteins. *Sci. Sin.* 11:1535–1558.

UCKERT, W.; WUNDERLICH, V.; GHYSDAEL, J.; PORTETELLE, D.; and BURNEY, A. 1984. Bovine leukemia virus (BLV): A structural model based on chemical cross-linking studies. *Virology* 133:386–392.

VANIN, E. F., and JI, T. H. 1981. Synthesis and application of cleavable photoactivable heterobifunctional reagents. *Biochemistry* 20:6754–6760.

VAS, M., and CSANADY, G. 1987. The two fast-reacting thiols of 3-phosphoglycerate kinase are structurally juxtaposed: Chemical modification with bifunctional reagents. *Eur. J. Biochem.* 163:365–368.

WAGNER, R., and GARRETT, R. A. 1978. A new RNA-RNA cross-linking reagent and its application to ribosomal 5S RNA. *Nucleic Acid Res.* 5:4065–4075.

WEBB, R. R., and KANEKO, E. 1990. Synthesis of 1-(aminooxy)-4-[(3-nitro-2-pyridyl)dithio]butane hydrochloride and 1-(aminooxy)-4-[(3-nitro-2-pyridyl)dithio]but-2-ene. Novel heterobifunctional cross-linking reagents. *Bioconjugate Chem.* 1:96–99.

WELLS, J. A.; KNOEBER, C.; SHELDON, M. C.; WERBER, M. M.; and YOUNT, R. G. 1980. Cross-linking of myosin subfragment. 1. Nucleotide-enhanced modification by a variety of bifunctional reagents. *J. Biol. Chem.* 255:11135–11140.

WHITE, F. H. 1972. Thiolation. *Methods Enzymol.* 25:541–546.

WOLD, F. 1972. Bifunctional reagents. *Methods Enzymol.* 25:623–651.

WOLF, B.; LESNAW, J. A.; and REICHMANN, M. E. 1970. Mechanism of the irreversible inactivation of bovine pancreatic ribonuclease by diethylpyrocarbonate. General reaction of diethylpyrocarbonate with proteins. *Eur. J. Biochem.* 13:519–525.

WONG, S. S. 1993. *Chemistry of Protein Conjugation and Cross-Linking.* CRC Press, Boca Raton, FL.

WOODWARD, R. B., and OLOFSON, R. A. 1961. The reaction of isoxazolium salts with bases. *J. Am. Chem. Soc.* 83:1007–1009.

WOODWARD, R. B.; OLOFSON, R. A.; and MAYER, H. 1961. A new synthesis of peptides. *J. Am. Chem. Soc.* 83:1010–1012.

5
Antibody Production

INTRODUCTION

The production and large scale availability of antibody of desired affinity and specificity and/or cross-reactivity are fundamental to the process of developing sensitive, precise, and specific immunoassay products for use in the diagnostics industry. Although the primary objectives of this phase of immunoassay product development appear to be quite simple, i.e., to elicit an immune response, and to select and purify antibodies that have the necessary specificity and sensitivity required for the end-use application, a bewildering variety of approaches and choices may complicate the process of antibody production. Irrespective of the strategy chosen, the process needs to yield an antibody having all the desired properties. Although there are no definitive guidelines or procedures that guarantee an ideal product to meet all requirements, certain principles nevertheless must be followed to produce an effective and workable antibody. Factors that primarily govern the production of a desirable antibody are described in this chapter.

An ideal antiserum should have a high titer coupled with high affinity and specificity. Because immunoassays utilize the primary antigen:antibody reactions that obey the Law of Mass Action, and because the antigen:antibody complex is also capable of dissociation during the incubation and separation stages of an immunoassay, it is the affinity of the antibody that primarily determines the sensitivity (or the minimum detection limit) of an assay. This follows from the fact that for the measurement of a given concentration of unlabeled ligand (e.g., drug levels in serum) the concentration of the enzyme or tracer label and the antibody should be comparable or even smaller. If the target concentration is low, as often is the case with several hormones and therapeutic drug levels, the concentration of the label and antibody must be correspondingly low. Thus, for any reaction to take place at these sub-nano- or pico-molar concentrations, the affinity constant of the antibody, as expressed by its K_a value, must be high.

The affinity or K_a value of a good antibody preparation may be as high as 10^9–10^{12} l/mol. Thus, if sensitivity is broadly defined as equivalent to $1/K_a$, the detection range will range from 10^{-9}–10^{-12} mol/l, or 1 nmol–1 pmol/l. This potential sensitivity is generally more than adequate for the majority of substances of biological interest.

The desired specificity and/or cross-reactivity of the antibody preparation also need to be considered prior to the preparation of the immunogen. If the intended product is supposed to determine the presence of different compounds belonging to an entire family of drugs such as β-lactams or sulfonamides, the antibody then needs to have a broad range of cross-reactivity. Therefore, it must be able to recognize the antigenic determinants that are common to that family of drugs. In some instances, the assay must distinguish a particular drug of interest from a variety of structurally related substances. The development of an immunoassay for the asthma drug, theophylline, which distinguishes it from a dazzling array of substituted xanthines and hypoxanthines is a classic example of a highly specific antiserum preparation. Thus, one needs to consider the specificity and/or cross-reactivity in the end-use application of the intended antibody preparation. Some selected examples of monitoring these two aspects will be described later in this chapter.

Large scale availability of the antibody is primarily a practical problem that, although it does not influence the characteristics of an assay, may nevertheless determine the choice between alternative systems of similar characteristics. Factors that need to be considered in this regard include the difficulty of the initial preparation of the antibody, the reproducibility of different preparations, its stability on storage, the cost, and the amount required in terms of the number of assays likely to be performed.

Generally, a specific antibody, once produced, is usually available in quantities sufficient for a very large number of assays. For example, a single large animal such as a goat or sheep may yield enough antibody to last for several years. Furthermore, with materials that are good immunogens, the production of new antiserum presents little or no difficulty. However, problems may arise with materials that are poor immunogens, or that require an elaborate and complicated immunogen preparation. Thus, if their antisera represent only a few bleeds from a small animal, supplies will be limited.

Polyspecific as well as specific monovalent antisera are commercially available for a wide variety of substances of biological importance. However, these antisera are expensive, and are often diluted to a low titer. Furthermore, the desired antibody may not be offered commercially. It is therefore ideal to have an in-house program to prepare desired antibodies. Moreover, the consistency in the quality of the antiserum forms the essential basis of all immunoassay products, and can be best controlled by such an in-house production program.

Almost all aspects of an antibody production program are, at best, arbitrary. In principle, this reflects the fact that the same end can be achieved by a variety of means, although in practice, it is very difficult to show that any one scheme or strategy is superior to any other, because the variable nature of the antibody response would demand large groups of animals and appropriate controls. Despite these uncertainties, there are several specific factors that can be discussed in relation to the success or failure of an antibody development program.

To accomplish the goal of producing a desirable antibody, one must decide the following:

- Whether to develop monoclonal or polyclonal antibodies,
- The type of immunogen and host animals to use,
- The type of immunization schedule to follow,
- How to set up appropriate antibody screening assays,
- The techniques to use for purifying and storing the antibody, and
- How to establish a quality control program to assess the performance of the product.

As an example, a general protocol for polyclonal antibody production and quality control is presented in Table 5.1. It is impossible to discuss all the relevant aspects in detail. Some excellent general treatises were edited by Williams and Chase (1967), Sela (1973–1982), Chard (1978), Steward (1984), Tijssen (1985), Clausen (1988), Catty (1988), Harlow and Lane (1988), Zola (1987), and Roitt et al. (1989). Hence, only the practical aspects and guidelines for the important stages in the production of antibodies are described here.

NATURE AND PREPARATION OF IMMUNOGEN

Large Molecular Weight Compounds

The immunogenicity of a molecule, i.e., its property to elicit an immune response, is directly related to its molecular weight (MW). In contrast, antigens are compounds that will react with an antibody, but are not necessarily immunogenic on their own (e.g., most haptens). Because of their high degree of conformation and structural rigidity, molecules of MW > 5000 are generally good immunogens as long as they are foreign to the recipient animal. Foreignness to the recipient host animal is absolutely essential to obtain a good immune response. For example, several mammalian serum proteins that exhibit structural homology are poor immunogens in closely related species, and therefore, will not elicit a strong immune response.

TABLE 5.1.   Strategy for Antibody Production and Quality Control[a]

1. **Preparation of Immunogen**
   - Purify
   - Test purity
   - Conjugate to carrier if less than 5,000-10,000 molecular weight

2. **Immunization**
   - Select an appropriate species as recipient
   - Decide on the form of immunogen presentation (e.g., with adjuvant), route of administration, dose and timing; allow as long as possible between injections—up to several months (advance planning is essential)
   - Prepare the immunogen in final format for injection (emulsion, precipitate, etc.), enough for a single dose

3. **Test Procedure**
   - Test bleed and separate the serum
   - Check specificity, titer, and binding properties; include tests under same conditions as planned for final application
   - Do trial absorptions on the basis of any detected cross-reactivities—prepare insoluble adsorbents
   - Retest absorbed serum. If satisfactory, prepare for scale-up absorption
   - Decide on boosting strategy on the basis of interim tests
   - Repeat all of the above as required

4. **Harvesting, Storage, and Absorption**
   - The volume required determines staged harvesting or immediate sacrifice of animals
   - Store in aliquots labeled according to batch/date, store at -20°C or preferably lower
   - Test and trial-absorb subsamples, then bulk-absorb aliquots as required

5. **Immunoglobulin Fractionation**
   - Prepare by the two-stage method of salt precipitation and anion-exchange chromatography
   - Test the purity of the fraction
   - Check the activity of the fraction
   - Conditions for storage vary by species

6. **Affinity Purification of Antibodies**
   - Check the purity of the antigen
   - Prepare the antigen affinity column by appropriate coupling chemistry to selected matrix.
   - Check antigenic activity of the column, loading capacity, and conditions for antibody elution
   - Perform trial purification, testing purity, activity, and stability of eluted product
   - Perform bulk purification with appropriate eluate storage in aliquots

TABLE 5.1. (*continued*)

---

7. Quality Control
- Assess the performance of the product across a range of applications, use antigen standards where available
- Calibrate against a reference reagent where applicable
- Quantitate antibody concentration (to soluble single antigens)
- Determine isotype(s) of antibody

---

[a]From Catty and Raykundalia (1988).

Most large MW proteins and polypeptides, therefore, are naturally immunogenic, and may be injected without further modification. They also possess several antigenic determinants. Certain complex polysaccharides also behave in a similar manner, although their natural immunogenicity is often due to small amounts of protein frequently associated with their structures (Burrin and Newman 1991).

Because of their tertiary and quaternary conformations, the antigenic determinants of these molecules are often discontinuous in sequence, and hence, reagents such as dithiothreitol (DTT or Cleland's reagent), mercaptoethanol, and strong chaotropic reagents such as urea and guanidine hydrochloride that disrupt their native conformation should not be used during the preparation and purification of such immunogens.

Not all proteins are naturally immunogenic. For example, human prolactin (MW ~ 25, 000) on its own is only very weakly immunogenic (Burrin and Newman 1991). In such instances, these molecules should be treated as haptens and coupled to a carrier protein prior to immunization. Although not necessary, the use of adjuvants is also often recommended for large MW compounds to obtain the best results (Catty and Raykundalia 1988). To raise antisera against biological fluids of low antigen content, e.g., tissue culture media, urine, or cerebrospinal fluids, a concentration or lyophilization step is often required to increase the concentration of proteins or other macromolecular antigens (Clausen 1988).

In certain instances, large molecules share common sequences with other members of the same family of compounds. For example, an antibody raised against the α-subunit of the glycoprotein pituitary hormones will react nonspecifically with other members of the family containing the conserved subunit. Thus, the antibody specificity to the individual hormones could be markedly improved by using the characteristic β-subunit of the protein as an immunogen.

Although antiserum to whole organ antigens can be produced successfully, such approaches are often limited by three factors (Warr 1982; Clausen 1978). First, the lipid content of the organ may mask certain protein antigens to a degree that makes it impossible to elicit an immune response. The lipids, therefore, need to be removed either by repeated freeze-thaw or by acetone extraction (Laterre and Heremans 1963). These procedures, however, may cause protein denaturation with possible alteration of antigenic specificity. The second difficulty in preparing antiserum against organ antigens is the autoimmune reaction in the recipient animal provoked by the introduction of antigens possessing partial immunological identity with proteins from corresponding organs in the recipient animals, thereby leading to its death. Finally, antigenic similarities between different animal species may give rise to immunotolerance (described in Chapter 2) towards some antigens, thereby causing an unsuccessful immune response.

Small Molecular Weight Compounds

Small peptides such as CTH (MW 4500) and the posterior pituitary hormones (MW 1000) do not elicit a strong immune response. Similarly, compounds of MW < 1000 (e.g., steroid hormones, therapeutic drugs) are practically nonimmunogenic. The immunochemistry of such low molecular weight molecules was very elegantly elucidated by Landsteiner (1945). He first demonstrated that antibodies can be raised to small molecules when they are covalently linked to a naturally immunogenic carrier protein as "haptens." The ability to couple many different structures to macromolecules, the high degree of antigenicity of such conjugates, the development of sensitive methods of detecting and quantitating reactions between antibody and hapten, and the perfection of techniques for obtaining highly purified preparations of antihapten antibodies have contributed significantly to the development of several of our modern immunological concepts, as well as to the popularity of immunodiagnostics as an analytical tool in the clinical field.

Proper point of attachment of the haptenic molecule to the carrier protein is absolutely essential in the production of an antibody of desired specificity and cross-reactivity. This is governed by Landsteiner's principle (1945), which states that antibody specificity is directed primarily at that position of the hapten farthest removed from the functional group used to link it to the carrier protein. Thus, antibody specificity is directed towards that position of the hapten molecule that is most accessible to the circulating macrophages and lymphocytes. Therefore, depending on the specificity required in the end product, great care must be taken to ensure that those groups that distinguish a molecule from its precursors, metabolites, and/or analogs are not used for conjugating to carrier protein.

Landsteiner's prediction is particularly applicable to molecules possessing ring structures. Several examples in the literature bear out this fact; some selected ones are briefly described below.

1. Opiate Immunoassays.   The basic structures of the two related opiates, viz., morphine and codeine, are shown in Figure 5.1. One of the first immunoassays for morphine was based on an antibody raised against the derivative, 3-*O*-carboxymethyl morphine (Figure 5.2A), covalently linked to bovine serum albumin (Spector and Parker 1970). Antibodies generated in this manner recognized

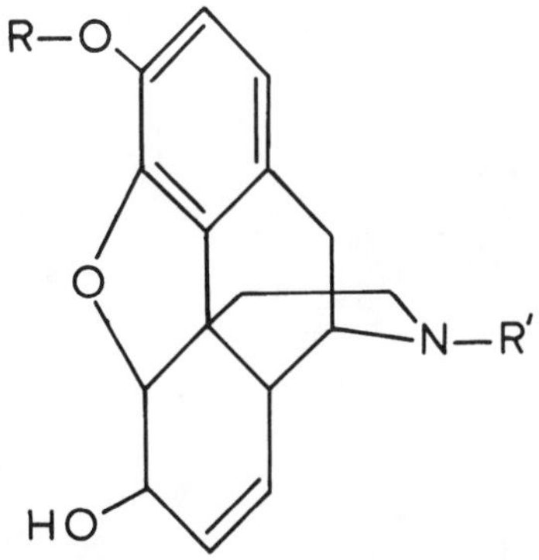

Figure 5.1. Basic structures of opiates

A: Morphine   R = H   R' = CH$_3$
B: Codeine   R = CH$_3$ R' = CH$_3$
C: Normorphine R = H   R' = H
D: Norcodeine   R = CH$_3$ R' = H

codeine (Figure 5.1B) more strongly than morphine (Figure 5.1A). Rubenstein et al. (1972) developed a homogeneous enzyme immunoassay for detecting opiate abuse. These researchers used nonspecific antibodies generated by this derivative, because the purpose of the assay was to detect all abused opiates in urine and to measure morphine glucuronide.

Subsequent researchers designed opiate immunoassays for a variety of purposes. For studies on the metabolism of codeine to morphine, specific immunoassays were required. Similarly, specificity is also required for in vivo pharmacological or pharmacokinetic studies. To achieve these goals, various morphine derivatives that have been used to stimulate antibody production illustrate the different approaches to the problem of distinguishing morphine from codeine and other opiates.

Spector et al. (1973) used morphine-6-hemisuccinate (Figure 5.2B) and another derivative (Figure 5.2C1) to generate antibodies and compared them with antibodies raised against the 3-*O*-carboxymethyl morphine prepared in their original work (Spector and Parker 1970). Gross et al. (1974) synthesized the axomorphine derivative (Figure 5.2C2) for antibody production. The functionalized *N*-normethyl compounds at the alkaloid nitrogen to generate the derivatives D1, D2, and D3 (Figure 5.2) were used by Findlay et al. (1976, 1977) and Morris et al. (1974).

The relative affinities of these various antibodies for codeine and morphine were evaluated in eight different immunoassay systems by these researchers (Table 5.2). Substitutions on the 3-oxygen (Figure 5.2A) and the 6-oxygen (Figure 5.2B) provided no selectivity between morphine and codeine. The discrimination of antisera raised against C2 was considered significant by Gross et al. (1974), yet

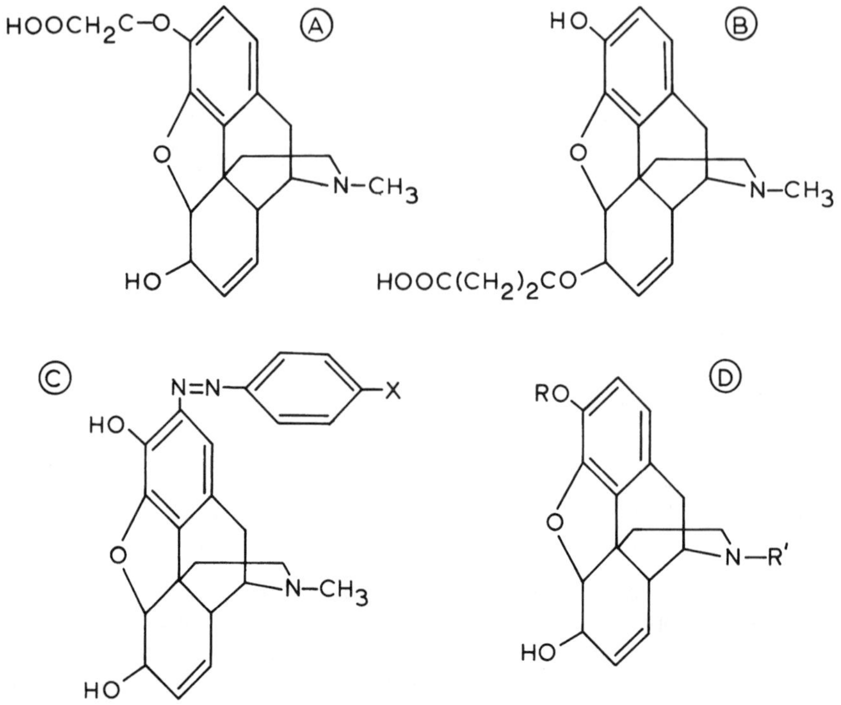

Figure 5.2. Haptens used for the production of antibodies with varying specificities and cross-reactivities to morphine derivatives
A: 3-O-Carboxymethyl morphine
B: Morphine-6-hemisuccinate
C: C1: X = NH$_2$
    C2: X = COOH
D: N-Normethyl morphine derivatives
D1: R = H   R' = (CH$_2$)$_3$COOH
D2: R = CH$_3$ R' = (CH$_2$)$_3$COOH
D3: R = H   R' = CO(CH$_2$)$_3$COOH

antisera raised against C1 showed poor selectivity (Spector et al. 1973). A highly specific and superior response was achieved using the haptens substituted on nitrogen (Findlay et al. 1976, 1977; Morris et al. 1974). Because this site is distal to the position where morphine and codeine differ, maximum opportunity for selectivity is provided.

2. Theophylline Immunoassay.   Theophylline immunoassay is yet another example of how hapten conjugation to carrier protein affects the specificity of the resulting antibody. Theophylline is considered the most useful and primary drug of choice in the treatment and prevention of asthmatic symptoms in children and

TABLE 5.2. Percent Cross-Reactivity of Opiate Immunoassays Based on Different Immunogens

Immunogen[a]	Morphine	Codeine	Reference
A	100	100	Spector et al. (1973)
A	100	140	Schneider et al. (1973)
B	100	100	Spector et al. (1973)
C1	100	83	Spector et al. (1973)
C2	100	10	Gross et al. (1974)
D1	10	0.03	Findlay et al. (1977)
D2	0.1	100	Findlay et al. (1977)
D3	100	4.8	Morris et al. (1974)

[a]Structures of the hapten derivatives used for immunogen preparation and antibody production are shown in Figure 5.2.

adults. The relationship between theophylline dosage and the likelihood of achieving both therapeutic effect and toxicity is well known. However, because there are remarkable variations in the rate of individual metabolism of theophylline in both children and adults, large percentages of patients in any population receiving the useful dosage will either not achieve a therapeutic effect or be at the risk of toxicity. This is particularly true in the case of infants, who are very susceptible to seizure, while other adverse side effects are usually difficult to recognize. It is now a well-recognized fact that safe, effective use of theophylline in adults and children requires careful adjustment of the dose based on serum measurements of theophylline concentration. Theophylline dosages that achieve the serum concentration ranges of 10–20 µg/ml are usually effective in suppressing chronic asthmatic symptoms. The development of a homogeneous enzyme immunoassay for theophylline was an important factor in the treatment of this disease (Gushaw et al. 1977).

The variety of structurally related substances that are important in an assay for theophylline (1,3-dimethylxanthine) is shown in Figure 5.3. Variations in the alkyl substituents on the nitrogen atoms provide a dazzling array of substituted xanthine and hypoxanthine derivatives. In order for an assay to be widely useful, the antisera must differentiate theophylline from other common methylated xanthines including the two most important ones, viz., caffeine (1,3,7-trimethylxanthine) and theobromine (3,7-dimethylxanthine).

Three different immunochemical approaches were used to solve the problem of antibody specificity. Cook et al. (1976) synthesized 8-(3-carboxypropyl)-theophylline (Figure 5.4A) as the hapten derivative. The rationale was to leave positions 1, 3, 7, and 9 all exposed to allow the antibody to recognize differences at these positions. In a second approach, Neese and Soyka (1977) prepared

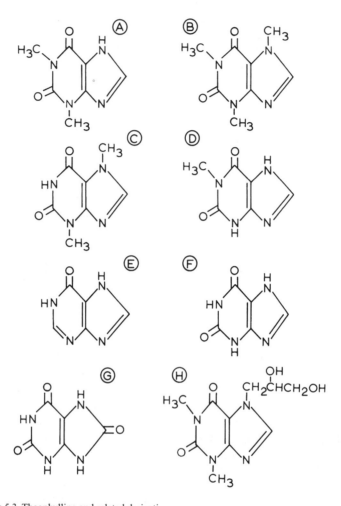

Figure 5.3. Theophylline and related derivatives
A: Theophylline   B: Caffeine
C: Theobromine   D: 1-Methyl xanthine
E: Hypoxanthine   F: Xanthine
G: Uric acid   H: Dyphylline

8-carboxytheophylline (Figure 5.4B), thus choosing the 8-position for linking to carrier protein and to maximize specificity at the 1-, 3-, and 7-positions. In the third approach, Gushaw et al. (1977) used (3-carboxypropyl)-theophylline (Figure 5.4C) as the hapten derivative. Their strategy was to emphasize differences between theophylline, caffeine, and theobromine. Position 1, where theobromine differs from the other two, and position 7, where theophylline differs from caf-

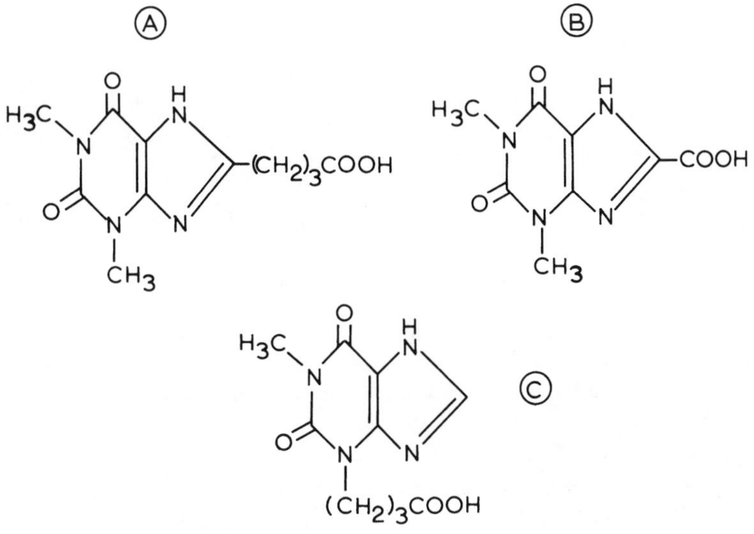

Figure 5.4. Theophylline derivatives used for haptenic immunogen preparations to produce antibody of desired specificity
A: 8-(3-Carboxypropyl)-theophylline
B: 8-Carboxytheophylline
C: (3-Carboxypropyl)-theophylline

feine and theobromine, were both left exposed. These researchers selected position 3, which is common to all three, as the linking point.

The relative responses of theophylline and its related derivatives to these three different antibody preparations and as evaluated in three different immunoassays are summarized in Table 5.3. The comparison is somewhat affected by differences in the definition of cross-reactivity between the systems, and by the fact that 1-methylxanthine is added to the antibody to desensitize the enzyme immunoassay to this compound (EMIT® Theophylline Assay 1978). Nonetheless, it is quite evident that conjugation of theophylline to carrier protein at the 3-position yields better sensitivity against caffeine than attachment at the 8-position. Conversely, the relative responses of 1-methylxanthine indicate that there is less cross-reactivity to this compound, but not to caffeine, in the two systems where the 8-conjugate was used.

3. Drug Abuse (Urine and Serum Toxicology) Assays.   Unlike the two examples discussed above, wherein the antibody specificity is essential to detect only the drug of interest, drug abuse assays are usually designed to detect an entire class of structurally similar drugs and their metabolites. The emphasis here is to detect the presence of drug rather than quantify its concentration. As an example, the structural similarities among benzodiazepines commonly encountered are

TABLE 5.3.   Relative Cross-Reactivity of Theophylline Immunoassays

Derivative used for immunogen[a] (Immunoassay)	Theo-phylline	Caffeine	Theobro-mine	1-Methyl-xanthine	3-Methyl-xanthine
8-(3-Carboxypropyl)-theophylline (RIA)[b]	100.0	4.2	0.09	0.08	2.0
8-Carboxytheo phyllinec (RIA)[c]	100.0	16.0	1.1	0.9	2.0
(3-Carboxypropyl)-theophylline (EIA)[d]	100.0	0.6	< 0.1	~ 5.0	~ 0.25

[a]See Figures 5.2, 5.3, and 5.4 for related chemical structures.
[b]From Cook et al. (1976), evaluated at 50% inhibition.
[c]From Neese and Soyka (1977), evaluated at 50% inhibition.
[d]EMIT® Theophylline Assay (1978), evaluated as the quantity necessary to give a response equivalent to 1 μg/ml theophylline in the EMIT® Theophylline Assay, Syva Co., San Jose, CA.

shown in Figure 5.5. The EMIT® assays, developed by Syva Co., San Jose, CA, are constructed in such a manner that any of these compounds will generate approximately the same response (Jaklitsch 1985). For example, the assay reports as positive 2 μg/ml flurazepam, as well as its metabolite, desalkylflurazepam. This wide cross-reactivity was achieved by coupling a synthetic drug derivative (Figure 5.5G) to carrier protein by the functional group that differs among compounds of this class. Thus, antibodies were raised to that part of the benzodiazepine nucleus that is common to members of this class.

This type of immunoassay, which detects a class of drug compounds, is particularly useful in the hospital emergency room where knowledge of the total drug load is critical to saving a patient's life. Typically, an estimate of the total drug load is made by comparing the response of the sample to the response of a representative member of the drug class. The unknown is then quantified in total drug equivalents.

The nature of hapten linkage to carrier protein is also critical to the specificity and sensitivity of an immunoassay. Immunogens prepared by haptenation of proteins often generate antibodies against the linkage. For example, Eisen and Siskind (1964) observed that antibodies raised to 2,4-dinitrophenol conjugated to lysyl groups of carrier proteins were more sensitive to ε-DNP-lysine than to dinitrophenol. Van Weeman and Schuurs (1971) also observed that the enzyme immunoassays developed for estrogens were more sensitive to the hemisuccinate derivatives than to the parent estrogens. These researchers further investigated the effects of heterology (the use of different hapten derivatives to prepare the enzyme conjugate and the immunogen) on the sensitivity and specificity of estrogen as-

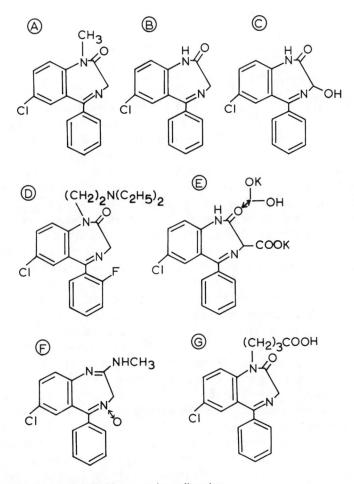

Figure 5.5. Structural similarities among benzodiazepines
A: Diazepam   B: N-Desmethyl diazepam
C: Oxazepam   D: Flurazepam
E: Chlorazepate   F: Chlordiazepoxide
G: N-Carboxypropyl diazepam, a synthetic hapten used to obtain an antibody of desired cross-reactivity to various benzodiazepines

says (van Weeman and Schuurs 1975). They defined three types of heterologies, any of which improves detectability:

1. Hapten heterology, where different but related haptens are attached through the same site by the same linkage. Van Weeman and Schuurs (1975) used an antibody to estradiol and an estrone-enzyme conjugate, both linked at the 11-OH position and using the same succinyl linking group in their study.

2. Bridge heterology, where different cross-linkers are used for coupling the haptens to carrier protein and the enzyme label, e.g., a succinyl linking group for the immunogen and a glutaryl side chain for the enzyme conjugate, and

3. Site heterology, where the same linking group is attached to different sites (e.g., 11- and 17-hydroxyl of the hormone) of the hapten.

It is critical to use at least one of the three approaches mentioned above while screening the antisera for the presence of and the determination of antibody titer. Van Weeman and Schuurs (1975) observed a 100–200-fold increase in assay sensitivity when any of the heterologous combinations of antibody and enzyme conjugate was used. This was explained by a reduction in the difference between the antibody affinity for the enzyme conjugate and the free estrogen in the mismatched system, thereby allowing increased competition by free estrogen.

Another important factor in preparing the haptenic immunogen relates to the optimal number of haptens bound to the carrier protein. The incorporation of too many hapten molecules as well as too few in a hapten-protein conjugate generally leads to a poor antibody response. Because there are no definitive rules, it may be beneficial to prepare the conjugates both with several carriers as well as with a range of hapten-carrier protein molar ratios to determine the best immunogen.

Generally, the number of hapten molecules conjugated to the carrier protein depends on the nature of carrier protein and the number of functional reactive groups available for conjugation. A ratio of 10:1 to 15:1 incorporation of hapten into the protein molecule appears optimal when bovine serum albumin (BSA) is the choice of protein, whereas Erlanger (1980) suggests an optimal density anywhere between 8 to 25 to obtain good antibody titers. With aromatic and hydrophobic haptens, lower ratios are sometimes preferred to avoid the problems of precipitation during immunogen preparation. Similarly, with protein carriers of low MW, such as ovalbumin, a lower degree of hapten incorporation is preferred. In any event, it is generally a good practice to determine the molar incorporation of the drug prior to immunization. This number can be easily estimated from the absorbance spectrum of the conjugate, or by determining the number of amino groups remaining on the carrier protein after conjugation with reagents such as trinitrobenzene sulfonic acid.

In summary, the examples from the work on opiates, theophylline, benzodiazepines, and steroid hormones emphasize the need and importance of considering the critical antigenic determinants of a hapten both during the preparation of immunogen as well as in screening, developing, and optimizing the immunoassay. Various techniques and approaches available for conjugating haptens to carrier proteins and enzymes were described in Chapter 4.

Choice of Carrier Proteins

No realistic studies were ever carried out regarding whether or not the choice of a particular carrier significantly influences the immune response. The choice of carrier protein often lies in personal preferences, and sometimes with availability and cost. Factors that primarily determine the choice of carrier include its immunogenicity and solubility and whether or not adequate conjugation can be achieved. The chosen carrier should also be irrelevant to all aspects of the future assays.

The commonly used protein carriers for haptenation include globulin fractions, the serum albumins of various species, ovalbumin, thyroglobulin, and keyhole limpet hemocyanin (KLH). Synthetic polypeptide carriers such as poly (L-lysine or poly(L-glutamic acid), although used sometimes, are generally not advisable choices.

Because "foreignness" of a protein to the recipient host animal induces a strong immune response, KLH which is isolated from the mollusc *Megathura crenulata,* is a widely used carrier protein. It has a large molecular mass (MW 4.5 $\times$ 10^5–1.3×10^7) and several available lysine groups. However, because of its large size, KLH often suffers from poor water solubility. Moreover, it also has several antigenic determinants that may compete with the hapten.

Antigenic competition is particularly important with hormones and drugs that are weakly immunogenic as epitopes. In such instances, a moderately immunogenic protein such as BSA should be used as a carrier protein (Burrin and Newman 1991). BSA in fact has been the most popular choice with laboratories developing immunoassays. It is inexpensive and readily available, has good immunogenicity and a number of functional groups, resists denaturation, and is very soluble.

Recently, Pierce Chemical Co. (1994) has introduced a new Imject® Supercarrier® system for use as a carrier protein. It uses a cationized BSA (cBSA) prepared by coupling a diamine to the carboxyl groups on BSA via EDC-mediated reaction. The highly cationized molecule (pI 11–12 as compared to 4.5 for native BSA), when coupled with haptens or other proteins, is believed to be more effective in presenting the immunogen to the cells for processing and ultimate antibody production. When normally immunogenic proteins are coupled to cBSA, the antibody response is claimed to be 2–3-fold greater.

The degree of purity of an immunogen is not necessarily related to the specificity of the antibody produced. However, traces of very immunogenic impurities may sometimes overwhelm the response to principle antigen. Generally, some purification of the hapten-protein conjugates prior to immunization to remove unreacted hapten and adsorbed material, either by dialysis or gel filtration, is desirable. This practice may not only minimize unwanted biological or pharmacological effects of the injected conjugate, but also improve the specificity of the resultant antibodies. The purification of larger particles such as viruses generally poses fewer problems. However, the complexity of these particles may in turn require their dissociation and fractionation to decrease the number of antigenic epitopes.

Adjuvants

Adjuvants are substances that nonspecifically enhance the immune response to immunogens. Several properties and functions of adjuvants in antibody production are summarized in Table 5.4. An excellent monograph on several aspects of adjuvants used in biotechnology is also available (Spier and Griffiths 1985). Generally, antibody responses to antigens in adjuvants are greater, more prolonged, and frequently consist of different classes to the response obtained without adjuvant (Roitt et al. 1989).

A range of adjuvants is available commercially; these adjuvants are differentiated into the following five classes.

1. Water-in-oil emulsions, e.g., Freund's complete and incomplete adjuvants, respectively containing and lacking heat-killed *Mycobacterium tuberculosis* bacteria, Pierce's AdjuPrime™ Immune Modulator containing branched glucose polymers, methylated BSA

2. Minerals, e.g., antigen absorbed onto aluminum hydroxide (alumina), quaternary ammonium salts, bentonite

3. Bacterial, e.g., Bacille Calmette-Guérin (BCG), *Bordetella pertussis,* and *Corynebacterium parvum* organisms

4. Bacterial products, e.g., endotoxins, muramyl dipeptide fragment of BCG, lipopolysaccharides, liposomes, and

5. Polynucleotides, e.g., poly I-Poly C, Poly (A-U)

The most popular adjuvants are the water-in-oil type with the immunogen in the aqueous phase. Of these, Freund's complete adjuvant is by far the most com-

TABLE 5.4.   Properties and Functions of Adjuvants[a]

- Stimulate immune response and antibody production nonspecifically by increasing the efficiency of antigen presentation and the number of collaborating and secreting cells involved
- Effectively reduce the required immunogen dose
- Enhance immunogenicity by allowing antibody response to molecules with weak immunogenic properties
- May change the mode of immune response, e.g., tolerance vs immunity
- Alter isotype pattern of antibody responses
- Act as immunogen depot for prolonged antigen stimulation
- Protect the immunogen from rapid removal and breakdown, thereby reducing the need for repeated injections
- Increase the average avidity and affinity of the antibody response

---

[a]Compiled from Catty (1988), Roitt et al. (1989), and Spier and Griffiths (1985).

monly used for the primary immunization. For subsequent boostings, Freund's incomplete adjuvant without the bacteria is used. A "water-in-oil" emulsion is more effective than an "oil-in-water" emulsion, and to ensure this, the proportion of adjuvant to antigen solution should always be 2:1 or greater by volume. The resulting emulsion should also be tested by dispersion on a water surface.

IMMUNIZATION

Choice of Animal

The choice of animal depends mostly on convenience, cost, the species from which the antigen is isolated and its availability, ease of handling, and volume of antisera required, rather than on necessity. Consistent differences between species in their response to a given immunogen have rarely been shown, although inbred strains may occasionally show such a difference. A notable exception is the production of anti-insulin antibodies, for which guinea pigs should be used (Burrin and Newman 1991).

Excellent polyclonal antibodies can be produced in many animals, although mice, rats, rabbits, guinea pigs, goats, and sheep are most frequently used. Mice and rats are the preferred choice for the production of monoclonal antibodies. The use of large animals such as sheep and goats enables large quantities of sera to be collected quickly with a minimal amount of pooling. These animals are often chosen for producing commercial quantities of polyclonal antisera. For antiglobulin antisera, both rabbit and sheep make excellent responses to human immunoglobulins of all classes. They also make similarly good antisera to rat and mouse isotypes. In addition, being phylogenetically distant, they make excellent antiglobulins to each other's molecules (Catty 1988).

The age of animals is an important consideration, because newborn mammals are virtually incapable of synthesizing antibody, whereas older animals show a reduced response. Factors such as poor nutrition, disease, and stress conditions also affect antibody response. Mice and rats of 2–4 months age and rabbits of 3–6 months are ideal. Similarly, a number of animals should be used for immunization with any given immunogen. This takes into account variation between animals, and allows pooling to provide maximum diversity where required or selection of the highest avidity/affinity response.

Finally, it must be remembered that the use of animals for laboratory experiments is controlled by specific legal regulations in most countries. Although these regulations vary, they are designed to ensure the welfare of animals and to ensure that the operator is skilled and the manipulation justified. Individuals, therefore, must take personal responsibility for the welfare of animals.

Routes of Immunization

The immune response is systemic and not local. Therefore, immunization by different routes seldom makes any difference in the end product. Nevertheless, the route of immunization may affect the immune response as well as the isotypes of the responding antibodies. For example, oral administration of immunogens such as liposomes is associated with a largely IgA response (Bienenstock and Befus 1980; Gregory et al. 1986).

The route of immunization is primarily governed by three factors: the volume to be delivered, the buffers and other components that will be injected with the immunogen, and how quickly the immunogen should be released into the lymphatics or circulation (Allen 1967; Harlow and Lane 1988). Whereas large volume injections are normally given at multiple subcutaneous sites for rabbits, only intraperitoneal injections are possible with mice. Similarly, immunogens containing adjuvants or particulate matters should not be delivered intravenously, although this method is often preferred for an immediate release of the immunogen. In contrast, intramuscular or intradermal injections are used if a slow release of the immunogen is desired.

Immunogens can be administered in several ways including intradermally, subcutaneously, intramuscularly, intraperitoneally, and intravenously; the choice is often influenced by the physical nature of the injection, the species, and the stage of the immunization schedule. The intradermal route, although technically difficult to administer with small animals such as rodents, is commonly used for the immunization of larger animals. It is often the preferred choice if small volumes of scarce immunogen are available. Intradermal injections can be used with particulate and adjuvant-containing inocula. They give a very slow rate of release of the immunogen.

The subcutaneous and intramuscular routes are most commonly used for immunogens in complete Freund's adjuvants. Although both routes favor granuloma formation around depot sites and a slow immunogen release to local lymph nodes, the immune response with the subcutaneous route is often slower. This route is nevertheless preferred for booster injections, because it decreases the chances of an anaphylactic shock.

Direct injection into the peritoneal cavity is particularly easy and appropriate in small rodents. It is, however, not recommended for inoculating rabbits. Particulate antigens or antigens in adjuvant can both be given safely by this route which involves the spleen in response. The intraperitoneal route may also prevent acute hypersensitivity death in repeated injections of immunogen in solution.

Intravenous inoculations are preferentially used for soluble or particulate immunogens. Although the immune response is rapid, it is not sustained over a long period. Although seldom used for primary immunization, the intravenous route is a practical method of delivering a secondary or later boosts for most laboratory animals. Intravenous injections, however, carry a serious risk to the host animal, as they can readily induce pulmonary embolisms or lethal anaphylactic shock in sensitized animals. Any toxic material present in the inoculum, such as bacterial endotoxins, is particularly dangerous when delivered intravenously. It is not safe or appropriate to use Freund's adjuvant with intravenous injections. Similarly, if possible, physiological buffers and salt conditions should be used when immunizing by this route. Detergent concentration should never be allowed to exceed 0.1% for ionic or 0.2% for nonionic detergents.

As far as the host species are concerned, intraperitoneal injections are easy to perform in mice, although subcutaneous, intravenous or even injections into the spleen can be good alternatives. Subcutaneous or intradermal routes in multiple sites are most commonly used with rabbits, while for larger animals, such as sheep, hogs, goats, and horses, intradermal or intramuscular routes are often preferred.

Several reviews and monographs are available on routes and techniques of immunization (Catty and Raykundalia 1988; Herbert and Fristensen 1986; Poole 1987; Tijssen 1985).

Immunogen Dose

The amount of immunogen capable of inducing an immune response depends on the nature of the antigen and on the host. The typical immunogen dose-response relationship is sigmoidal in nature (Figure 5.6). Therefore, very large doses of immunogen may result in "high-dose tolerance" and antiserum of rela-

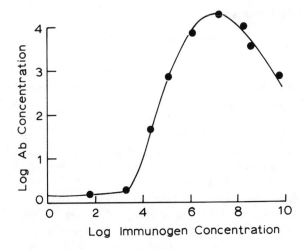

Figure 5.6. Immunogen dose-response relationship
From: Burrin and Newman (1991)

tively low avidity (Kohler and Milstein 1975; Harlow and Lane 1988; Burrin and Newman 1991). Because immunodiagnostic assays require high affinity and avidity antibodies to achieve the desired sensitivity, it is essential to use the lowest effective dose of the immunogen for the production of antiserum.

To determine the appropriate dose of the immunogen, two criteria need to be considered: the optimum dose to achieve the strongest response and the minimum dose likely to induce the production of a useful polyclonal antiserum or donor animals for hybridoma fusions (Harlow and Lane 1988). Much of the injected material will be catabolized and cleaved before reaching an appropriate target cell. The efficiency of this process varies with host factors, the route of immunization, the use of adjuvants, and the intrinsic nature of the antigen. Thus, the effective dose delivered to the immune system may bear little relationship to the introduced dose. Nevertheless, some general guidelines do need to be followed.

A suitable primary dose for rabbits and guinea pigs is in the range of 100 µg, although amounts as high as 500–1000 µg have been reported in the literature. Doses of up to 500–1000 µg are generally used for larger animals. For second and subsequent booster injections, traditional suggestions recommend 10–50% of the amount of the primary dose. However, it is not uncommon to use similar amounts for both primary and subsequent booster injections. Most of these recommendations are based on good immunogens that are available in large supplies. For rare antigens, one or two injections with a low dose to prime the animals followed by a larger secondary boost have often been shown to be effective for antibody production (Harlow and Lane 1988).

Immune Response

A typical antibody response following primary immunization and subsequent boosts is shown in Figure 5.7. An increase in B cells bearing surface antibodies specific for the antigen is first detected 5–6 days after the primary injection of the immunogen. The specific antibody is usually detected in the serum from around 7 days after the injection. It persists at a low level for a few days, typically reaching peak titer around day 10. The first antibodies produced are of the IgM type, characterized by high avidity but low affinity for the antigen. This is followed by a peak of IgG class antibodies as a rule higher than the IgM one (Figure 5.7). The IgG class antibodies display lower avidity, but are characterized by much higher affinity for the antigen than the IgM type.

The primary antibody responses often are very weak, especially for readily catabolizable, soluble antigens. Effective priming, however, can take place even in the absence of detectable antibody production. Most animals will remain effectively primed for at least a year after receiving the first injection. Because the response is generally low or weak, it is not necessary to monitor the process for antibody production at this stage.

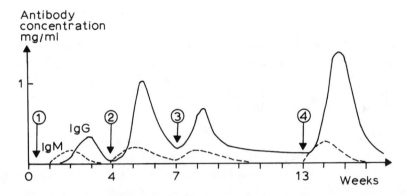

Figure 5.7. Immunogenicity and antibody production
1. A primary immune response to a soluble antigen induces first an IgM antibody production
followed by IgG-specific antibodies
2. A booster injection induces a marked increase in the IgG antibody level
3. A second booster injection soon after the previous one (3 weeks) is less effective in IgG antibody
production
4. An increased delay (6 weeks) between two successive boosts yields more antibodies
From: Paraf and Peltre (1991)

The response to a second antigen or booster injection is characterized by a rapid increase in the antibody titers (Figure 5.7). At this stage, the IgG class antibodies predominate. As a general rule, a minimum of 3–4 weeks delay following the primary inoculation is recommended for the second injection. Antibodies in the serum generally can be detected after 3–4 days, while peak titers are usually observed around days 10–14. The peak antibody titers also increase exponentially after a secondary challenge. Typically high titers of antibody persist for about 2–4 weeks after the second injection.

The immune response to the third and subsequent booster injections typically mirrors that of the secondary injection. High antibody titers can be repeatedly induced following booster injections provided they are properly spaced throughout the lifetime of the immunized animal. At this stage of the immunization schedule, both the nature and quality of the antibodies produced also change because of the maturation of the immune response, thus providing high affinity antibody preparations.

Although the intervals between booster injections can be varied, sufficient time must be allowed for the circulating level of the antibody to drop enough to prevent rapid clearance of the newly injected immunogen. The example shown in Figure 5.7 illustrates the importance of spacing the booster injections. The secondary immune response in this case led to a peak of more than 1 mg precipitating antibody per ml serum. The third injection performed 3 weeks after the previous one, in contrast, induced a peak of only 0.6 mg antibody per ml serum, whereas the fourth injection 6 weeks later resulted in the production of more than twice the amount of antibody produced during the previous boost. As a general rule, booster injections, therefore, should normally be given every 4–6 weeks to obtain the desired titers of the antibody in the serum.

ANTIBODY SCREENING

The quality of antiserum is more important in immunoassays than in other qualitative immunological techniques. The antiserum needs to be screened for antibody titer, i.e., the concentration of antibody present, its affinity and avidity, and its specificity for the target analyte.

Serum samples should be taken 7–14 days after the booster injections. This timing approximately corresponds with the peak antibody titers for most routes of immunization. Generally, smaller samples are collected until the desired antibodies are detected and the titers have reached acceptable levels.

Test bleeds normally are screened against the haptenic immunogen itself by one of several techniques, the most common being immunoassays and immunoblots. In this regard, antibody capture assays are the most versatile. If a pure or nearly pure antigen is available, the presence of antibody and its level can be easily monitored by using a variation of this type of immunoassay. In this format, the antigen preparation is bound to a solid support. The test solution containing an unknown amount of antibody is allowed to bind to the antigen-matrix. The amount of antibody bound to the solid phase is then determined by incubation with a labeled secondary reagent such as an anti-immunoglobulin antibody.

Although, when selecting the appropriate antisera from a number of animals used in the production, the one with the highest titer is often chosen for further evaluation, this may not necessarily be the most appropriate way of screening for the desired antibody. Using such an approach, one is quite likely to miss an antiserum of exceptionally high affinity which may not be of quite such high titer as the others. Hence, a greater emphasis should be placed on monitoring the affinity of the antibody of a workable titer, than on the titer itself of the antiserum.

Selecting an antibody of high affinity also allows one to lower the detection limit of the assay being developed. Moreover, assay incubation times tend to be shorter. In addition, for heterogeneous immunoassays where a phase separation of the free and bound forms is required, a high affinity antibody tends to perform much better than one with low affinity for the target analyte.

As described earlier in this chapter, whichever screening method is used for monitoring the antibody titer, the importance of and the need for using at least one of the three heterologous combinations (i.e., hapten, site, or bridge heterology) must be kept in mind to improve the sensitivity of the detection method. Ideally, to assess the quality of sera, the screening method used should resemble as much as possible the techniques for which the antibodies are being raised. At this stage, the test bleeds should also be screened against a panel of target and nontarget compounds to identify the presence and quantity of antibody specific to the target compound.

Finally, one should also determine the possibility of the selected antibody yielding a false positive result when no specific antigen is present in the assay system. This phenomenon is usually caused by "noise factors" in the antiserum (Morris 1985). It is possible to check for their presence by assaying samples known to contain no, or negligible, amounts of the specific target antigen. Any spurious false positive results at this stage are likely to be due to the antibody recognizing and cross-reacting with a component of the matrix other than the target analyte.

It is often possible to obtain an indirect assessment of the antibody affinity and cross-reactivity at the same time that one is monitoring the titer of the antiserum. One experiment can be set up to derive the antiserum dilution or titer curve (Figure 5.8, Curve A). In a second set of experiments under identical conditions, a standard solution of the unlabeled antigen, similar to its expected detection limit in the assay to be developed, can be added to the test system. This experiment should also yield a sigmoidal curve similar to that of antibody titer, with its linear

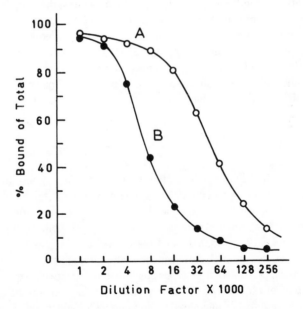

Figure 5.8. Antiserum titration curves for an antitriiodothyronine (T_3) antiserum.
Curve A: Antiserum dilution curve
Curve B: Antiserum displacement curve in the presence of 4 ng/ml of added T_3
From: Morris (1985)

portion parallel to that of the antiserum dilution curve (Figure 5.8, Curve B). However, this second curve will be displaced to the left of the dilution curve, the horizontal degree of displacement being an indirect measure of the affinity of the antibody (Morris 1985). The greater the displacement, the greater the affinity of the antibody.

Potential cross-reactivity of the antiserum to the most likely interfering compounds can also be measured in a similar way. A third curve can be generated under similar conditions, except that instead of the standard unlabeled antigen, the cross-reacting substances are added to the test system. The closer this curve is to the antiserum dilution curve (Figure 5.8, Curve A), the more specific the antibody being screened for the target analyte. Ideally, an antibody exhibiting no cross-reactivity should yield a curve that is superimposable with the antiserum dilution curve.

During the initial screening process, the serum samples from individual animals should not be mixed. The sera should be compared to other test bleeds in titrations to determine the strength of the antibody response. Comparing titrations of various test bleeds generally gives an accurate measure of the course of the immunization process, and provides a good measure of the correct time to begin collecting large volumes of serum.

Once a good titer and sensitivity have developed against the antigen of interest, regular boosts and bleeds are performed to collect the maximum amount of serum. The production bleeds from rabbits generally amount to 30–50 ml, whereas larger animals such as sheep and goats may yield up to 500 ml of antiserum per bleed. At this stage, depending on the required sensitivity and cross-reactivity of the antibodies, antisera from different animals may be pooled to form a large quantity of uniform polyclonal antisera.

MONOCLONAL ANTIBODIES

The ability of each B lymphocyte of the humoral immune system to make only one particular antibody molecule or clonotype expressed at the cell surface as a receptor for antigen was well recognized by the late 1950s. An immunized animal generally produces a random number of clonotypes to yield a polyclonal antiserum. Klinman (1972) estimated that $10–40 \times 10^6$ distinct clonotypes can theoretically be generated by a BALBc mouse that contains about 2×10^8 B cells. About 1 out of every 10,000 clonotypes appears to recognize a given epitope with varying degrees of affinity. Thus, for any given epitope, several different clonotypes could be produced by an animal (Klinman 1972; Kohler and Milstein 1976). In practice, however, only a random few (up to about 10) B cell clones are activated and only few distinct antibodies generated out of this huge repertoire of randomly formed specificities (Briles and Davie 1980; Tijssen 1985; Harlow and

Lane 1988; Roitt et al. 1989). It is, therefore, practically impossible to make reproducible reagents against any given epitope. Moreover, even antisera from the same animal taken at different times differ in their properties.

In this regard, monoclonal antibodies produced by a single clone of B cells offer four distinct advantages: (1) uniform affinity and specificity of binding, (2) homogeneity (they constitute a well-defined reagent), (3) the ability to be produced in unlimited quantities of the same homogeneous reagent, and (4) obviation of the need for use of a pure immunogen due to cloning and selection during the production process. A monoclonal antibody, therefore, may be defined as a uniform homogeneous antibody directed against a single epitope or antigenic determinant, and produced continuously from one cell clone.

Monoclonal antibodies were first isolated as a homogeneous population of antibodies from B cell tumors. Clonal populations of these cells can be propagated as tumors in animals and grown in tissue culture. Unfortunately, B cell tumors secreting antibodies of a predefined specificity cannot be isolated conveniently. Furthermore, plasma cells secreting the antibodies from an immunized animal could not be grown successfully in tissue culture to produce large amounts of monoclonal antibody. This was first achieved by Kohler and Milstein (1975), who developed a technique that allows the growth of clonal populations of cells secreting antibodies of a predefined specificity. In this technique, a specific antibody-producing cell isolated from an immunized animal is fused with a myeloma cell line to produce hybrid cells or "hybridomas." Myeloma cells are derived from a mutant cell line of a tumor of B lymphocytes. These mutant cells are unable to produce immunoglobulins. The hybridomas can be maintained in vitro and will continue to secrete antibodies of a defined specificity.

Production

Detailed descriptions of the process and techniques involved in the production of monoclonal antibodies are beyond the scope of this chapter. Numerous reviews and monographs are available on this topic (Zola 1987; Harlow and Lane 1988; Brown and Ling 1988; Gordon 1988; Goding 1986; Nakamura 1983; Galfre and Milstein 1981).

The basic principles involved in the production of monoclonal antibodies are illustrated schematically in Figure 5.9. The process involves five major steps: immunization, fusion, selection, cloning, and production. Mice and rats are the preferred animals for the production of monoclonal antibodies.

The animals are first injected with an immunogen preparation, and once a good humoral response is observed, an appropriate procedure is developed for screening the desired antibodies. The sera from test bleeds are used to develop and validate the screening procedure. When an appropriate screen has been established, the actual process of producing the hybridomas can begin. Several days prior to

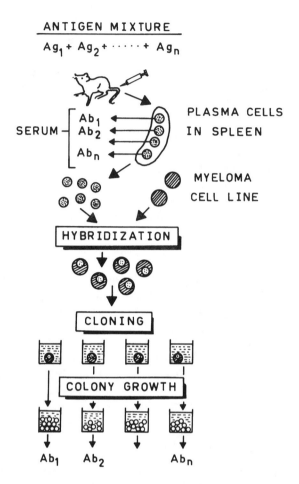

ANTIGEN MIXTURE

$Ag_1 + Ag_2 + \cdots\cdots + Ag_n$

SERUM — $\begin{bmatrix} Ab_1 \\ Ab_2 \\ Ab_n \end{bmatrix}$

PLASMA CELLS
IN SPLEEN

MYELOMA
CELL LINE

HYBRIDIZATION

CLONING

COLONY GROWTH

Ab_1 Ab_2 Ab_n

Figure 5.9. A schematic representation of the process of immortalizing an antibody-producing clone by hybridization, cloning, and selection of clones producing the desired antibodies. From: Zola (1987)

the fusion, the animals are boosted with a sample of the immunogen. The antibody-secreting cells are then prepared from the spleen of the immunized animals, mixed with the myeloma cells, and fused. The hybridomas are diluted in selective medium and plated in multiwell tissue culture plates. After about one week after the fusion, the hybridomas are ready for screening. Cells from positive wells are grown and then a single cell is cloned. The entire process seldom takes less than 2 months from start to finish, and sometimes may require up to a year.

The positive clones generated from individual cells then can be produced in large quantities either by in vivo ascites production in mice or by in vitro tissue culture techniques. Ascites is intraperitoneal fluid normally extracted from mice that have developed peritoneal tumors. These tumors are induced by injecting hybridoma cells into the peritoneum, which serves as a growth chamber for the cells. The hybridoma cells grow to high densities and continue to secrete the antibody of interest, creating a high-titered solution of antibodies for collection. Ascites contain up to 5–15 mg/ml of the monoclonal antibody, and as much as 40 ml of fluid may be obtained per mouse over a period of one week or more. This process can yield approximately 50–100 mg of antibody per mouse.

In contrast, the in vitro production of monoclonal antibodies by tissue culture techniques enables easy expansion of the scale of production; the processes can be engineered to be highly reproducible, and the rate of contamination is much reduced. Two approaches may be used, the first requiring the immobilization and entrapment of hybridoma cells into a solid matrix, and the second involving culturing of the cells in a homogeneous suspension. High-titer supernatants containing the monoclonal antibody obtained from these approaches may be used as such or after further purification.

Hybridomas and monoclonal antibodies can be characterized in a number of ways as follows (Nakamura 1992):

1. Fusion partners: a description of myeloma cells, antibody-producing cells, culture media, passage level, and cloning,
2. Immunogen used and schedule of immunization,
3. Antibody type: immunoglobulin type, including isotype and idiotype when applicable, and
4. Specificity of the antibody, including antigenic determinant reactivity, affinity constants for a series of defined antigens, and yields and titers of antibody produced in vitro or in vivo.

Advantages and Limitations

As mentioned earlier, the usefulness of monoclonal antibodies is primarily due to their high specificity of binding, their homogeneity, and their ability to be produced in unlimited quantities. The use of monoclonal antibodies in clinical laboratories has been established in numerous assays in the fields of clinical chemistry, microbiology, immunology, hematology, surgical pathology, and cytopathology (Nakamura 1992). Their use in immunodiagnostics has simplified the immunoassay formats, in addition to greatly increasing the assay specificity.

Monoclonal antibodies have allowed the development of a one-step, enzyme-labeled, two-site immunoenzymometric (or sandwich) assay. The enzyme-labeled antibody can be simultaneously incubated with a solid-phase immobilized antibody and with the test analyte; the requirement is that the monoclonal antibody that is used for the solid phase must react with a different determinant of the poly-

TABLE 5.5.   Advantages and Limitations of Using Monoclonal Antibodies

---

Advantages
- Homogeneous reagent of well-defined specificity for a single antigenic determinant
- No variability in specificity and affinity
- Purity of immunogen is not important
- Very little presence of irrelevant immunoglobulins
- Much less likely to cross-react

Limitations
- Time-consuming and expensive preparation
- Fixed affinity, which, if low, may make the antibody unsuitable for use in a very sensitive assay system
- Limited sensitivity as compared to polyclonal antisera unless a mixture of monoclonal antibodies is used
- Unlike polyclonal antisera, cross-reactivity, if present, is not easily removable
- Limited biological activity or unusual physical characteristics
- Lack of precipitant or agglutinating properties
- Limited stability to changes in pH or salt concentration
- Discernment of only a small portion of the total antigenicity of the immunogen, resulting in performance inferior to that of polyclonal antibodies in certain assay systems
- Difficult to label

---

valent antigen than does the specific enzyme-labeled antibody (Siddle 1985). The most important advantage of monoclonal antibodies in enzyme immunoassays, however, appears to be the possibility to standardize the assay methods, specificities, detectabilities, and sensitivities. The major advantages and limitations of monoclonal antibodies in the field of immunodiagnostics are summarized in Table 5.5.

PURIFICATION

Once produced, both polyclonal and monoclonal antibodies may require further purification to immunoglobulin fractions or purified immunoglobulins, respectively. Although crude antisera can be used in immunoassay systems, purification generally helps to improve assay sensitivity and reproducibility, reduce assay times and backgrounds, and aid in the standardization of the system.

Several other techniques also rely on the use of purified antibodies at least in some steps. These include cell staining, immunoblots, and immunoaffinity techniques, which often use labeled purified antibodies to determine the presence of an antigen or another antibody.

In choosing a purification strategy, the source material, nature of contaminants, and the final application of the end product, as well as the relevant regulatory aspects if applicable, must be considered. To develop an efficient purification strategy, the antibody concentration in the source material must be known, as it can dictate the purification method. The most widely used antibody quantitation techniques include ELISA, radial immunodiffusion, and RIA. These methods depend on two factors: the analyte concentration (epitope density) and the binding affinity of the capture reagent for the analyte.

Antibodies can be purified from a number of sources (Table 5.6). Although the total immunoglobulin concentration in the serum for a polyclonal antibody preparation may be as high as 10 mg/ml, the desired antibody of interest may constitute only about 10% of this fraction. In contrast, monoclonal antibodies in ascites preparation can be up to 90% pure. Compared to these two sources, monoclonal antibodies in the supernatant obtained from cell culture techniques are present in very dilute amounts, and may, therefore, require a concentration step prior to purification.

Polyclonal antisera as well as monoclonal antibody preparations from both cell culture supernatant and ascites contain a number of nonantibody contaminants at varying levels. These include serum proteins/antibodies, secreted proteins/proteases, serum lipids, endotoxins, viruses, and nucleic acids (Grandics 1994a). The ratio of desired antibody relative to the contaminating antibodies is unpredictable and varies from animal to animal. Therefore, obtaining monospecific an-

TABLE 5.6. Sources for Purifying Antibodies[a]

Source	Antibody type	Total antibody (mg/ml)	Specific antibody (mg/ml)	Contaminating antibodies	Possible purity of specific antibody
Serum	Polyclonal	10	1 at least (10% max)	Other serum antibodies	10% at best[b]
Tissue Culture (in vitro techniques)					
With 10% FBS[c]	Monoclonal	1	0.05 (5%)	Calf antibody	> 95%
Serum-free media	Monoclonal	0.05	0.05 (100%)	None	> 95%
Tissue Culture (in vivo techniques)					
Ascites	Monoclonal	1-10	0.9-9 (90%)	Mouse antibody	90%

[a]Modified from Harlow and Lane (1988).
[b]Except for antigen-affinity column.
[c]Fetal bovine serum.

tibody from the source is often very difficult unless antigen-immunoaffinity chromatography purification is used.

The desired purity of an antibody preparation depends on its end use. For many in vitro research and diagnostic applications, ~ 95% purity can be satisfactory. For in vivo applications of monoclonal antibodies, purity of > 99.99% may be required (FDA 1987). The amount and biochemical heterogeneity in the antibody preparations from different sources, therefore, make purification methods diverse and complex.

It is recommended that the source material containing the desired antibody be purified as early as possible. However, in practice, this is often not feasible. It is, therefore, important to store, characterize, and prepare the starting material properly in order to maximize the yield, activity, stability, and integrity. Antibodies can be stored conveniently at –20°C in the same material.

The tissue culture supernatants can be stored for a few days at 2–5°C, but the best method of storage is freezing (–20°C to –70°C) after 10 times concentration (Table 5.7). It is also desirable to buffer the medium by the addition of 1/20 volume of 1 M Tris (pH 8.0) prior to freezing (Harlow and Lane 1988). Concentration of the culture supernatant prior to freezing helps to control proteases and has a cryoprotective effect. For the concentration of small volumes (up to 50 ml), a centrifugal concentrator is suitable, while larger volumes (up to 1 liter) can be concentrated in a hollow-fiber type concentrator, as it is the most gentle on the antibody (Grandics 1994a). Stirred cell type laboratory concentrators, in which the driving gas (usually nitrogen) is mixed/dissolved into liquid phase, are not recommended. The microscopic bubbles that are formed upon releasing the pressure often promote antibody aggregation and denaturation, increased turbidity of the protein solution, and formation of a precipitate. Once concentrated, the material should be sterile filtered, and dispensed in convenient volumes in sterile culture flasks for storage.

TABLE 5.7.   Recommended Storage Conditions for Antibody Sources

Source	Preparation	Storage	Stability
Polyclonal antiserum	Centrifuged, delipidated	2–5°C	Days
		–20°C	≥ one year
		–70°C	4–5 years
Cell culture supernatant	10× concentrated[a], sterile filtered in sterile containers	2–5°C	Days
		–20°C	≥ one year
		–70°C	4–5 years
Ascites	Centrifuged, delipidated, sterile filtered	2–5°C	Days
		–20°C	≥ one year
		–70°C	4–5 years

[a]Only centrifugal or hollow-fiber concentrators should be used.

Serum or ascitic fluid are already well buffered. However, they often need to be clarified by centrifugation and then delipidated prior to freezing. Depending on the storage conditions, the shelf life of the source material varies from a few days at 2–5°C to up to 4–5 years when kept frozen at –70°C (Table 5.7).

Similar purification approaches can be used for both polyclonal and monoclonal antibodies. To effectively resolve a crude mixture of substances, it may be necessary to use a combination of techniques. A wide variety of methods is available to purify antibodies (Table 5.8). These methods include classical and affinity methods, both of which may be employed in a given process. The choice of the purification method is often governed by a number of variables, including the species in which it was raised, its class and subclass if it is a monoclonal antibody,

TABLE 5.8.   Methods for the Purification of Antibodies

Method	Examples
Fractional Precipitation	
Neutral salts	Ammonium sulfate, sodium sulfate, magnesium sulfate
Organic solvents	Ethanol, polyethylene glycol (PEG)
Metal ions	Zinc
Short-chain fatty acids	Caprylic acid
Organic cations	Rivanol
Electrophoretic Separation	
Carrier free media	Free boundary electrophoresis
Solid support	Zone electrophoresis on cellulose, acrylamide, or starch gel
Isoelectric focusing	Liquid media (sucrose gradient), gel media (acrylamide)
Ion-Exchange and Gel Filtration Chromatography	
Anion exchangers	Aminoethyl (AE-), diethylaminoethyl (DEAE-) and quarternary aminoethyl (QAE-) cellulose, Sephadex®, or Sepharose®
Cation exchangers	Carboxymethyl (CM-) cellulose, Sephadex®, or Sepharose®
Gel filtration	Sephadex® G series, Sepharose® 2B, 4B, 6B or CL 2B, CL 4B, CL 6B, Sephacryl® S series, Biogel® agarose or polyacrylamide beads, Ultrogel® polyacrylamide-agarose beads
Preparative Ultracentrifugation	
	Sucrose density gradients, salt gradients
Affinity Chromatography	
	Protein A, protein G, protein A/G, thiophilic adsorption, avidin-biotin systems, antigen affinity column, anti-immunoglobulin affinity column, hydroxylapatite

the source that will serve as the starting material for the purification, sample size, antibody quantity and purity required, and the intended use of the antibody.

A classical scheme for the purification of antibody typically employs fractional precipitation as the initial separation from the other proteins present in the serum and ascites, followed by ion-exchange and hydrophobic interaction chromatography (Table 5.8). A final size exclusion or gel filtration chromatography step allows aggregate removal, polishing purifications, and buffer exchange. These techniques exploit the chemical and physicochemical differences between various classes of immunoglobulins. These classical methods are labor intensive, and require time-consuming process optimization. Such lengthy applications also encourage undesirable antibody aggregation and loss of yield and activity.

Recently, Blizzard and Garramone (1995) have described the E-Z-SEP® antibody purification kit, marketed and sold by Pharmacia Biotech, Uppsala, Sweden. The kit contains solutions of nonionic linear polymers and buffers designed to selectively precipitate immunoglobulins from heterogeneous protein-containing solutions by the volume exclusion effect, also referred to as "crowding." The addition of polymer reagents to a protein solution increases electrostatic interaction, which enhances protein aggregation and thus separation. The selectivity of immunoglobulin separation is based upon the size, structure, and concentration of polymers; size, shape, charge, and composition of the proteins; pH, temperature, and ionic strength of the reaction; and nonspecific protein-protein interactions.

The E-Z-SEP® antibody purification kits are optimized to precipitate immunoglobulins from ascites fluid, cell culture supernatant, bioreactor fluid, and animal sera, while contaminating proteins and lipids remain solubilized. The polymer reagents isolate immunoglobulins independent of species, subclass, or isoelectric point. Precipitated antibodies are pelleted by centrifugation, while contaminants remain solubilized and are decanted to waste.

Comparative immunological activity studies of E-Z-SEP® processed immunoglobulins have been shown to have similar activity levels compared to conventional ammonium sulfate precipitation followed by ion-exchange chromatography (Blizzard and Garramone 1995).

To isolate group-specific antibodies, particularly IgGs, ligand affinity chromatographic techniques are widely used. These methods use protein A or protein G as affinity ligands. These, often one-step procedures, yield > 95% purification of most antibodies. Because of their lower concentration of the specific antibodies, affinity methods are particularly useful for purifying antibodies from tissue culture supernatants. Affinity-purified antibody may be further fractionated by classical chromatography methods. In affinity-based separations, the need for process optimization is considerably reduced compared with classical methods of antibody purification.

As a general rule, one should avoid precipitation methods such as ammonium sulfate or polyethylene glycol (PEG), because precipitation may reduce the activity and stability of the antibody. Antibodies purified or concentrated using precipitation tend to form aggregates more readily, particularly during lyophilization or wet storage (Grandics 1994b).

If small amounts of antibody (milligrams to grams) need to be purified from small samples, purification method development may become a lengthy and expensive process. Therefore, the simpler affinity purification methods are more advantageous to use on a small scale. Large scale production of antibodies intended for in vivo use requires particularly careful method development and validation. In these cases, the cost of method development can be easily recovered in the long run from savings in production costs. Two excellent review articles present information on process development for purification of large amounts of monoclonal antibodies intended for therapeutic use (Schmidt 1989; Gagnon et al. 1993).

Retaining immunoreactivity of the antibody during purification is critical. Activity may be compromised at several stages, including purification, buffer exchange, concentration, and storage. During purification, the antibody may be exposed to pH extremes, chaotropic reagents, denaturants, detergent, or hydrophobic surfaces, all of which may affect integrity and activity. As a general rule, methods that avoid low-pH elution tend to reduce antibody activity, while neutral elution conditions maximize it.

STABILIZATION AND STORAGE

Stabilizing antibodies/proteins to prevent irreversible changes is a major concern in the diagnostics industry. Minimizing protein inactivation, therefore, is crucial in any successful purification, storage, or isolation procedure. Historically, protein stabilization has been considered more of an art than a science. This is understandable because protein inactivation depends not only on external factors but also on the nature of the protein itself. Consequently, it is impossible to recommend generic recipes to stabilize a particular protein.

External factors commonly responsible for protein inactivation include heat, extremes of pH, surfactants and detergents, high levels of salt, chelating agents, mechanical forces, repeated freezing and thawing, dehydration, and radiation. Furthermore, the degradation of proteins can be divided into two areas: one involves a covalent bond and is usually irreversible, and the other does not and is often referred to as denaturation.

Covalent degradation of proteins includes hydrolysis, imide formation, deamidation, cross-linking, isomerization, oxidation, and disulfide exchange. Although these issues need to be thoroughly investigated with pharmaceutical proteins, they are less important in research because this type of degradation develops gradually over time and may have only a small effect on protein activity. Such changes in proteins can be monitored by techniques such as SDS-PAGE, isoelectric focusing, and reversed-phase HPLC.

Noncovalent degradation or denaturation of proteins involves aggregation, precipitation, and adsorption. Proteins often exist in partially unfolded states which, in some cases, may lead to aggregation and loss of partial or full activity. Aggregation can be caused by moderate amounts of denaturant, such as guanidine-HCl or urea, in the solution which can generate partially unfolded states, or by heat and incorrect storage pH. The latter two are possibly the most common cause of the loss of protein activity.

Precipitation is the macroscopic equivalent of aggregation and a result of denaturation when the proteins fall out of solution. Adsorption of proteins to surfaces of glass, plastics, and delivery pumps can also lower the overall protein activity.

Several approaches, including addition of stabilizers, freeze drying, and cold storage, are used for stabilizing antibodies/proteins. Depending on the type of protein, one needs to use these techniques individually or in combination.

Various types of molecules such as proteins, sugars, amino acids, surfactants, and fatty acids have been used to stabilize proteins (Grandics 1994c). Serum albumin (human or animal) has been extensively used as a stabilizer of proteins. The concentration required varies from 0.1% to 1%. The same range also applies to lyophilized preparations. Protein aggregation can also be effectively prevented by using amino acids such as glutamic acid, lysine, and glycine.

Other stabilizers include carbohydrates, fatty acids, phospholipids, and polyols such as polyhydric alcohols (Harlow and Lane 1988; Grandics 1994c). Alcohols such as sorbitol, mannitol, and glycerol have been used widely in reagent formulations in the diagnostics industry to prevent aggregation, along with sugars such as maltose, mannose, trehalose, sucrose, and glucose. Polyols also protect proteins from oxidation, strengthen hydrophobic bonds, and are thought to have an effect on protein hydration.

Cold storage is the most common way to reduce reactions related to protein unfolding or inactivation. Antibodies can be stored conveniently in buffer solutions of neutral pH and containing 0 to 150 mM salt concentrations at 4–5°C. Phosphate-buffered saline (PBS) or similar isotonic solutions are commonly used buffers for storing purified antibodies. Solutions of purified antibody should be stored at relatively high concentrations of $\geq$ 1 mg/ml. Concentrations of up to 10 mg/ml are normally recommended. Dilute solutions, therefore, need to be concentrated prior to freezing.

There is often a choice between lyophilized storage and storage in a liquid solution. The optimum choice is to keep the protein in a solution form. For most IgG antibodies, sterile storage at concentrations mentioned above in a physiological saline at 5°C allows a relatively long shelf life.

For more delicate antibodies, storage in solid phase is recommended. Recently, Grandics (1994c) has described a promising novel approach in which proteins are stored in solid phase after entrapment into sugar crystals. The antibody solution is made to 50–60% in sucrose and dried at 25–30°C under vacuum. The resulting crystals are recovered and stored as a powder desiccated at 2–8°C. Using this approach, the shelf life of the antibody can be extended to several years. An alternative approach is to dilute the antibody 1:1 with glycerol and store at –30°C to –70°C.

Preservatives such as thiomersal (Merthiolate®) at 1: 10, 000 or 0.02–0.1% sodium azide are recommended for long term storage of sera. Sodium azide inhibits a number of biological assays and also interferes with some coupling reactions. Hence, it needs to be removed by prolonged dialysis or gel filtration. Preservatives are unnecessary with sterile sera or sera stored at –70°C.

Antibody solutions should not be frozen and thawed repeatedly. This can lead to aggregation and thus a potential loss of activity. Hence, solutions should be stored in convenient aliquots. Many antibody solutions also generate an insoluble lipid component with prolonged storage that needs to be separated from the aqueous phase.

The antibody may also be stored lyophilized. This depends on the successful development of a drying cycle, which in itself may not be an easy task. Because antisera vary in solubility after lyophilization, preservation by this method can be best reserved for small samples of valuable reagents pretested for solubility and activity after reconstitution, for samples that need to be shipped over long distances, and for cases where refrigeration is unavailable. Lyophilization is widely employed in the pharmaceutical industry for protein stabilization.

REFERENCES

ALLEN, L. 1967. Lymphatics and lymphoid tissues. *Ann. Rev. Physiol.* 29:197–224.

BIENENSTOCK, J., and BEFUS, A. D. 1980. Mucosal immunity. *Immunology* 41:249–270.

BLIZZARD, C. D., and GARRAMONE, S. 1995. Antibody purification by selective precipitation with polymers. *Amer. Biotech. Lab.* 13(3):71–72.

BRILES, D. E., and DAVIE, J. M. 1980. Clonal nature of the immune response. II. The effect of immunization on clonal commitment. *J. Exp. Med.* 152:151–160.

BROWN, G., and LING, N. R. 1988. Murine monoclonal antibodies. In *Antibodies. Vol. I. A Practical Approach,* ed., D. Catty, pp. 81–104, IRL Press, Oxford.

BURRIN, J., and NEWMAN, D. J. 1991. Production and assessment of antibodies. In *Principles and Practice of Immunoassay,* eds. C. P. Price and D. J. Newman, pp. 19–52, Stockton Press, New York.

CATTY, D. 1988. *Antibodies. Vol. I. A Practical Approach.* IRL Press, Oxford.

CATTY, D., and RAYKUNDALIA, C. 1988. Production and quality control of polyclonal antibodies. In *Antibodies. Vol. I. A Practical Approach,* ed., D. Catty, pp. 19–80, IRL Press, Oxford.

CHARD, T. 1978. *An Introduction to Radioimmunoassay and Related Techniques.* North-Holland, Amsterdam.

CLAUSEN, J. 1988. *Immunochemical Techniques for the Identification* and Estimation of Macromolecules. *Elsevier, Amsterdam.*

COOK, C. E.; TWINE, M. E.; MYERS, M.; AMERSON, E.; KEPLER, J. A.; and TAYLOR, G. F. 1976. Theophylline radioimmunoassay. Synthesis of antigen and characterization of antiserum. *Res. Commun. Chem. Pathol. Pharmacol.* 13:497–504.

EISEN, H. N., and SISKIND, G. W. 1964. Variations in affinities of antibodies during the immune response. *Biochemistry* 3:996–1008.

EMIT® THEOPHYLLINE ASSAY. 1978. Package insert, Syva Co., Palo Alto, CA.

ERLANGER, B. F. 1980. The preparation of antigenic hapten-carrier conjugates. A survey. *Methods Enzymol.* 70:85–104.

FDA. 1987. *Points to Consider in the Manufacture and Testing of Monoclonal Antibody Products for Human Use.* Office of Biologics Research and Review, Center for Drugs and Biologics, Food and Drug Administration, Bethesda, MD.

FINDLAY, J. W. A.; BUTZ, R. F.; and WELCH, R. M. 1976. A codeine radioimmunoassay exhibiting insignificant cross-reactivity with morphine. *Life Sci.* 19:389–393.

FINDLAY, J. W. A.; BUTZ, R. F.; and WELCH, R. M. 1977. Specific radioimmunoassays for codeine and morphine. Metabolism of codeine to morphine in the rat. *Res. Commun. Chem. Pathol. Pharmacol.* 17:595–603.

GAGNON, P.; CARTIER, P. G.; MAIKNER, J. J.; EKSTEEN, R.; and KRAUS, M. 1993. A systematic approach to the purification of monoclonal antibodies. *LCGC* 11(1):26–34.

GALFRE, G., and MILSTEIN, C. 1981. Preparation of monoclonal antibodies. Strategies and procedures. *Methods Enzymol.* 73:3–46.

GODING, J. W. 1986. *Monoclonal Antibodies. Principles and Practice.* Academic Press, London.

GORDON, J. 1988. Human monoclonal antibodies. In *Antibodies. Vol. I. A Practical Approach,* ed. D. Catty, pp. 105–112, IRL Press, Oxford.

GRANDICS, P. 1994a. Monoclonal antibody purification guide. Part 1. *Amer. Biotech. Lab.* 12(6):58,60,62.

GRANDICS, P. 1994b. Monoclonal antibody purification guide. Part 2. *Amer. Biotech. Lab.* 12(7):12–14.

GRANDICS, P. 1994c. Monoclonal antibody purification guide. Part 3. *Amer. Biotech. Lab.* 12(8):16,18.

GREGORY, R. L.; MICHALEK, S. M.; RICHARDSON, G.; HARMON, C. C.; HILTON, T.; and MCGHEE, J. R. 1986. Characterization of immune response to oral administration of *Streptococcus sobrinus* ribosomal preparations in liposomes. *Infect. Immunol.* 54:780–786.

GROSS, S. J.; GRANT, J. D.; WONG, S. R.; SCHUSTER, R.; LOMAX, P.; and CAMPBELL, D. H. 1974. Critical antigenic determinants for production of antibody to distinguish morphine from heroin, codeine and dextromethorphan. *Immunochemistry* 11:453–456.

GUSHAW, J. B.; HU, M. W.; SINGH, P.; MILLER, J. G.; and SCHNEIDER, R. S. 1977. Homogeneous enzyme immunoassay for theophylline in serum. *Clin. Chem.* 23:1144 (Abstr.).

HARLOW, E., and LANE, D. 1988. *Antibodies. A Laboratory Manual.* Cold Spring Harbor Laboratory Press, New York.

HERBERT, W. J., and FRISTENSEN, F. 1986. Laboratory animal techniques for immunology. In *Handbook of Experimental Immunology,* eds. D.M. Weir, L.A. Herzenberg, C. Blackwell, and L.A. Herzenberg, pp. 133.1–133.36, Blackwell Scientific, Oxford.

JAKLITSCH, A. 1985. Separation-free enzyme immunoassay for haptens. In *Enzyme-Mediated Immunoassay,* eds. T. T. Ngo and H. M. Lenhoff, pp. 33–56, Plenum Press, New York.

KLINMAN, N. R. 1972. Mechanism of antigenic stimulation of primary and secondary clonal precursor cells. *J. Exp. Med.* 136:241–260.

KOHLER, G., and MILSTEIN, C. 1975. Continuous cultures of fused cells secreting antibody of predefined specificity. *Nature* (London) 256:495–497.

KOHLER, G., and MILSTEIN, C. 1976. Derivation of specific antibody-producing tissue culture and tumor lines by cell fusion. *Eur. J. Immunol.* 6:511–519.

LANDSTEINER, K. 1945. *The Specificity of Serological Reactions.* Harvard University Press, Boston, MA.

LATERRE, E. C., and HEREMANS, J. F. 1963. A note on proteins aparently specific for cerebrospinal fluid (CSF). *Clin. Chim. Acta* 8:220–226.

MORRIS, B. A. 1985. Principles of immunoassay. In *Immunoassays in Food Analysis,* eds. B. A. Morris and M. N. Clifford, pp. 21–52, Elsevier Applied Science Publishers, London.

MORRIS, B. A.; ROBINSON, J. D.; PIALL, E.; AHERNE, G. W.; and MARKS, V. 1974. Development of a radioimmunoassay for morphine having minimal cross-reactivity with codeine. *J. Endocrinol.* 64:6P–7P.

NAKAMURA, R. M. 1983. Monoclonal antibodies. Methods and applications. *Clin. Physiol. Biochem.* 1:160–172.

NAKAMURA, R. M. 1992. General principles of immunoassays. In *Immunochemical Assays and Biosensor Technology for the 1990s,* eds. R. M. Nakamura, Y. Kasahara, and G. A. Rechnitz, pp. 3–22, *Am. Soc. Microbiol.,* Washington, D.C.

NEESE, A. L., and SOYKA, L. F. 1977. Development of a radioimmunoassay for theophylline. Application to studies in premature infants. *Clin. Pharmacol. Ther.* 21:633–641.

PARAF, A., and PELTRE, G. 1991. *Immunoassays in Food and Agriculture.* Kluwer Academic Publishers, Dordrecht, The Netherlands.

PIERCE CHEMICAL CO. 1994. *Life Science and Analytical Research Products Catalog and Handbook.* Rockford, IL.

POOLE, T. B. 1987. *The UFAW Handbook on the Care and Management of Laboratory Animals.* 6th ed., Longman, London.

ROITT, I. M., Brostoff, J.; and MALE, D. 1989. *Immunology.* Gower Medical Publishing, London.

RUBENSTEIN, K. E.; SCHNEIDER, R. S.; and ULLMAN, E. F. 1972. "Homogeneous" enzyme immunoassay. A new immunochemical technique. *Biochem. Biophys. Res. Commun.* 47:846–851.

SCHMIDT, C. 1989. The purification of large amounts of monoclonal antibodies. *J. Biotech.* 11:235–252.

SCHNEIDER, R. S.; LINDQUIST, P.; WONG, E. T.; RUBENSTEIN, K. E.; and ULLMAN, E. F. 1973. Homogeneous enzyme immunoassay for opiates in urine. *Clin. Chem.* 19:821–825.

SELA, M. 1973–1982. *The Antigens.* Vols. 1–6, Academic Press, New York.

SIDDLE, K. 1985. Properties and applications of monoclonal antibodies. In *Alternative Immunoassays,* ed. W. P. Collins, pp. 13–33, John Wiley, New York.

SPECTOR, S., and PARKER, C. W. 1970. Morphine radioimmunoassay. *Science* 168: 1347–1348.

SPECTOR, S.; BERKOWITZ, B.; FLYNN, E. J.; and PESKAR, B. 1973. Antibodies to morphine, barbiturates and serotonin. *Pharmacol. Rev.* 25:281–291.

SPIER, R., and GRIFFITHS, B. 1985. *Adjuvants in Animal Cell Biotechnology.* Vols. 1 and 2, Academic Press, New York.

STEWARD, M. W. 1984. *Antibodies. Their Structure and Function.* Chapman & Hall, New York.

TIJSSEN, P. 1985. *Practice and Theory of Enzyme Immunoassays.* Elsevier, Amsterdam.

VAN WEEMAN, B. K., and SCHUURS, A. H. W. M. 1971. Immunoassay using hapten-enzyme conjugates. *FEBS Lett.* 15:232–236.

VAN WEEMAN, B. K. and SCHUURS, A. H. W. M. 1975. The influence of heterologous combinations of antiserum and enzyme-labeled estrogen on the characteristics of estrogen enzyme immunoassays. *Immunochemistry* 12:667–670.

Warr, G. W. 1982. Preparation of antigens and principles of immunization. In *Antibody as a Tool,* eds. J. J. Marchalonis and G. W. Warr, pp. 21–58, John Wiley, New York.

WILLIAMS, C. A., and CHASE, M. A. 1967. *Methods in Immunology and Immunochemistry,* vols. I and II, Academic Press, New York.

ZOLA, H. 1987. *Monoclonal Antibodies. A Manual of Techniques.* CRC Press, Boca Raton, FL.

6

Enzymes and Signal Amplification Systems

INTRODUCTION

Since the introduction of radioimmunoassays in the late 1950s and the widespread reliance on radioisotopic labels in various fields of clinical and medical science and practice during the 1960s and 1970s, rapid changes have occurred in label technology for immunoassays. Even though radioimmunoassays have their unquestionable advantages, certain of their limitations have created a strong demand to develop nonradioisotopic alternatives. Thus, alternative labels and techniques have been adopted primarily in order to obtain simpler or more sensitive assays that do not involve radioisotopes.

Almost any compound that can be detected with high sensitivity and that can be firmly attached to the ligand without grossly altering its binding or antigenic properties can be used as an alternative to radioisotopes. An ideal label should cover most of the applications and the wide concentration range of analytes present in biological samples. The following are some general requirements for an ideal label in immunoassays.

Sensitivity

In order to exceed the sensitivity of radiolabels, an alternative label should enable the labeled reagent to be detected at concentrations down to 10^{-15}–10^{-18} mols. The detection system should be relatively simple, rapid, and inexpensive. However, the sensitivity of immunoassays is not solely determined by the detection sensitivity of the label, but rather by the properties of the immunoreagents, i.e., the relative affinity and low nonspecific binding of the labeled antibodies.

Coupling Properties

The label used needs to be coupled to immunological reagents in order to be used for assays of many different compounds. The coupling should not alter the immunological properties of labeled antigens, affect the affinity or specificity of labeled antibodies, or decrease the signal produced.

Homogeneity

In homogeneous types of immunoassays, the extent of immunoreaction is monitored without physical separation of the reaction components. The label used needs to enable the monitoring either via a change in signal level upon binding, via susceptibility to signal modulation, or via physical or optical properties.

Stability

The alternative label should enable storage of labeled reagent for extended periods of time. Stable signal levels from assay to assay and from day to day are required for automated or mechanized immunoassay systems to decrease the need for frequent standardization. The ideal label should also be inert against effectors derived from samples, e.g., color compounds, quenching agents, endogenous enzymes, or enzyme inhibitors.

Stability to Environmental Factors

Radioisotopes offer robust detection that is not affected either by any compounds present in samples or assay buffers, or by physical conditions like pH or temperature. Alternative labels are often quite sensitive to these factors, and this must be taken into consideration in choosing the label and detection system.

Clinical Applicability

To be useful in routine clinical use, or preferable even at doctors' offices or at patients' homes, the label introduced should provide robust practical detection and be amenable to automation.

Safety

Unlike the radiolabels, an alternative label should be nontoxic and safe. It should not involve dangerous reagents or cause waste problems.

Availability

To be widely accepted both in research and in routine clinical use, the labeling reagent should be generally available. The labeling reaction should be mild and easily optimized for each particular immunoreagent. The labeling should also be highly reproducible.

Several different labels, alone or in combination, can be used in immunoassays (Table 6.1). Important factors that determine the suitability of a labeling substance in immunoassay include its specific activity, ease of labeling, ease of endpoint determination, associated hazards, and possibilities for convenient assay formulation or homogeneous operation (Gosling 1990; Deshpande and Sharma 1993). High specific activity of the label is essential for the development of assays requiring low detection limits. The specific activity of an immunoassay label is related to three main considerations: the fraction available for detection, the degree of amplification, and the efficiency of detection.

The relative specific activity (defined as the number of detectable events per labeled molecule per unit time) varies with the type of label. In the case of ^{125}I radioactivity, only a very small fraction of the total events generated by the ^{125}I label are available for use during the counting times normally employed in immunoassays. Most of the potentially detectable events are usually lost during the storage of the label and are never utilized for assay purpose.

The chemiluminescent labels produce at least one detectable event per labeled molecule during the measurement. In contrast, the enzyme labels can produce many detectable events per enzyme molecule depending on the turnover number of the enzyme used. The fluorescent labels can also produce many detectable events per labeled molecule because one molecule can cycle many times through its excited and ground states during a short measurement period. These apparent advantages in specific activity of nonisotopic labels are, however, offset by background, matrix, quenching, light screening, or other effects that limit the signal-to-noise ratio (Diamandis 1988; Deshpande 1994).

The popularity of the most important types of labels has varied considerably in the last decade (Figure 6.1). The use of radioisotopes has declined from about 50% to an apparently stable 25% of new assays. Because they are totally impervious to normal environmental changes, radioimmunoassay methods are inherently stable with low between-assay variability. Hence, they continue to be popular in large-scale clinical analytical service laboratories (Gosling 1990). Because of the relative convenience of gamma counting (especially with multiheaded, computer-assisted modern counters), tritiated labels are becoming less popular and ^{125}I is the predominant choice for commercial RIAs, even where high specific activity is unnecessary.

The popularity of fluorescent and luminescent labels has remained relatively constant, at about 15% and 5%, respectively (Gosling 1990). However, this does not indicate the recent trend from more traditional labeling compounds to lanthanide ion-chelates and acridinium esters. Enzymes as labels in immunoassays have continued to increase in popularity, from use in about 20% of new assays to about 35%.

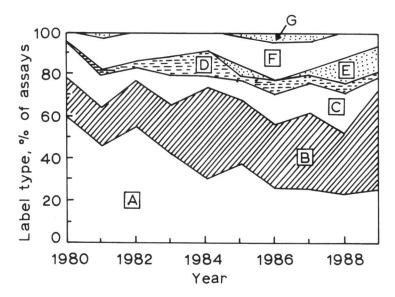

Figure 6.1. Trends in use of immunoassay labels.
A: Radioisotopes   B: Enzymes
C: Fluorescent compounds   D: Luminescent compounds
E: Biotin
F: No label of latex assays (nephelometric, turbidimetric, latex agglutination, and particle-counting assays)
G: Miscellaneous labels (liposomes, FAD^+, hematoxylin-stained bacterial cells, etc.)
From: Gosling (1990)

This chapter summarizes the current information on various enzymes commonly used in commercial immunoassays. Although a large number of enzymes have been tried and found suitable as labels, only those important in commercial diagnostic products are described in detail. A recurring approach to maximizing the final signal obtained in an immunoassay is to attempt to attach multiple molecules of the final labeling substance to each immune complex or component to be detected. Such signal amplification systems commonly used in these products are also described.

ENZYMES

The use of enzymes, because of their amplifying effect, as immunoassay labels has steadily increased during the past two decades. Theoretically, any enzyme can be used as a label in enzyme immunoassays. However, for commercial product development, an ideal enzyme label should have the following characteristics:

1. High specific activity at low substrate concentrations, i.e., a high turnover number, a low K_m for substrate, but a high K_m for product.

2. Compatibility with a wide range of sample media and assay conditions including pH, ionic strength, detergents, and buffers. It should be stable at the pH required for good antibody-antigen binding (generally neutral pH) and should have adequate activity at this pH, particularly for homogeneous enzyme immunoassays.

3. Utility in inexpensive, accurate, easy, nontoxic, and sensitive assay methods, preferably with a spectrophotometric endpoint.

4. Presence of reactive groups through which enzymes can be covalently linked to antibody, antigen, or hapten with minimum loss of enzyme or immune activities.

5. Stability under routine storage and assay conditions both in free form and when conjugated to required compounds. It should have a long shelf life.

6. Easy availability in a soluble, purified form at low cost.

7. Absence of health hazards attributable to enzyme, substrates, and cofactors.

8. Absence of enzyme activity and factors affecting enzyme activity from the test fluid, particularly for homogeneous enzyme immunoassays.

9. Availability of inexpensive and stable substrates, and easily and precisely detected product over a wide range.

10. Low nonspecific binding to antibody and solid phase.

The use of enzymes in immunoassays, however, is not without limitations. Being proteins, enzymes are denatured by known protein denaturing agents such as temperature extremes, acids, bases, organic solvents, detergents, salts, and chaotropic agents. Denatured enzymes generally lack catalytic activities.

More than 25 different enzymes including acetylcholinestarase, adenosine deaminase, amine oxidase, carbonic anhydrase, catalase, glucoamylase, penicillinase, and urease have been used in immunoassays. However, only three enzymes, viz., horseradish peroxidase (HRP), alkaline phosphatase (ALP), and *E. coli* β-D-galactosidase (BG) have been extensively used in both research applications and in commercially available diagnostic kits (Figure 6.2). For new enzyme-labeled assays, HRP is used about 50% of the time and ALP about 25% of the time (Gosling 1990). At present, no other enzyme looks set to seriously rival either of these two with respect to widespread general use.

In solid-phase immunoassays, the solid phase should have only a minimal influence on the enzyme activity. In contrast, in homogeneous immunoassays, the reaction of hapten- or antigen-labeled enzyme with antibody should strongly affect the enzyme activity, e.g., through steric inhibition of the substrate at the catalytic site of the enzyme.

It is impossible to provide information on all the enzymes used in immunoassays; therefore, only the enzymes most commonly used in commercial diagnostic products are described below.

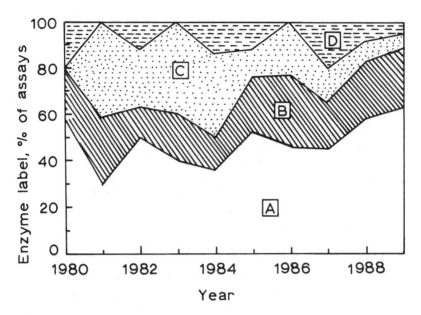

Figure 6.2. Major enzymes used in labeling
A: Horseradish peroxidase (HRP)
B: Alkaline phosphatase (ALP)
C: β-D-galactosidase (BG)
D: Other enzymes (β-lactamase, urease, glucose oxidase, and melittin)
From: Gosling (1990)

Horseradish Peroxidase (HRP)

HRP (hydrogen peroxide oxidoreductase, E.C. 1.11.1.7) is the most widely used enzyme in the enzyme immunodiagnostic products. It belongs to the ferro-protoporphyrin group of peroxidases and is isolated from horseradish roots (*Armoracia rusticana*).

1. Physicochemical Properties.   HRP is a holoenzyme of molecular weight 40,200 and contains one ferriprotoporphyrin III (also referred to in the literature as "ferriprotoporphyrin IX" and as "protohemin") group per molecule. The apoenzyme is a glycoprotein of 308 amino acid residues and 8 neutral carbohydrate side chains attached through asparagine residues. The molecular weight of the polypeptide chain alone is 33,890. It has 4 disulfide linkages, 1 tryptophan, 3 histidines, 6 lysines, and 21 arginines per molecule. Of the 6 lysine residues, 4 are available for conjugation without any loss of enzyme activity.

The covalent structure of the enzyme consists of two compact domains between which the hemin group is sandwiched (Welinder 1979). In protohemin, four of the six coordination bonds of iron interact with the pyrrole ring nitrogens. The other two are occupied by water molecules or OH⁻, depending on the pH. In HRP, one of these two remaining coordination bonds is to a carboxyl group of the protein, while the other is coordinated to an amino group or to a water molecule.

Seven isozymes of HRP have been isolated by ion-exchange chromatography. These are designated as A-1, A-2, and A-3 for the anionic forms, and as B, C, D, and E for the cationic forms. The acidic or anionic isozymes are characterized by a very high carbohydrate content, which gradually decreases with the more basic or cationic isozymes. Isozyme C (pI 8.7–9.0, apoprotein pI 6.8) is the predominant cationic form, constituting more than 50% of the total HRP in horseradish, and contributing the bulk of commercially available highly purified HRP (Tijssen 1985; Ngo 1991).

Because of protohemin, HRP has absorption maxima not only at 275 nm (tyrosine and tryptophan residues of the apoprotein), but also at 403, 497, and 641.5 nm (Whitaker 1994). The extinction coefficient of a 1% solution at 403 nm is 22.5. A commonly used criterion for purity of HRP is the RZ (for *Reinheits Zahl)* or purity number. It is the ratio of maximum absorbance in the Soret region (403 nm) to the absorbance at 275 or 280 nm. Completely pure HRP has a RZ value of 3.04. Although it is a common practice to express the purity of HRP in terms of RZ values, the method has several disadvantages. Some of these drawbacks include the following.

1. RZ refers only to purity as a protein and not as an enzyme.
2. Absorption at 275 nm is influenced by the presence of inorganic iron (Fe^{3+}); that at 403 nm by other hemoprotein contaminants.
3. The RZ value is different for peroxidases from different sources and for the different isozymes of HRP, because of their variable aromatic amino acid and carbohydrate contents. The different isozymes of HRP have RZ values ranging from 2.50 to 4.19 (Table 6.2).

TABLE 6.2. RZ Values for Isozymes of HRP[a]

Isozyme	RZ
A-1	4.19
A-2	4.12
A-3	3.71
B	3.37
C	3.42
D	2.57
E	2.50

[a]From Shannon et al. (1966).

enzymatically less active, and are thought not be involved directly in the mechanism of action of HRP. The k_1 also is much smaller than k_3 in the overall reaction mechanism.

There are marked changes in the spectra of the intermediates formed in the enzymatic reactions. When compound I is formed, there is a marked decrease in the absorbance at 403 and 497 nm and an increase in absorbance above 550 nm (broad peak at 575 nm and a peak at 650 nm). This accounts for the change in color from brown to green on formation of compound I. There are also correspondingly distinctive spectral changes in the formation of compounds II, III, and IV.

Cyanide or sulfide reversibly inhibits HRP at concentrations of 10^{-5} to 10^{-6} M, while azide, fluoride, hydrazine, hydroxamic acid, or hydroxylamine inhibits only at concentrations higher than 10^{-3} M (Saunders et al. 1964). Similar to other enzymes, HRP activity is also dependent on the temperature. In the absence of nonionic detergents, the highest activity is observed at 15°C. At higher temperatures, a higher initial rate for up to 10 min is followed by an inactivation that is stronger for free than for immobilized HRP. Depending on the hydrogen donor, Tween®-20 and Triton® X-100 delay HRP inactivation (Porstmann et al. 1981). However, an incorrect H_2O_2 concentration is probably the single most important factor in interassay variations and enzyme inactivation in the commercial enzyme immunodiagnostic products (Tijssen 1985).

Detailed reviews on the kinetics and mechanism of HRP-catalyzed reactions are available (Saunders et al. 1964; Dunford and Stillman 1976).

3. Hydrogen Donors.   HRP activity is almost always measured indirectly by the rate of transformation of the hydrogen donor. Hydrogen donors should have the following properties to be suitable for enzyme immunoassay (Tijssen 1985).

1. There should be negligible oxidation rate in the absence of enzyme,
2. The product should not undergo secondary reactions unless yielding another single product with a rate constant much faster than k_4,
3. There should be much slower reaction rate between the donor and enzyme than k_1,
4. There should be little pH dependence of the oxidation reactions,
5. Oxidation of donor must produce a stable change in a physical property in a single form that can be quantitated, e.g., spectrophotometrically,
6. Both the reduced and the oxidized states of the donor should be sufficiently soluble for most enzyme immunoassay applications; whereas in products where a precipitating reaction needs to be observed, the oxidized donor should be highly insoluble, and
7. The donor should not be toxic.

A number of aromatic phenols and amines serve as hydrogen donors and are used as chromogenic, fluorogenic, and chemiluminogenic substrates for HRP.

These reactions are carried out at neutral or slightly acidic pH. Chromogenic substrates commonly used in the diagnostic products in conjunction with HRP include: *o*-phenylenediamine (OPD); 2,2'-azino-di (3-ethyl-benzthiazoline) sulfonic acid (ABTS); *o*-dianisidine (ODIA); 5-aminosalicylic acid (5-AS); dicarboxidine; 3,3',5,5'-tetramethylbenzidine (TMB); phenol and aminoantipyrine (Trinder-Emerson reagent); and 3-methyl-2-benzothiazolinone hydrazone (MBTH) and 3-dimethyl-aminobenzoic acid (DMAB) (Ngo-Lenhoff reagent) (Ngo 1991; Ngo and Lenhoff 1980; Tijssen 1985; Geoghegan 1985). Chloronaphthol and diaminobenzidine (DAB)/cobalt are popular reagents where insoluble oxidation products are used as a means of detection.

Among these substrates, ABTS, OPD, and TMB are more widely used hydrogen donors. Each of these substrates also requires the presence of H_2O_2. The specific activity of HRP with ABTS is about 1000 U/mg at 25°C and that of the other substrates a few-fold higher. ABTS is a good all-purpose reagent with a wide working range. TMB is often preferred, as it gives the highest absorbance values, low backgrounds, and is not mutagenic.

The practical detection limit of HRP using these substrates is in the region of 10^{-14} to 10^{-17} moles. This is comparable to that of the commonly employed radioisotope [125]I. All these substrates are commercially available.

HRP-based assays are generally capable of greater sensitivity than alkaline phosphatase with colorimetric substances, as the signal intensity under identical conditions is an order of magnitude greater.

Detailed descriptions of preparation and assay conditions for the substrate systems described above are provided by Tijssen (1985) and Porstmann and Porstmann (1988).

Well-known luminogenic substrates for HRP include luminol, isoluminol, and acridinium ester; while 3-(4-hydroxyphenyl)propionic acid, 4-hydroxyphenyl acetic acid, homovanillic acid, tyramine, and 3',6'-diacetyl-2',7'-dichlorofluorescein are widely used as fluorogenic substrates (Whitehead et al. 1983; Wood et al. 1988; Thorpe et al. 1984; Guilbault et al. 1968a,b).

A number of widely used HRP substrates, including ABTS, OPD, *o*-toluidine, and benzidine, are known mutagens and carcinogens. Hence, caution should be exercised in handling these compounds.

4. Conjugation Methods. HRP has been conjugated to a variety of antibodies, antigens, and haptens. Some of the more widely used techniques for coupling HRP to these reagents are briefly described below.

CONJUGATION WITH AMINO AND THIOL DIRECTED CROSS-LINKERS. Among the heterobifunctional reagents selective toward amino and thiol groups, *N*-succinimidyl 4-(*N*-maleimidomethyl)cyclohexane-1-carboxylate (SMCC) and *N*-succinimidyl 6-maleimidohexanoate (SMH) have been most favorably used to cross-link the amino group of HRP with the thiol group of Fab', reduced IgG, or thiolated F(ab')$_2$ (Wong 1993). Coupling of labeled HRP to these reagents can be achieved by incubating the components in 1:1 molar ratio. Because the incorporation of cross-linker into HRP is limited due to few available lysyl chains on HRP, as well as the small number of thiols in immunoglobulins, this method affords 1:1 conjugates.

Detailed experimental protocols using such an approach are available in the literature (Ishikawa et al. 1983; Imagawa et al. 1982; Hasida et al. 1984; Weiss and Van Regenmortel 1989).

CONJUGATION THROUGH DISULFIDE FORMATION.   HRP and IgG or its fragments can be conjugated through a disulfide bond (Ishikawa et al. 1983). The amino groups of HRP are activated by reaction with SPDP through labeling with pyridyl disulfide, which can react further with free thiol groups of Fab' or thiolated IgG and F(ab')$_2$. Alternatively, HRP can be thiolated with *S*-acetylmercaptosuccinic anhydride followed by hydroxylamine to generate a free thiol that can react with dithiopyridine introduced into Fab' or thiolated IgG. These disulfide interchange methods allow one to prepare HRP-labeled protein conjugates with a more defined ratio of HRP to the protein.

CONJUGATION WITH GLUTARALDEHYDE.   Glutaraldehyde-treated HRP can be coupled to proteins with (two-step method) or without (one-step method) prior removal of unreacted glutaraldehyde (Avrameas 1969; van Weeman and Schuurs 1974). Conjugates prepared by this method often give a wider range of assay linearity, which is useful for quantitative measurement purposes. Nonetheless, the recovery of both immunological and enzyme activities by these methods is low.

Immunoconjugates prepared by the direct one-step coupling of HRP and antibodies with glutaraldehyde often result in heterogeneous, high molecular weight aggregates with less than 10% recovery of enzyme activity (Avrameas et al. 1978). Hence, the two-step procedure is often followed.

CONJUGATION USING PERIODATE OXIDATION.   Because HRP is a glycoprotein, oxidation of its carbohydrate moiety with sodium periodate results in the formation of aldehyde groups. The latter will form Schiff bases with amino groups of ligands and immunoglobulins. Stable conjugates then can be obtained after reduction with sodium borohydride or other suitable reducing agents (Wilson and Nakane 1979; Tijssen 1985; Tijssen and Kurstak 1984).

To prevent self-coupling, the reactive amino groups of HRP can be first irreversibly blocked by alkylation with fluorodinitrobenzene. Alternatively, HRP can be oxidized under acidic conditions (pH 4 to 5) without prior blocking of its amino groups. In acid conditions, the formation of Schiff's base is not favored, and therefore, the extent of self-aggregation is minimal (Wilson and Nakane 1979).

Using the periodate oxidation method, a maximum of 5 to 6 mol of HRP can be coupled per mole of IgG (Nakane and Kawaoi 1974; Nakane 1975). Conjugates prepared by this method are generally capable of providing very high sensitivity detection but with narrow assay linearity range. Hence, they are often useful for the "plus-minus" type of immunoassays.

HRP fulfills most of the desired criteria of an ideal enzyme for use in immuno-diagnostic products. It is the smallest of the three most popular enzyme labels, and glycosylation by eight neutral carbohydrate residues renders it less sticky. Thus, it results in fewer nonspecific signals than other enzymes in the immunoassays. It was also one of the first enzymes used in developing ELISAs for low molecular weight haptens, antigens, and infectious agents. HRP-labeled antibodies are often the reagent of choice for visualizing antigens in dot-blot format assays, and are often superior to alkaline phosphatase- or β-galactosidase-labeled conjugates in terms of assay sensitivity and time.

HRP stability in solution is rather limited, but can be greatly improved by storing the conjugates at higher concentrations and by inclusion of certain substrates. Single-component substrate solutions are also available commercially, but in order to obtain the best stability of reagents and optimal sensitivity, it is necessary to use a two-component substrate to be mixed immediately prior to use. Another disadvantage of HRP is product inhibition, which limits the effective length of incubation with substrate. The enzyme is also less robust than alkaline phosphatase, and unless well protected, can be irreversibly inactivated by certain samples such as urine. HRP is also very sensitive to metal ions.

Alkaline Phosphatase (ALP)

ALP (orthophosphoric monoester phosphohydrolase, alkaline optimum, E.C. 3.1.3.1) is the second most widely used enzyme in commercial enzyme immuno-diagnostic products. It is found primarily in animal tissues and microorganisms. ALP is present in particularly high concentrations in the lactating mammary gland and in the kidney.

ALP used in enzyme immunoassays generally comes from bovine intestinal mucosa or from *E. coli*. ALPs from these two sources differ considerably in their properties and should not, as often practiced in the industry, be assayed under identical conditions.

1. Physicochemical Properties. ALPs are dimeric glycoproteins, the properties of which vary according to the origin of the enzyme (Table 6.3) They are all zinc metalloenzymes, containing at least two Zn (II) per molecule. Zinc binds to three essential histidyl residues per subunit, of which at least one seems to be in the active site (McCracken and Meighen 1981). ALP from *E. coli* contains four Zn (II) (two active site and two structural) and forms a covalent intermediate that has two phosphate groups attached to several histidyl groups. Its low pI suggests a high percentage of acidic amino acids.

TABLE 6.3.   Physicochemical Properties of Alkaline Phosphatase[a]

Property	Source	
	Bovine mucosa	*E. coli*
Molecular weight	140,000	80,000
pH optimum	10.3	8.0
Sedimentation coefficient (S, pH 6.0)	6.0	5.2
Partial specific volume (ml/g)	0.73	0.73
pI	5.7	4.5
Extinction coefficient (1%, 278 nm, 1 cm)	7.8	7.2

[a]Compiled from Tijssen (1985) and Ngo (1991).

ALP contains approximately 50 lysyl residues per molecule of the enzyme (Ngo 1991). The catalytically nonessential lysyl residues are available for conjugation purposes without significantly affecting the enzyme activity. ALP contains no detectable free thiol groups.

ALP is stable at room temperature in solutions of slightly alkaline or neutral pH. It is unstable in acid pH and is irreversibly inactivated at pH less than 3. Being a metalloenzyme, it is inhibited by metal chelators such as EDTA, cysteine, and thioglycolic acid.

2. Reaction Mechanism.   ALP hydrolyzes numerous phosphate esters including those of primary and secondary alcohols, phenols, and amines. It has a rather broad specificity in that it has primary specificity for the monophosphate group only, and the alcohol moiety of the substrate is of secondary importance. The enzyme transfers the phosphoryl residue via a phosphoryl-enzyme intermediate, which can be repressed by inorganic phosphate, P_i (Figure 6.4). ALP is able to hydrolyze P-F, P-O-C, P-O-P, P-S, and P-N bonds but not the phosphonate P-C bond. Excellent reviews of the mechanism of action of ALP are available (Fernley 1971; Reid and Wilson 1971).

As mentioned earlier, the microbial and bovine ALPs differ in their properties. For example, there seems to be only one active site per dimer for the microbial enzyme at low substrate concentrations ($< 10^{-4}$ M), whereas both sites are active at higher concentrations (Heppel et al. 1962). In contrast, the pH optimum of the bovine ALP is shifted to higher values with increasing substrate concentrations. The pH optimum tends to shift to neutrality with increasing pK values of the substrate. Addition of ions such as Mg^{2+} may significantly enhance enzyme activity in some buffers (e.g., glycine buffers, which have insufficient buffering capacity for ALP) but not in others (e.g., diethanolamine) (Tijssen 1985; Fernley 1971; Bergmeyer 1974). Tris buffers produce a sharp increase in enzyme activity (Neumann et al. 1975).

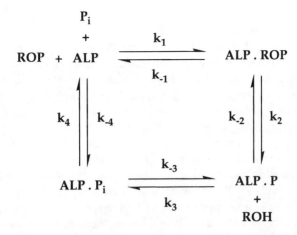

Figure 6.4. A schematic representation of hydrolysis of a phosphate ester (ROP) by alkaline phosphatase (ALP). Pi represents inorganic phosphate, ROP is the substrate, and ALP.P is phosphoryl enzyme.
From: Tijssen (1985)

3. Substrates. ALP hydrolyzes numerous phosphate esters, but not phosphodiesters. By far the most commonly used chromogenic substrate system is *p*-nitrophenyl phosphate (PNPP) in 1 *M* diethanolamine containing 0.5 mM $MgCl_2$ at pH 9.8. Under these conditions, the specific activity at 25°C is in the range 1000–2000 U/mg. Because ALP is larger than HRP and the ε_{mM} of the nitrophenyl product (18.3) is lower than that of the acidified TMB product, the detectability of ALP with PNPP substrate is not as good as that of HRP, unless incubation times are significantly increased.

PNPP, however, is still a popular substrate for ALP, because its spontaneous hydrolysis is low below 30°C; it is soluble to a concentration 100 times higher than the K_M value, thereby establishing zero-order kinetics; and its hydrolysis product *p*-nitrophenol absorbs strongly at 405 nM (Tijssen 1985).

Other commonly used chromogenic substrates for ALP include phenolphthalein monophosphate, thymophthalein monophosphate, β-glycerophosphate, and uridine phosphate.

Fluorogenic substrates are also utilized in conjunction with ALP. β-Naphthyl phosphate, 4-methylumbelliferyl phosphate, and 3-o-methylfluorescein monophosphate are widely used (Ngo 1991).

The choice of buffer is an important consideration when working with ALP. Different buffers produce different pH-rate profiles. Tris, diethyl barbiturate, glycine, borate, ethanolamine, and carbonate with pK_a values ranging from 8.1 to 9.5 are commonly used in ALP assays. In general, both the K_M and V_{max} of the enzyme increase with increasing pH and reach a limiting value at pH greater than

9.0. The addition of zinc and magnesium ions to buffers increases both the enzyme activity and the reaction rate.

While working with ALP, one should avoid the use of phosphate buffers at all stages of the immunoassay. Even though the substrate is present in another buffer, incubations and washings with phosphate-buffered saline (PBS), as is often practiced in the industry, should not be performed. Pi is a competitive inhibitor of ALP and forms an intermediate with ALP that is indistinguishable from the intermediate formed during catalysis of the hydrolysis of phosphate esters (Caswell and Caplow 1980). PBS contains a relatively high amount of P_i (15 mM) and behaves as a highly concentrated solution of inhibitor for ALP.

Similarly, spontaneous substrate hydrolysis, producing P_i, needs to be avoided. The presence of only 1% P_i in the commonly used substrate PNPP produces a significant inhibition of the enzyme (Jung and Pergande 1980). Commercially available substrates may contain up to 6% P_i. Hence, they need to be tested, if necessary, purified, and stored at low temperatures.

4. Conjugation Methods

CONJUGATION WITH GLUTARALDEHYDE.   Cross-linking of ALP to haptens or antibodies by using either a one-step or two-step glutaraldehyde method was one of the early conjugation methods used in the industry. Generally, a weight ratio of 2:1 (4:1 for Fab') is used for the enzyme and the IgG (Avrameas 1969; Ford et al. 1978). However, conjugation by this method often leads to the formation of heterogeneous aggregates. It also severely diminishes the enzyme activity. These methods are now largely replaced by newer techniques since the introduction of several heterobifunctional cross-linkers.

CONJUGATION WITH AMINO OR THIOL DIRECTED REAGENTS.   Similar to HRP, ALP can be coupled to antibodies by SMCC, SMH, or SPDP. In a two-step reaction at neutral pH, the succinimidyl part of SMCC reacts with the amino groups of the enzyme, leaving the maleimido group of SMCC free to react with thiol groups. At neutral pH, the maleimido group of the cross-linker is not reactive enough to react with the amino groups of ALP. In the second step, it reacts with the thiol group of the IgG or its fragments in a more controlled manner (Ishikawa et al. 1983,1988).

For labeling ALP to a protein without a free thiol group, the latter must first be introduced to the protein by reacting it with SPDP. The pyridyldithiopropionylated protein is then treated with DTT to generate the desired free thiol groups that can react with maleimido labeled ALP (Ngo 1991).

CONJUGATION WITH PERIODATE OXIDATION.   The procedure for coupling ALP to IgG with sodium periodate is similar to that for HRP. The enzyme is first oxidized, and then after the removal of free reagents, mixed with IgG to obtain the desired conjugates. This approach, however, is not widely followed.

Although ALP is considerably more expensive than HRP, it offers several advantages in immunoassay products. It has much better liquid and high temperature stability than HRP, it is less sensitive to inhibition by bacteria and other sample and environmental factors, and it affords alternate detection methods of high specific activity such as enzyme amplification, chemiluminescence, or fluorometry. However, in some pathological conditions the ALP activity is increased; therefore one must guard against obtaining false-positive results while working with serum samples. In this regard, HRP, which is abundant in plant but not in animal tissues, is popular for diagnostic products intended to test samples from human or animal origin. ALP, however, is widely used in a number of commercially available in-office and in-home self-testing ELISA kits for the detection of pregnancy and ovulation.

β-D-Galactosidase (BG)

BG (β-D-galactoside galactohydrolase, or β-galactosidase, E.C. 3.2.1.23) is found in microorganisms, animals, and plants. It is also known as "lactase." BG from E. coli is the most popular choice for enzyme immunoassays.

1. Physicochemical Properties. BG is a tetramer of molecular weight 465,000 and a pI of 4.6 (Fowler and Zabin 1977). It dissociates into inactive monomers at pH < 3.5 or > 11.5, or when treated with chaotropic reagents and heavy metal ions. It contains no carbohydrate or phosphate groups. BG contains over 20% acidic amino acids, 2.5% lysine, and 3–4% cysteine. The lysine and cysteine residues, about 100 and 60 mol per mole of the enzyme, respectively, can be conveniently used for cross-linking the enzyme with haptens, antibodies, or antigens.

BG is also an excellent immunogen. Anti-BG antibodies do not inhibit enzymatic activity (Cohn and Torriani 1952). The antibodies to the wild-type enzyme can also activate an inactive mutant enzyme (Rotman and Celada 1968).

BG is stable for at least 30 min at 40°C at pH 6–8 (Wallenfels and Weil 1972). However, its stability at 40°C falls sharply below pH 6.0 and slowly above pH 8. Thiogalactosides, heavy metal ions, organomercurials, metal chelators, and some alcohols and amines are potent inhibitors of the enzyme.

2. Reaction Mechanism. BG catalyzes the hydrolysis of terminal nonreducing β-D-galactose residues in β-galactosides. It has a specific requirement for D-galactopyranoside moiety and β-glycosidic linkage. Substitution at C-2, C-3, C-4, and C-6 by groups that are bulkier than hydroxyls prevents binding of the substrate to the enzyme, whereas the replacement of oxygen by a sulfur atom in the glycosidic bond decreases the rate of enzymatic hydrolysis (Ngo 1991). BG also catalyzes the hydrolysis of C-N and C-F bonds.

Wallenfels and Malhotra (1960) have proposed a three-step minimal kinetic mechanism for BG reaction. The sulfhydryl group of the active site cysteine of the enzyme acts as a general acid to protonate the glycosidic oxygen atom, while the imidazole group of histidyl residue acts as a nucleophile. The latter attacks the nucleophilic center at C-1 of the glycone. A covalent intermediate involving a carbon-nitrogen bond is proposed. In removal of the galactosyl group, the sulfhydryl anion (S^-) acts as a general base to abstract a proton from water which assists in the attack of OH^- at the C-1 position. The proposed mechanism thus encompasses (a) a stereospecific solvolysis of a carbonium ion transition state in the active site that allows only one-sided attack by the solvent and (b) a double displacement mechanism that involves protonation of the glycosidic oxygen and attack by a nucleophilic group of the enzyme to form a galactosyl-enzyme intermediate and is then followed by a second displacement reaction by a nucleophile. These two consecutive displacement reactions result in the retention of configuration at the anomeric carbon atom.

3. Substrates. The most commonly used chromogenic substrate for BG is o-nitrophenyl-β-D-galactopyranoside (ONPG), which is converted to o-nitrophenol. Its relative catalytic rate is considerably higher and less solid-phase interference is expected. ONPG is hydrolyzed at a rate of about 500 U/mg.

BG, however, can tolerate a great number of variations in the structure of the aglycon part of β-galactosides. Thus, p-nitrophenyl-β-D-galactopyranoside (PNPG), 6-bromo-2-naphthyl-β-D-galactoside, indolyl-β-D-galactoside, methyl-umbelliferyl-β-D-galactoside, 2-naphthyl-β-D-galactoside, and fluorescein di-β-D-galactoside have all been used as substrates for the enzyme.

BG can be detected fluorometrically with greater sensitivity. Methylumbelliferyl-β-D-galactoside is the most widely used fluorogenic substrate.

The commonly used assay conditions include 0.1 M Tris or phosphate buffer pH 7.0–7.5 containing 1 mM $MgCl_2$, 0.1 M reducing agent such as 2-mercaptoethanol (2-ME) and 5 mM of substrate such as ONPG. Acceptor alcohols such as 2-ME stimulate the rate of substrate hydrolysis.

4. Conjugation Methods. Similar to HRP and ALP, BG can be cross-linked to haptens and IgG or its fragments via amino and thiol directed reagents such as SMCC, SMH, MBS, or GMBS. Dimaleimide bifunctional reagents such as *N,N'*-o-phenylenedimaleimide or *N,N'*-oxydimethylenedimaleimide can also be used to couple the thiol groups of BG and Fab', reduced or thiolated IgG or F(ab')$_2$ (Kato et al. 1975a,b). Because of its large size, BG cannot be easily separated from the conjugate by gel filtration; thus, it is essential to use excess Ig to completely convert BG to the conjugate.

Cross-linkers containing diazo and maleimido functional groups that react selectively with tyrosyl and histidyl residues and with thiol groups, respectively, have also been used to prepare BG conjugates (Fujiwara et al. 1988). Glutaraldehyde coupling methods are not used because of the heterogeneous cross-linking nature of the procedure.

BG has been used as an enzyme label in both nonseparation (homogeneous) and separation-required (heterogeneous) enzyme immunoassays. The enzyme and its substrate, ONPG, have a long shelf life. Using this enzyme as the label, a number of ELISAs with a sensitivity comparable to radioimmunoassays (RIAs) have been developed for antigens and low molecular weight compounds. BG appears to be ideally suited as a label for developing highly sensitive ELISAs and thus holds great promise for the eventual development of extremely sensitive and simple homogeneous immunoassays for proteins. Single molecules of this enzyme have been successfully detected, albeit in very small volume droplets, by the use of fluorogenic substrates.

BG does not appear to offer any advantage over ALP. However, as the two enzymes function under similar conditions, this offers the possibility of designing assays that measure two analytes simultaneously .

Glucose Oxidase (GO)

GO (β-D-glucose:oxygen 1-oxidoreductase, E.C. 1.1.3.4) is frequently used in activity-amplification type homogeneous enzyme immunoassays. The enzyme is also known as penicillium B, notatin, and glucose aerodehydrogenase. GO used in ELISAs is obtained from *Aspergillus niger.*

1. Physicochemical Properties. GO consists of two identical polypeptide subunits of molecular weight 80, 000 each that are covalently linked by disulfide bonds (O'Malley and Weaver 1972). Each subunit contains one mole of Fe and one mole of FAD (flavine adenine dinucleotide). The FAD is replaceable with FHD (flavine hypoxanthine dinucleotide) without loss of activity. The molecule contains approximately 74% protein, 16% neutral sugar (mostly mannose) and 2% amino sugars (Tsuge et al. 1975).

GO has a pI of 4.2–4.4, and an optimum activity at pH 5.5. Oxidation of the carbohydrate with periodate does not affect the enzyme activity, immunological properties or antigenicity, or heat stability (Tijssen 1985). The oxidized enzyme, however, is more susceptible to detergents than the native enzyme. Its activity is inhibited by Ag^+, Hg^{2+}, and Cu^{2+}.

Crystalline GO is stable for at least two years at $0°C$ and 8 years at $-15°C$ (Bentley 1963). Aqueous solutions of the enzyme are stable for a week at $5°C$. It is, however, unstable above $40°C$ or at pH values above 8, although the presence of glucose may exert a protective action. GO is inhibited to varying degrees by 2-deoxy-D-glucose, *p*-chloromercuricbenzoate, sodium nitrite, D-arabinose, and 8-hydroxyquinoline.

2. Reaction Mechanism. GO catalyzes the oxidation of β-D-glucose to δ-D-gluconolactone in the presence of molecular oxygen. It has a high specificity for β-D-glucopyranose. There is an absolute requirement for a hydroxyl group at C-1, and the activity is about 160 times higher if the hydroxyl group is in the β-position. Changes in the substrate at C-2 through C-6, except for L-glucose and 2-*o*-methyl-D-glucose, do not completely prevent the compounds from serving as substrates, but the enzyme activity is considerably reduced (Whitaker 1994).

During substrate hydrolysis, the oxidized form of the enzyme (EFAD) functions as a dehydrogenase to extract two hydrogen atoms from β-D-glucose to form the reduced enzyme, EFADH$_2$ and δ-D-gluconolactone. In a subsequent step, δ-D-gluconolactone is hydrolyzed nonenzymatically to D-gluconic acid and the reduced enzyme is reoxidized by molecular oxygen (enzymatic step).

3. Substrates. GO activity can be determined by colorimetric, fluorometric, manometric, or electrochemical methods. The colorimetric assay is based on a peroxidase indicator reaction to measure the amount of H$_2$O$_2$ liberated. For this reaction, an excess of peroxidase and hydrogen donor has to be used. Thus, GO activity can be determined by incorporation of peroxidase and a chromogen such as *o*-dianisidine into the reaction mixture. Peroxidase utilizes the H$_2$O$_2$ produced to oxidize the chromogen to a colored compound, which can be determined colorimetrically.

Homovanillic acid (4-hydroxy-3-methoxy-phenylacetic acid, or HVA) is the popular fluorogenic substrate for GO. It is converted to a highly fluorescent, stable compound with an excitation wavelength of 315 nm and an emission wavelength of 425 nm.

GO assays can be carried out in 100 mM phosphate buffer, pH 6–7, containing 0.17 mM D-glucose, 82 mM *o*-dianisidine, and 1.2 U/ml peroxidase. Commercial sources of GO, however, may contain some catalase activity which interferes in the assay.

4. Conjugation Methods. GO can be coupled to IgG through the activation of its carboxyl group with EEDQ (*N*-ethoxycarbonyl-2-ethoxy-1,2-dihydroquinoline) after its amino groups are first blocked with FDNB or by biotinylation. The labeled enzyme is then coupled to IgG at pH 8.0 (Guesdon 1988). Heterobifunctional cross-linkers such as SMCC, MBS, SMH, and SPDP can also be used to couple GO to Fab', reduced IgG, or thiolated antibody. Periodate oxidation of the carbohydrate moiety of GO to aldehydes for cross-linking with amino groups of IgG is also feasible.

Lysozyme

Lysozyme (*N*-acetylmuramide glycanohydrolase, E.C. 3.2.1.17), also known as muramidase, was the first enzyme used in activity modulation type homogeneous enzyme immunoassays (Rubenstein et al. 1972). Since then, assays using this enzyme have been developed for a large number of haptens, and are commercially employed in the detection of urinary metabolites of abused drugs. Lysozyme is isolated from hen egg white.

1. Physicochemical Properties. Lysozyme is a small protein of molecular weight 14, 000 and has an unusually stable three-dimensional structure. It is a very basic protein of pI 11, and has optimal activity around pH 9.2. Lysozyme is inhibited by surface-active reagents such as dodecyl sulfate, alcohols, and fatty acids. Imidazole and indole derivatives are inhibitors via formation of charge-transfer complexes.

The enzyme has a deep cleft on one side, lined mostly with hydrophilic amino acid residues to which the substrate (6 sugar residues) is bound. It contains 6 lysyl residues of which one is near the active site. Conjugation of a hapten to this group does not affect the activity of the enzyme, However, incubation of the enzyme-hapten conjugate with the specific antihapten antibody prior to the addition of the bulky substrate produces serious inhibition. The small Fab' is not as effective for inhibition as complete IgG.

Acetylation of lysyl residues, which removes positive charges from the enzyme, affects enzyme activity (Tijssen 1985). Similarly, modification of the carboxyl groups of glutaryl and aspartyl residues, which are involved in substrate binding, has a profound influence on the activity of lysozyme.

Lysozyme stored as a dry lyophilized or crystalline powder at 5°C is stable for years. Solutions at pH 4–5 are stable for several weeks refrigerated and for days at ambient temperatures.

2. Reaction Mechanism. Lysozyme catalyzes the hydrolysis of N-acetylmuramyl (NAM) bonds of the alternating N-acetylglucosamine (NAG)-N-acetylmuramic acid copolymer found in the cell walls of certain bacteria. The enzyme, which possesses β-$(1\rightarrow4)$-glucosaminidase activity, hydrolyzes the bond between the C-1 of NAM and the C-4 of NAG.

The side chains of Asp 32 and Glu 35 are directly involved in the hydrolysis of the substrate (Phillips 1967). The carboxyl group of the glutamyl side chain is nonionized and located about 0.3 nm from the glucosidic oxygen between the C-1 of NAM and the C-4 of NAG. It donates a proton to this oxygen and produces a C-1-OH in NAM and a C-4^+-H carbonium ion intermediate in NAG. This enzymatic reaction is promoted by the presence of a negatively charged group on the aspartyl side chain 0.3 nm away from the carbonium ion intermediate and by the distortion of the NAG ring into a half-chair form. During binding, the substrate is forced to assume the geometry of the transition state, which is similar to that of the carbonium ion. This mechanism explains the strong pH and ionic strength dependence of lysozyme action (Chang and Carr 1971; Tijssen 1985). Derivation of these groups inactivates the enzyme.

3. Substrates and Assay Conditions. Low molecular weight synthetic substrates are not suitable for measuring lysozyme activity. The reaction kinetics with these substrates are complicated by their nonproductive binding outside of the catalytic site. Hence, cell wall fragments or whole cells of *Micrococcus luteus* are routinely used as substrates. The reaction is carried out in 0.05–0.1 M phosphate buffer, pH 7.4, containing 0.1% sodium azide, 1 mg BSA/ml, and 0.1 ml of lysozyme sample in 1 ml buffer (Tijssen 1985). In the homogeneous immunoassays, the activity is monitored based on the principles of turbidimetry. Hydrolysis of the cell walls and subsequent cell lysis results in a decrease in light scattering, which is followed by absorbance measurement at 436 nm.

4. Conjugation Methods.   Water-soluble carbodiimides such as 1-ethyl-3(3-dimethylaminopropyl)-carbodiimide (EDC) or N-benzyl-N^1-3-dimethylaminopropyl carbodimide have been used for the modification of lysozyme carboxyl groups for coupling to amino-group containing haptens (Hoare and Koshland 1966; Atassi et al. 1973; Lin and Koshland 1969; Yamada et al. 1981). As mentioned earlier, coupling of haptens to the lysyl side chain near the active site of lysozyme does not affect the enzyme activity. Hence, this group can also be selectively used for conjugating haptens to lysozyme.

Malate Dehydrogenase (MDH)

MDH (L-malate:NAD$^+$ oxidoreductase, E.C. 1.1.1.37) is widely used in homogeneous enzyme immunoassays because of its ability to form hapten conjugates, which in some cases are inhibited while in others are activated by the appropriate antihapten antibodies. It is widely used for the detection of morphine, thyroxine, triiodothyronine, and tetrahydrocannabinol. MDH is isolated from pig heart mitochondria.

1. Physicochemical Properties.   As an enzyme of the citric acid or Kreb's cycle, MDH is ubiquitous in aerobic organisms. It is composed of two identical subunits of molecular weight 35, 000 each. These subunits are held together by secondary interactions. Each subunit has an active site for the coenzyme NAD$^+$ (Holbrook and Wolfe 1972). The two subunits of MDH can dissociate without losing catalytic activity and reassemble in the presence of the substrate (Shore and Chakrabarti 1976). MDH contains 14 sulfhydryl groups, two of which are required to bind substrate (Sequin and Kosicki 1967).

The optimal pH for MDH activity is 7.4. Several iodinated agents, thyroxine, iodine cyanide, and molecular iodine, inactivate the enzyme by oxidizing the sulfhydryl groups. Phosphate, arsenate, and zinc ions are stimulatory. Various organic compounds seem to have a significant stabilizing effect on MDH.

MDH is stable for one year when stored at 5°C as a suspension in ammonium sulfate.

2. Reaction Mechanism and Assay Conditions.   MDH catalyzes the following reversible reaction:

$$\text{Malate} + \text{NAD}^+ \rightleftharpoons \text{Oxaloacetate} + \text{NADH} + \text{H}^+$$

Similar to most dehydrogenases, MDH functions by a compulsory ordered mechanism (Harada and Wolfe 1968). Oxaloacetate inhibits substrate catalysis by feedback inhibition. Activation of MDH by malate is seen as a binding of malate to some effector site other than the active site, thereby resulting in a tighter binding of NAD^+.

MDH can be irreversibly inhibited by modification of about two histidines or two arginines, and can be either activated or inhibited upon modification of specific sulfhydryl groups (Harada and Wolfe 1968).

MDH activity is monitored by the change in optical density at 340 nm per min due to the disappearance of NADH according to the reaction

$$\text{Oxaloacetate} + \text{NADH} + H^+ \rightarrow \text{Malate} + NAD^+$$

The reaction is carried out in 0.1 M phosphate buffer of pH 7.4 and containing 0.5 mM oxaloacetate and 0.2 mM NADH.

3. Conjugation Methods. Various haptens and/or their derivatives can be conjugated to MDH via amino, tyrosyl, sulfhydryl, and carboxyl groups using the approaches described earlier for other enzymes. However, one needs to carefully assess the influence of conjugation procedures, the reactive groups involved, and the degree of hapten substitution on the activity or inhibition of the enzyme. For example, various derivatives of morphine have been successfully conjugated to lysyl, tyrosyl, and cysteinyl side chains of MDH. The effect of antimorphine antibodies on the activity of these conjugates has been evaluated (Rowley et al., 1975).

Enzyme inhibition occurs only when the antibody is bound to morphines that are attached through amino groups of the enzyme. Removal of the tyrosyl-bound groups with hydroxylamine has no effect on instability, while enzyme conjugates that are labeled on cysteine are not further inhibited by the antibody. Furthermore, conjugation in the presence of substrates yields relatively noninhibitable enzyme, suggesting that it is the derivatization of only one or a few specific amino groups on lysyl side chains that produces the inhibition effect required in a homogeneous immunoassay.

When morphine occupies all 14 cysteine side chains, the enzyme activity is practically inhibited. In contrast, a much more gradual decrease in activity is observed on labeling tyrosines and lysines.

Similar behavior is generally seen with other haptens conjugated to MDH or to other enzymes used in homogeneous enzyme immunoassays. Thus, each system needs to be optimized to obtain the best results and assay performance.

Glucose-6-Phosphate Dehydrogenase (G6PDH)

G6PDH (D-glucose-6-phosphate:$NADP^+$ oxidoreductase, E.C. 1.1.1.49) has been employed for a wide range of homogeneous enzyme immunoassays. It is ideally suited for the assay of substances in human serum.

G6PDH is widely distributed, however, the one from *Leuconostoc mesenteroides* is used in practically all immunoassays. Unlike the mammalian enzyme, which is specific for NADP, the bacterial enzyme can employ either NAD or NADP (DeMoss et al. 1953). Thus, by using NAD^+ as the cofactor in G6PDH-based immunoassays, interference from the presence of the mammalian enzyme in the serum sample being assayed can be avoided.

1. Physicochemical Properties. G6PDH has a molecular weight of 104,000. It consists of two identical subunits, each containing a reactive lysyl residue near the active site (Milhausen and Levy 1975). The bacterial enzyme contains no cysteine and is stable for up to one year at 4°C.

The optimal pH for its activity is 7.8 in 0.05 M Tris buffer and with NAD as coenzyme. Pyridoxal 5'-phosphate inhibits competitively with respect to G6PDH and noncompetitively with respect to NAD or NADP. FDNB inhibits the enzyme irreversibly via its blocking of active site lysyl residue of the enzyme.

2. Reaction Mechanism and Assay Conditions. G6PDH catalyzes the following reaction:

$$\text{Glucose-6-phosphate} + NAD^+ \rightarrow \text{6-Phospho-D-gluconate} + NADH + H^+$$

Similar to other dehydrogenases, G6PDH probably operates according to an ordered pathway.

G6PDH activity may be determined by several methods; spectrophotometric determination of NADH at 340 nm is the most convenient. Mg^{2+} stimulates the enzyme up to 10 mM concentration, but inhibits its activity at higher concentrations. Other bivalent cations are inhibitory. Phosphate is an important inhibitor of the enzyme, and hence, the use of phosphate buffers should be avoided in the assay. Typically, G6PDH assays are carried out in 0.1 M triethanolamine buffer, pH 7.8, and containing < 10 mM $MgCl_2$, 1.2 mM glucose-6-phosphate, and 0.3–0.4 mM NAD^+.

3. Conjugation Methods. Haptens or their derivatives can be coupled to the lysyl or tyrosyl side chains of G6PDH by one of the several methods described earlier. Similar to MDH conjugates, the activity of G6PDH conjugates is also inhibited by excess of antihapten antibodies. The reduction in total enzyme activity upon antibody binding appears to involve conformational changes in the enzyme.

SIGNAL AMPLIFICATION SYSTEMS

In a conventional enzyme immunoassay, the enzyme label is responsible for the conversion of a substrate into a product that can be readily detected, either by its color, or by some other property such as fluorescence or luminescence. At very low enzyme concentrations, hardly any product is generated, resulting in a weak signal that is difficult to measure because of background noise. To overcome these problems, a variety of methods have been used or advocated as amplification systems to improve the sensitivity of enzyme immunoassays. Some of the commonly used signal amplification systems are described below.

Soluble Enzyme: Antienzyme Complexes

One approach to increasing the sensitivity of detection using labeled antibodies is to increase the size of the immune complex. Modest signal amplification can thus be obtained by direct coupling of a number of enzyme molecules to the antibody, or through indirect coupling via enzyme:antienzyme complexes.

In the latter case, complexes of enzyme and antienzyme antibodies are prepared. If the two molecules are added at nearly equal molar ratios, multimeric interactions occur quickly, allowing the formation of large enzyme:antienzyme complexes. These large complexes are linked to the primary antibody by using a bridging anti-immunoglobulin antibody. For example, a mouse monoclonal antibody bound to an antigen could be detected using a rabbit antimouse immunoglobulin antibody to bridge it to a complex of peroxidase:antiperoxidase, if the antiperoxidase antibodies are mouse. The most common example of this method is the peroxidase:antiperoxidase complexes (PAP) (Mason et al. 1969; Sternberger et al. 1970). These complexes can be purchased already in their large aggregated forms.

Soluble PAP complexes are useful as tracers in enzyme immunoassays. In general, their activity is superior, sometimes by several orders of magnitude, to that of chemically linked conjugates. Their use, however, may lead to increased nonspecific binding of enzyme and thus an increase in background noise. Thus, a suitable compromise must be made between signal-to-noise ratio and increased assay sensitivity.

Coupled Enzyme Cascade Systems

The degree of signal amplification can be greatly increased by using an enzyme label to produce a catalyst for a second reaction or a series of reactions leading to a detectable product. In fact, several metabolic pathways are controlled in this fashion where the final response is amplified by several orders of magnitude over the initial signal. Two well-known examples include the kinase/phosphatase cycles of phosphorylase control in the glycogen metabolic pathway and the mechanism of fibrin formation in the blood clotting cascade.

In order to have widespread utility for enzyme immunoassay, an amplification system must be made up of components that have good stability and be simple and reproducible in use. The more complex the system, the more difficult it becomes to satisfy these criteria, because several enzymes, substrates, and activators need to be stable in a single compatible solution (Johannsson 1991).

The first amplification system based on cofactor cycling fulfilling the above criteria for enzyme immunoassay was described by Self (1982) and later in detail by Johannsson and Bates (1988). In this system, the dephosphorylation of one substrate by ALP results in one product capable of initiating cycling reactions, therefore amplifying the formation of highly colored products.

A practical example of this system, which has been widely applied for quantitative competitive assays and both quantitative and qualitative homogeneous type assays, uses NADP as the substrate for ALP (Figure 6.5). The dephosphorylation of NADP by ALP to NAD activates an alcohol dehydrogenase (ADH)-diaphorase redox cycle. A consequence of ethanol oxidation by ADH is the reduction of NAD to form NADH. The latter in the presence of diaphorase reduces the colorless iodonitrotetrazolium violet (INT-violet) to highly colored formazan, and concomitantly is itself oxidized to NAD. NAD can again be used in the ADH-catalyzed reaction, thereby repeating the cycle. The net result is an accumulation of formazan (amplification) by cycling of NAD and NADH in the presence of excess ADH, diaphorase, and their substrates.

In the above reaction, each ALP label is capable of producing approximately 60,000 NAD molecules per min, and in turn, each NAD molecule will initiate the production of 60 molecules of the colored formazan product. This highly sensitive assay can detect 0.011 amol of ALP or about 1500-fold lower than the 1-hr non-amplified PNPP system (Johannsson et al. 1986).

This remarkable system has been used successfully commercially. It is a genuine biomolecular amplifier, resulting in as much as 1000-fold amplification of the signal. However, long term stability of the liquid reagents can only be achieved by storing them as two solutions, separating the NADP from the enzymes ADH and diaphorase, as only the slightest amount of contaminating phosphatase activity in a solution containing all the components would quickly lead to background color development. For this reason of lower backgrounds, the amplified detection of ALP is usually carried out in two sequential incubations.

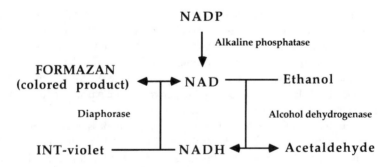

Figure 6.5. Combined enzyme cascade system for measuring alkaline phosphatase using secondary enzyme redox amplifying system.

Other enzyme cascade detection systems for enzyme labels have been described based on the production of an activator for an enzyme or the destruction of an enzyme inhibitor (Mize et al. 1989).

Avidin-Biotin Systems

In recent years, the avidin-biotin complex has become very useful as an extremely versatile, general mediator in a wide variety of bioanalytical applications. The major distinguishing feature of the avidin-biotin system is the extraordinary affinity ($K_a = 10^{15}$ l/mol) that characterizes the complex formed between the water-soluble vitamin, biotin (molecular weight 244), and the egg white protein, avidin (or streptavidin, its bacterial relative from *Streptomyces avidinii*). Each molecule of avidin has four binding sites for biotin. The basic principle of this approach is that biotin, coupled to low or high molecular weight molecules, can still be recognized by avidin, either as the native protein or in derivatized form containing any one of a number of different labels.

The strength and speed of the binding between avidin and biotin is used to provide amplification of signal and as the basis of generic signal generation reagents. Streptavidin is generally preferred to avidin, as the former has a neutral isoelectric point and it does not contain carbohydrates. These properties make streptavidin more inert in assay systems, resulting in lower nonspecific binding and hence greater sensitivity.

In the avidin-biotin system, one participating component must always be biotinylated. Biotin-labeled compounds, including immunoglobulins, thus can be detected by avidin-conjugated or -complexed substances. Unlike other immunologic methods, this technique may not require an antigen-antibody reaction for the detection. In most immunoassay applications, biotin is introduced into detection antibodies at 10–20 biotin molecules per antibody by reacting the ε-amino groups of lysine with *N*-hydroxysuccinimide ester of biotin in a very simple one-step reaction. Depending on the detection systems, avidin can then be conjugated to different labels.

The avidin-biotin system can be used for detecting a diverse number of compounds with three different basic configurations: the labeled avidin-biotin (LAB) technique, the bridged avidin-biotin (BRAB) technique, and the avidin-biotin complex (ABC) technique (Figure 6.6).

In the LAB method, biotinylated antibody links the immobilized target molecule or antigen to avidin labeled with a detectable molecule such as an enzyme (Figure 6.6a). This detection format is used widely for immunoassays. In the BRAB method, unlabeled avidin acts as a link between two or more biotin-conjugated molecules (Figure 6.6b). It is based on the principle that avidin possesses four active sites, not all of which react with the biotin residues associated with the antigen:biotinylated antibody complex. Thus, the remaining free active sites can operate as

(a)

(b)

(c)

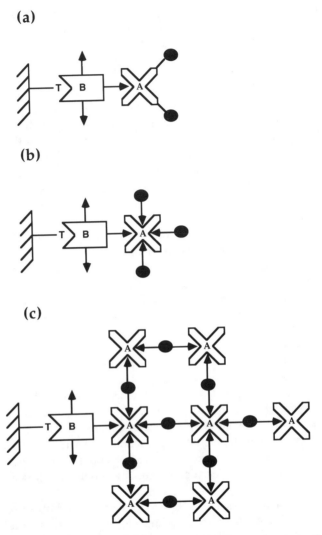

Figure 6.6. Basic configurations of the avidin-biotin system. In all cases, the target molecule (T) is shown immobilized on a solid support.

(a) Labeled avidin-biotin (LAB) technique: The specific binder (B) of the target is biotinylated (→), and avidin (A) carries a label, such as an enzyme (•).

(b) Bridged avidin-biotin (BRAB) technique: Same as (a) except unlabeled avidin is used to link the biotinylated binder and the biotinylated label.

(c) Avidin-biotin complex (ABC) technique: Unlabeled avidin is mixed with biotinylated label to form a polymeric complex with some residual biotin-binding activity. This polymer binds to the biotinylated binder as in (a).

From: Diamandis and Christopoulos (1991).

acceptors for another biotin-labeled protein (enzyme label) secondarily added to the system. This format is also used widely in immunoassays.

The ABC format (Figure 6.6c) combines effectively the principles of the above two methods to yield a significantly more sensitive system. It is based on the biochemical characteristics of avidin and biotin conjugates. Because it has four binding sites for biotin conjugates, avidin can form a link between two or more biotin conjugates. With this approach, the sensitivity of the detecting system is generally increased because an increased number of enzyme molecules are present on the complex.

Perhaps the greatest benefit of avidin-biotin systems has been for the speed and simplicity of assay development. Use of generic signal generation reagents labeled with avidin avoids the need to develop individual conjugation methods for each assay. The system is commercially utilized by a number of companies to develop generic capture systems. The solid-phase (e.g., polystyrene tubes) is coated with avidin, and the capture antibody is biotinylated. This minimizes the need for new coating methods and facilitates the use of antibodies with high affinities for analyte but poor coating properties.

Kits offering the avidin and biotinylated enzymes at optimized concentrations, so that one can form the complex by simply mixing the reagents, are commercially available and are known as ABC (avidin-biotin complex) kits (Diamandis and Christopoulos 1991).

REFERENCES

ATASSI, M. Z.; HABEEB, A. F. S. A.; and ANDO, K. 1973. Enzymic and immunochemical properties of lysozyme. VII. Location of all the antigenic reactive regions. A new approach to study immunochemistry of tight proteins. *Biochim. Biophys. Acta* 303:203–209.

AVRAMEAS, S. 1969. Coupling of enzymes to proteins with glutaraldehyde. Use of the conjugates for the detection of antigens and antibodies. *Immunochemistry* 6:43–52.

AVRAMEAS, S.; TERNYNCK, T.; and GUESDON, J. L. 1978. Coupling of enzyme to antibodies and antigens. In *Quantitative Enzyme Immunoassay,* eds. E. Engvall and A. J. Pesce, pp. 7–23, Blackwell Scientific, Oxford.

BENTLEY, R. 1963. Glucose oxidase In *The Enzymes,* vol. 7, 2d ed., eds. P. D. Boyer, H. Lardy, and K. Myrback, pp. 567–586, Academic Press, New York.

BERGMEYER, H. U. 1974. *Methods of Enzymatic Analysis.* Academic Press, New York.

BOVAIRD, J. H.; NGO, T. T.; and LENHOFF, H. M. 1982. Optimizing the *o*-phenylenediamine assay for horseradish peroxidase. Effects of phosphate and pH, substrate and enzyme concentrations and stopping reagents. *Clin. Chem.* 28:2423–2426.

CASWELL, M., and CAPLOW, M. 1980. Correlation of thermodynamic and kinetic properties of the phosphoryl-enzyme formed with alkaline phosphatase. *Biochemistry* 19:2907–2911.

CHANG, K. Y., and CARR, C. W. 1971. Structure and function of lysozyme. I. Effect of pH and cation concentration on lysozyme activity. *Biochim. Biophys. Acta* 229:496–503.

COHN, M., and TORRIANI, A. M. 1952. Immunochemical studies with β-galactosidase and structurally related proteins. *J. Immunol.* 69:471–480.

DEMOSS, R. D.; GUNSALUS, I. C.; and BARD, R. C. 1953. A glucose-6-phosphate dehydrogenase in *Leuconostoc mesenteroides. J. Bacteriol.* 66:10–16.

DESHPANDE, S. S. 1994. Immunodiagnostics in agricultural, food, and environmental quality control. *Food Technol.* 48(6):136–141.

DESHPANDE, S. S., and SHARMA, B. P. 1993. Immunoassays, nucleic acid probes, and biosensors. Two decades of development, current status and future projections in clinical, environmental, and agricultural applications. In *Diagnostics in the Year 2000,* eds., P. Singh, B. P. Sharma, and P. Tyle, pp. 459–525, Van Nostrand Reinhold, New York.

DIAMANDIS, E. P. 1988. Immunoassays with time-resolved fluorescence spectroscopy: Principles and applications. *Clin. Biochem.* 21:139–150.

DIAMANDIS, E. P., and CHRISTOPOULOS, T. K. 1991. The biotin-(strept)avidin system: principles and applications in biotechnology. *Clin. Chem.* 37:625–636.

DUNFORD, H. B., and STILLMAN, J. S. 1976. On the function and mechanism of action of peroxidases. *Coord. Chem. Rev.* 19:187–257.

FERNLEY, H. N. 1971. Mammalian alkaline phosphatases. In *The Enzymes,* vol. 4, 3d ed., ed. P. D. Boyer, pp. 417–447, Academic Press, New York.

FORD, D. J.; RADIN, R.; and PESCE, A. J. 1978. Characterization of glutaraldehyde coupled alkaline phosphatase-antibody and lactoperoxidase-antibody conjugates. *Immunochemistry* 15:237–243.

FOWLER, A. V., and ZABIN, I. 1977. The amino acid sequence of β-D-galactosidase of *Escherichia coli. Proc. Natl. Acad. Sci. (*USA) 74:1507–1510.

FUJIWARA, K.; SAITA, T.; and KITAGAWA, T. 1988. The use of *N*-[β-(4-diazophenyl)ethyl] maleimide as a coupling agent in the preparation of enzyme-antibody conjugate. *J. Immunol. Meth.* 110:47–53.

GEOGHEGAN, W. D. 1985. The Ngo-Lenhoff (MBTH-DMAB) peroxidase assay. In *Enzyme-Mediated Immunoassay,* eds. T. T. Ngo and H. M. Lenhoff, pp. 451–465, Plenum Press, New York.

GIBBONS, I.; SKOLD, C.; ROWLEY, G.L.; and ULLMAN, E. F. 1980. Homogeneous enzyme immunoassay for proteins employing β-galactosidase. *Anal. Biochem.* 102:167–170.

GOSLING, J. P. 1990. A decade of development in immunoassay methodology. *Clin. Chem.* 36:1408–1427.

GRIBNAU, T. C. J.; LEUVERING, J. H. W.; and VAN HELL, H. 1986. Particle-labeled immunoassays: a review. *J. Chromatogr.* 376:175–189.

GUESDON, J. L. 1988. Amplificaton systems for enzyme immunoassay. In *Nonisotopic Immunoassay,* ed. T. T. Ngo, pp. 85–106, Plenum Press, New York.

GUILBAULT, G. G.; BRIGNAC, P. J.; and JUNEAU, M. 1968a. New substrates for the fluorimetric determination of oxidative enzymes. *Anal. Chem.* 40:1256–1263.

GUILBAULT, G. G.; BRIGNAC, P. J.; and ZIMMER, M. 1968b. Homovanillic acid as a fluorimetric substrate for oxidative enzymes. Analytical applications of the peroxidase, glucose oxidase and xanthine oxidase systems. *Anal. Chem.* 40:190–196.

HARADA, K., and WOLFE, R. G. 1968. Malate dehydrogenase. VII. The catalytic mechanism and possible role of identical protein subunits. *J. Biol. Chem.* 243:4131–4137.

HASCHKE, R. H., and FRIEDHOFF, J. M. 1978. Calcium-related properties of horseradish peroxidase. *Biochem. Biophs. Res. Commun.* 80:1039–1042.

HASIDA, S.; IMAGAWA, M.; INOUE, S.; RUAN, K. H.; and ISHIKAWA, E. 1984. More useful maleimide compounds for the conjugate of Fab' to horseradish peroxidase through thiol groups in the hinge. *J. Appl. Biochem.* 6:56–63.

HEPPEL, L. A.; HARKNESS, D. R.; and HILMOE, R. J. 1962. A study of the substrate specificity and other properties of the alkaline phosphatase of *Escherichia coli. J. Biol. Chem.* 237:841–846.

HOARE, D. G., and KOSHLAND, D. E. 1966. A procedure for the selective modification of carboxyl group proteins. *J. Am. Chem. Soc.* 88:2057–2058.

HOLBROOK, J. J., and WOLFE, R. G. 1972. Malate dehydrogenase. X. Fluorescence microtitration studies of D-malate, hydroxymalonate, nicotinamide dinucleotide, and dihydronicotinamide-adenine dinucleotide binding by mitochondrial and supernatant porcine heart enzymes. *Biochemistry* 11:2499–2502.

IMAGAWA, M.; YOSHITAKE, S.; HAMAGUCHI, Y.; ISHIKAWA, E.; NITSU, Y.; URUSHIZAKI, I.; KANAZAWA, R.; TACHIBANA, S.; NAKAZAWA, N.; and OGAWA, H. 1982. Characteristics and evaluation of antibody-horseradish peroxidase conjugates prepared by using a maleimide compound, glutaraldehyde, and periodate. *J. Appl. Biochem.* 4: 41–57.

ISHIKAWA, E.; IMAGAWA, M.; HASHIDA, S.; YOSHITAKE, S.; HAMAGUCHI, Y.; and UENO, T. 1983. Enzyme-labeling of antibodies and their fragments for enzyme immunoassay and immunohistochemical staining. *J. Immunoassay* 4:209–327.

ISHIKAWA, E.; HASHIDA, S.; KOHNO, T.; and TANAKA, K. 1988. Methods for enzyme-labeling of antigens, antibodies and their fragments. In *Nonisotopic Immunoassay,* ed. T. T. Ngo, pp. 27–55, Plenum Press, New York.

JOHANNSSON, A. 1991. Heterogeneous enzyme immunoassay. In *Principles and Practice of Immunoassay,* eds, C. P. Price and D. J. Newman, pp. 295–325, Stockton Press, New York.

JOHANNSSON, A., and BATES, D. L. 1988. Amplification by second enzymes. In *ELISA and Other Solid Phase Immunoassays,* eds. D. M. Kemeny and S. J. Challacombe, pp. 85–106, John Wiley, Chichester, England.

JOHANNSSON, A.; ELLIS, D. H.; BATES, D. L.; PLUMB, A. M.; and STANLEY, C. J. 1986. Enzyme amplification for immunoassays. Detection limit of one hundredth of an attomole. *J. Immunol. Meth.* 87:7–11.

JUNG, K., and PERGANDE, M. 1980. Influence of inorganic phosphate on the activity determination of isoenzymes of alkaline phosphatase in various buffer systems. *Clin. Chim. Acta* 102:215–219.

KATO, K.; HAMAGUCHI, Y.; FUKUI, H.; and ISHIKAWA, E. 1975a. Enzyme-linked immunoassay. I. Novel method for synthesis of the insulin-β-D-galactosidase conjugate and its applicability for insulin assay. *J. Biochem.* 78:235–237.

KATO, K.; HAMAGUCHI, Y.; FUKUI, H.; and ISHIKAWA, E. 1975b. Enzyme-linked immunoassay. II. A simple method for synthesis of the rabbit antibody-β-D-galactosidase complex and its general applicability. *J. Biochem.* 78:423–425.

LIN, T. Y., and KOSHLAND, D. E. 1969. Carboxyl group modification and the activity of lysozyme. *J. Biol. Chem.* 244:505–508.

MASON, T. E.; PHIFER, R. F.; SPICER, S. S.; SWALLOW, R. A.; and DRESKIN, R. B. 1969. An immunoglobulin-enzyme bridge method for localizing tissue antigens. *J. Histochem. Cytochem.* 17:563–569.

MCCRACKEN, S., and MEIGHEN, E. A. 1981. Evidence for histidyl residues at the Zn^{2+} binding sites of monomeric and dimeric forms of alkaline phosphatase. *J. Biol. Chem.* 256:3945–3950.

MILHAUSEN, M., and LEVY, H. R. 1975. Evidence for an essential lysine in glucose-6-phosphate dehydrogenase from *Leuconostoc mesenteroides. Eur. J. Biochem.* 50:453–461.

MIZE, P. D.; HOKE, R. A.; LINN, C. P.; REARDON, J. E.; and SCHULTE, T. H. 1989. Dual-enzyme cascade: An amplified method for the detection of alkaline phosphatase. *Anal. Biochem.* 179:229–235.

MORISHIMA, I.; KURONO, M.; and SHIRO, Y. 1986. Presence of endogenous calcium ion in horseradish peroxidase. Elucidation of metal-binding site by substitutions of divalent and lanthanide ions for calcium and use of metal-induced NMR resonance. *J. Biol. Chem.* 261:9391–9399.

NAKAMURA, R. M.; VOLLER, A.; and BIDWELL, D. E. 1986. Enzyme immunoassays: Heterogeneous and homogeneous systems. In *Handbook of Experimental Immunology,* Vol. 1, Immunochemistry, 4th ed., ed. D. M. Weir, pp. 27.1–27.20, Blackwell Scientific, Oxford.

NAKANE, P. K. 1975. Recent progress in the peroxidase-labeled antibody method. *Ann. N.Y. Acad. Sci.* 254:203–228.

NAKANE, P. K., and KAWAOI, A. 1974. Peroxidase-labeled antibody, a new method of conjugation. *J. Histochem. Cytochem.* 22:1084–1091.

NEUMANN, H.; KEZDY, F.; HSU, J.; and ROSENBERG, I. H. 1975. *Biochim. Biophys. Acta* 391:292–300.

NGO, T. T. 1991. Enzyme systems and enzyme conjugates for solid phase ELISA. In *Immunochemistry of Solid-Phase Immunoassay,* ed. J. E. Butler, pp. 85–102, CRC Press, Boca Raton, FL.

NGO, T. T., and LENHOFF, H. M. 1980. A sensitive and versatile chromogenic assay for peroxidase and peroxidase-coupled reactions. *Anal. Biochem.* 105:389–397.

NGO, T. T., and LENHOFF, H. M. 1985. *Enzyme-Mediated Immunoassay.* Plenum Press, New York.

OELLERICH, M. 1984. Enzyme immunoassay: A review. *J. Clin. Chem. Clin. Biochem.* 22:895–904.

OGAWA, S.; SHIRO, Y.; and MORISHIMA, I. 1979. Calcium binding by horseradish peroxidase C and the heme environmental structure. *Biochem. Biophys. Res. Commun.* 80:674–678.

O'MALLEY, J. J., and WEAVER, J. L. 1972. Subunit structure of glucose oxidase from *Aspergillus niger. Biochemistry* 11:3527–3532.

PHILLIPS, D. C. 1967. The hen egg white lysozyme molecule. *Proc. Natl. Acad. Sci.* (USA) 57:484–495.

PORSTMANN, B., and PORSTMANN, T. 1988. Chromogenic substrates for enzyme immunoassay. In *Nonisotopic Immunoassay,* ed. T. T. Ngo, pp. 57–84, Plenum Press, New York.

PORSTMANN, B.; PORSTMANN, T.; GAEDE, D.; NUGEL, E.; and EGGER, E. 1981. Temperature dependent rise in activity of horseradish peroxidase caused by non-ionic detergents and its use in enzyme immunoassay. *Clin. Chim. Acta* 109:175–181.

REID, T. W., and WILSON, I.B. 1971. *Escherichia coli* alkaline phosphatase. In *The Enzymes,* vol. 4, 3d ed., ed. P. D. Boyer, pp. 373–415, Academic Press, New York.

ROTMAN, M. B., and CELADA, F. 1968. Antibody-mediate activation of a defective β-galactosidase extracted from an *Escherichia coli* mutant. *Proc. Natl. Acad. Sci.* (USA) 60:660–667.

ROWLEY, G. L.; RUBENSTEIN, K. E.; HUISJEN, J.; and ULLMAN, E. F. 1975. Mechanism by which antibodies inhibit hapten-malate dehydrogenase conjugates. *J. Biol. Chem.* 250:3759–3766.

RUBENSTEIN, K. E.; SCHNEIDER, R. S.; and ULLMAN, E. F. 1972. "Homogeneous" enzyme immunoassay. A new immunochemical technique. *Biochem. Biophys. Res. Commun.* 47:846–851.

SAUNDERS, B. C.; HOLMES-SIEDEL, A. G.; and STARK, B. P. 1964. *Peroxidase.* Butterworths, London.

SELF, C. H. 1982. European Patent Application No. 82301170.5 (EP 60123). *Chem. Abstr.* 97:212066D.

SEQUIN, R. J., and KOSICKI, G. W. 1967. Studies on the conformatinal changes of mitochondrial malate dehydrogenase in urea-phosphate solutions. *Can. J. Biochem.* 45:659–670.

SHANNON, L. M.; KAY, E.; and LEW, J. Y. 1966. Peroxidase isozymes from horseradish roots. I. Isolation and physical properties. *J. Biol. Chem.* 241:2166–2172.

SHIRO, Y.; KURONO, M.; and MORISHIMA, I. 1986. Presence of endogenous calcium ion and its functional and structural regulation in horseradish peroxidase. *J. Biol. Chem.* 261:9382–9390.

SHORE, J. D., and CHAKRABARTI, S. K. 1976. Subunit dissociation of mitochondrial malate dehydrogenase. *Biochemistry* 15:875–879.

STERNBERGER, L. A.; HARDY, P. H.; CUCULIS, J. J.; and MEYER, H. G. 1970. The unlabeled antibody enzyme method of immunohistochemistry: Preparation and properties of soluble antigen-antibody complex (horseradish peroxidase-antihorseradish peroxidase) and its use in identification of spirochetes. *J. Histochem. Cytochem.* 18:315–333.

THORPE, G. H. G.; HAGGART, R.; KRICKA, L. J.; and WHITEHEAD, T. P. 1984. Enhanced luminescent enzyme immunoassays for Rubella antibody, immunoglobulin E and digoxin. *Biochem. Biophys. Res. Commun.* 119:481–487.

TIJSSEN, P. 1985. *Practice and Theory of Enzyme Immunoassays.* Elsevier, Amsterdam.

TIJSSEN, P., and KURSTAK, E. 1984. High efficient and simple methods for the preparation of peroxidase and active peroxidase. Antibody conjugates for enzyme immunoassays. *Anal. Biochem.* 136:451–457.

TSUGE, H.; NATSUAKI, O.; and OHASHI, K. 1975. Purification, properties and molecular features of glucose oxidase from *Aspergillus niger. J. Biochem.* (Tokyo) 78:835–843.

VAN WEEMAN, B. K., and SCHUURS, A. H. W. M. 1974. Immunoassay using antibody-enzyme conjugates. *FEBS Lett.* 43:215–221.

WALLENFELS, K., and MALHOTRA, O. P. 1960. β-Galactosidase. In *The Enzymes,* vol. 4, 2d ed., eds. P. D. Boyer, H. Lardy, and K. Myrback, pp. 409–430, Academic Press, New York.

WALLENFELS, K., AND WEIL, R. 1972. β-Galactosidase. In *The Enzymes,* vol. 7, 3d ed., eds. P. D. Boyer, H. Lardy, and K. Myrback, pp. 617–663, Academic Press, New York.

WEISS, E., and VAN REGENMORTEL, M. H. 1989. Use of rabbit Fab'-peroxidase conjugates prepared by the maleimide method for detecting plant viruses by ELISA. *J. Virol. Methods* 21:11–16.

WELINDER, K. G. 1979. Amino acid sequence studies of horseradish peroxidase. *Eur. J. Biochem.* 96:483–502.

WHITAKER, J. R. 1994. *Principles of Enzymology for the Food Sciences.* 2d ed., Marcel Dekker, New York.

WHITEHEAD, T. P.; THORPE, G. H. G.; CARTER, T. J. N.; GROUCUTT, C.; and KRICKA, L. J. 1983. Enhanced luminescence procedure for sensitive determination of peroxidase-labeled conjugates in immunoassay. *Nature* (London) 305:158–159.

WILSON, M. B., and NAKANE, P. K. 1979. Preparation and standardization of enzyme-labeled conjugates. In *Immunoassays in Clinical Laboratory,* eds. R. M. Nakamura, W. R. Ditto, and E. S. Tucker, pp. 81–98, Alan R. Liss, New York.

WONG, S. S. 1993. *Chemistry of Protein Conjugation and Cross-Linking.* CRC Press, Boca Raton, FL.

WOOD, W. G.; FRICKE, H.; and STRASBURGER, C. J. 1988. Solid-phase luminescence imunoassays using kinase and aryl hydrazide labels. In *Nonisotopic Immunoassay,* ed. T. T. Ngo, pp. 257–270, Plenum Press, New York.

YAMADA, H.; IMOTO, T.; FUJITA, K.; OLCAZAKI, K.; and MOTOMURA, M. 1981. Selective modification of aspartic acid-101 in lysozyme by carbodiimide reaction. *Biochemistry* 20:4836–4842.

7
Separation and
Solid-Phase Systems

SEPARATION SYSTEMS

Introduction

To quantitate the amount of analyte present in an enzyme immunoassay (EIA), the extent of reaction of the enzyme-labeled ligand (or antibody) with antibody (or the ligand) must be determined. Since the reaction between the ligand and its specific antibody does not produce a precipitate, it is necessary to effect a physical separation of the two forms before either or both forms can be quantitated. By definition, in all heterogeneous EIAs, a physical separation of the free and antibody-bound fractions is required. The homogeneous EIAs, in contrast, are separation-free systems; they do not require separation of the free and bound forms.

Separation techniques have become the critical issue in the development and advancement of immunoassays. Numerous techniques have been developed to separate labeled antigen-antibody complexes from the unbound labeled fraction. Removal of one of these labeled fractions enables the development of quite sensitive and relatively precise immunoassays. Regardless of methodology, either competitive or immunometric; or of the label, radioisotopic, enzymic, or fluorometric, the more efficient the separation of the bound from the unbound fraction, the more the sensitivity of the assay will be enhanced. In exchange for the sensitivity and specificity benefits obtained, these separation steps can be slow, difficult to automate, and may require multiple operations. This in turn increases the technical requirements for performing the assay.

Of the new assays or assay systems described during the past two decades, only about 10–20% require no separation step (Figure 7.1). Therefore, most immuno-

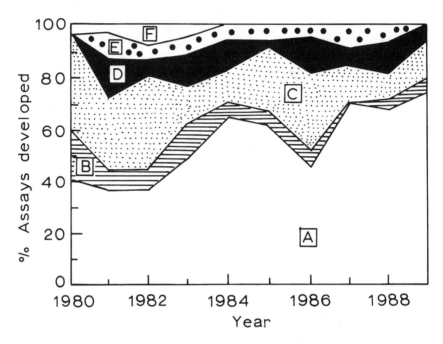

Figure 7.1. Trends over the past decade with respect to the use of different separation methods or no separation step (homogeneous, turbidimetric assays, etc.) in new immunoassays. The survey includes new immunoassays described in *Clinical Chemistry* from January 1980 to December 1989, and does not include articles on immunoassays concerned with immunoassay evaluation, comparison, improvement, automation, problems of interference, etc. The average sample size was 42 (range, 27 in 1980 to 57 in 1985). The adsorption methods used charcoal (a total of 28 assays over the decade) or Florisil® (one assay). The precipitation methods largely involved second antibody (52 assays) but also polyethylene glycol alone (31 assays), protein A (four assays), and ammonium sulfate (two assays).

A: Solid-phase methods   B: Adsorption methods
C: Precipitation methods   D: Homogeneous assays
E: Turbidimetric, agglutination, etc., assays
F: Miscellaneous assays
From: Gosling (1990)

assays, particularly those designed to operate in the picomole range or lower, involve a distinct separation step or even a series of such steps.

In EIA systems, the separation step serves the dual function of determining the proportion of enzyme label distributed between the free and antibody-bound fractions, and separating exogenous enzyme from sample components that might interfere with its measurement. The latter may include endogenous enzymes with similar substrate specificity, enzyme inhibitors or activators, and competing substrates. Generally, measurement of activity of the antibody-bound enzyme label is preferred. A failure to do so has in many cases led to matrix interferences and background problems with low dilution samples.

The separation of the free form varies with the type of assay. In a competitive assay, separation removes the unbound enzyme tracer and free analyte so that only the antibody-bound fraction remains. If a low affinity antibody is used in the assay, this step may occasionally also remove the bound fraction to a varying degree. In an immunometric or sandwich assay, the separation is responsible for removing the unbound labeled antibody. In both assay designs, the efficiency of the separation has direct effects on the quality of the assay (Wild and Davies 1994). In competitive assays, the efficiency is reduced by the following:

1. A failure to remove all of the free enzyme tracer. The residual unbound tracer causes high background signal,
2. Partial removal of enzyme tracer bound to antibody, especially when low affinity antibodies are used in the assay,
3. Interference in the reaction between antibody and analyte, and
4. Inconsistent separation due to variability in the sample matrix.

In immunometric assays, the major source of error in practice is failure to remove all of the unbound labeled antibody.

To consider a separation technique for a routine immunoassay, five potential characteristics must be considered: speed, simplicity, applicability, reproducibility, and cost. In order to maximize assay precision and sensitivity, an ideal separation technique must ensure complete separation of the free and bound fractions with relatively simple and foolproof manipulations. The technique should not modulate the primary ligand-antibody reaction. It should also be accomplished rapidly, preferably without elaborate or expensive equipment. Furthermore, an ideal separation method should be unaffected by the constituents of the sample (whole blood, plasma, serum, urine, saliva, cerebrospinal fluid, feces, milk, etc.), be generally applicable to a wide variety of analytes, and be amenable to automation.

The "efficiency" of a separation technique can be defined as the completeness with which the bound and free forms are separated. In theory, a perfect separation system should completely divide the two components of the assay (i.e., be 100% efficient). However, in practice, this is seldom achieved. An incomplete separation often causes "misclassification errors," thereby resulting in assay bias and imprecision. Conventionally, the efficiency of separation can be measured as "nonspecific binding" (NSB) of the enzyme label to the assay tube. This is so-called diluent or assay blank.

NSB may occur due to physical trapping of the free ligand in the bound complex, the presence of impurities in the enzyme label with properties similar to those of the bound complex, adsorption of the free ligand to the assay tube, or simply a failure to remove all the supernatant liquid (Newman and Price 1991; Chard 1978; Morris 1985).

NSB of the enzyme tracer is in fact the most common problem encountered in competitive EIAs. Such binding adds a level of background noise that can reduce precision, particularly at high concentrations. It can be minimized by using solid-phase or second antibody separation systems (described later in this section). In contrast, the separation in immunometric sandwich assays has a direct effect on assay sensitivity, i.e., the precision at very low and zero concentrations of the analyte. The separation in such assay designs can be greatly improved by repeatedly washing off the unbound, residual labeled antibody from the solid phase with wash solution.

Classification

A wide variety of separation techniques have been used in immunodiagnostic products. All such techniques exploit the physicochemical differences between the ligand in its free and bound forms. The pioneering radioligand assays of the early 1960s utilized several different separation methods. The chromatoelectrophoresis method used by Yalow and Berson (1960) in their first RIA took advantage of differences between free and antibody-bound insulin in two properties: adsorption to cellulose and charge. Electrophoretic separation was also used by Ekins (1960) in his thyroid-binding globulin (TBG)-based thyroxine assay, and by Hunter and Greenwood (1962) in their first growth hormone RIA. The double antibody methods were first developed by Hales and Randle (1962) and by Morgan and Lazarow (1962) for the insulin assay, and by Utiger and his coworkers (1962) for the growth hormone assay.

Since these early developments, a great variety of separation methods have been developed. This reflects the fact that immunoassays are so universally applicable that no single separation method fits the properties of the multiple combinations of ligand-antibody (or receptors) that can be used as a universal basis for an assay. Furthermore, the performance demanded of a method depends to an extent on the purpose of the procedure. For example, an assay may be optimized for speed and ease of performance, or sensitivity and reproducibility. These requirements in turn determine the choice of separation technique.

Several methods are available for separating the free and bound forms (Table 7.1) and these essentially represent a chronicle of assay development and sophistication in the immunodiagnostics field. Among the groups listed in Table 7.1, all but the first group have achieved widespread use in immunodiagnostics. This is primarily because there is no single ideal separation method applicable to all systems. The first four methods described in Table 7.1 are liquid-phase (sometimes also referred to as solution-phase) systems where separations occur in solution. The walls of the assay tube are not actively involved in these reactions and there

TABLE 7.1. Classification of Separation Systems

Principle	Examples	Advantages	Limitations
Physicochemical properties of antibody or ligand	Chromatoelectrophoresis (starch gel, cellulose acetate, polyacrylamide gel), gel filtration (column or batch)	Generally applicable, well researched	Slow, equipment intensive, requires high level skills, complex steps, very low throughput
Chemical/fractional precipitation	Ammonium or sodium sulfate[a], trichloroacetic acid, ethanol[b], dioxane, polyethylene glycol (PEG)[a], cross-linked dextran[a]	Inexpensive, fast, generally applicable, well researched	Multistep procedure, test performance suffers due to number of steps, messy reagents, requires a degree of laboratory skills
Adsorption	Coated charcoal[c], silicates, hydroxyapatite	Inexpensive, fast, generally applicable, well researched	Multistep procedure, test performance suffers because of a number of steps, messy reagents, requires a degree of laboratory skill
Immunology	Double or second antibody precipitation[d] (applicable to both soluble liquid and solid-phase systems), protein A	Specific, fast, generally applicable, well researched	Multistep, requires centrifugation
Solid phase	Coated particles, membranes, discs, magnetic particles, tubes, microtiter plates, paddles, etc.	Easy, eliminates steps, improves performance, requires fewer skills, easier to automate, many product configurations, some eliminate handling and allow faster kinetics because of increased protein binding	Higher cost per test, kinetics are generally slow, large sample volumes

[a] At concentrations of ≥ 200 g/l.
[b] Usually at 70% concentration at $-20°C$.
[c] Binds free fraction.
[d] With or without PEG.

197

are no antibody-coated particles. They are sometimes used in competitive immunoassays.

Many of the parameters of a suitable separation technique are shared by both EIAs and their analogous RIA procedures. However, only the last two groups, viz., immunological precipitation and solid-phase separation, have found significant applications in EIAs, and of these two, solid-phase separation is the most commonly employed in commercial EIA products.

The principles of various separation techniques are briefly described below.

1. Separations Based on Physicochemical Properties of Antibody and/or Ligand

CHROMATOELECTROPHORESIS. Paper chromatoelectrophoresis was the first method described for separation of the reactants in a RIA (Yalow and Berson 1960). It evolved directly from the techniques originally used for the identification of antibodies to insulin in patients' serum.

For the purposes of separation of bound and free hormone, the technique depends on the fact that free hormone is adsorbed to the paper at the site of application, while the antibody-bound fraction migrates towards the center of the strip. After drying, sections of the strip can be counted to determine the distribution of radioactivity between the two phases.

Electrophoresis, however, has many practical disadvantages including high NSB (> 20%), and is rarely used in modern commercial products. It is also expensive and time consuming for routine use.

GEL FILTRATION CHROMATOGRAPHY. Because the ligand-antibody complex is much larger than the free ligand, they can be separated by gel filtration chromatography using a cross-linked gel matrix in the form of small particles, such as Sephadex® or Biogel®. Two approaches have been used. In the first, the gel matrix is used in the form of a column. This procedure is much too complex and time consuming for routine application. In the second, more practical, approach, the gel is actually incorporated in the incubation medium. Low molecular weight ligand can then distribute freely both inside and outside the gel, whereas the ligand-antibody complex cannot enter the gel, thus effecting the separation of the two forms. Separation by batch addition of a gel or similar matrix was widely used in earlier commercial kits for the measurement of thyroid hormones.

Similar to electrophoretic methods, separation by gel filtration chromatography is expensive, time consuming for routine use, and gives high NSB.

2. Chemical/Fractional Precipitation. The above two techniques were soon superseded by fractional precipitation approaches, which were widely used in the 1970s and even the early 1980s. The general mechanism of fractional precipitation depends on the use of salts and solvents that reduce the amount of "free" water in a system. Immunoglobulins, which have the lowest solubility of all proteins, are precipitated in solvents and solutes such as ethanol, dioxane, and poly-

ethylene glycol (PEG), or are salted out by ammonium or sodium sulfate. Antibodies are almost completely precipitated in 33% saturated ammonium sulfate, in 70% ethanol, and in 15% PEG.

When the primary antibody-ligand reaction is completed, the separation material is added at a concentration in which the antibody and the antibody-ligand complex are insoluble and therefore precipitate, while the free fraction remains in solution. The precipitate is pelleted by centrifugation. The tracer activity is then determined in the pellet (bound fraction) or the supernatant (free fraction).

Precipitation methods have the advantage of being simple, rapid, and inexpensive, and usually do not require any special incubation period. They are also amenable to automated pipetting systems. Furthermore, they are highly reproducible and virtually devoid of lot-to-lot variation. However, NSB still tends to be high, in the range of 10–20%. These methods, however, are not suitable for EIAs.

Ethanol precipitation methods have been used in assays for insulin and hCG; ammonium sulfate for assays of angiotensin, vasopressin, cyclic AMP, steroids, and prostaglandins; and PEG in a variety of immunoassays including growth hormones, thyroid hormone, and steroids.

3. Adsorption Methods.   The nonspecific adsorption of biological molecules to particle surfaces is widely used as a method for the separation of bound and free ligand. Most such procedures depend on the fact that only the ligand and not the antibody or antibody-ligand complex have this property. Materials with a high adsorptive power include cellulose, glass powder, silica powder, talc, active charcoal, and fuller's earth.

Of the adsorption methods, the activated charcoal technique, first described by Herbert et al. (1965) for the vitamin B_{12} assay, is the most extensively used. It may be used coated or uncoated. The coating, particularly with dextran, has been assumed to work like a molecular sieve, permitting only the small molecules from coming into contact with the adsorptive surface. However, a number of proteins including albumin, gelatin, and even serum may work equally well to reduce the binding of the antibody or antibody-ligand complex to the charcoal.

Typically, after the primary ligand-antibody reaction is completed, the adsorptive material is added and the tubes are centrifuged. These separation methods are unusual in that the free fraction is pelleted and the bound fraction stays in solution. The supernatant is decanted into another tube for counting.

Charcoal adsorption techniques have been extensively used in assays of small molecules, such as steroids, drugs, and peptides up to a molecular weight of approximately 10,000. Time and temperature are critical factors during the separation and centrifugation, and great care needs to be taken to ensure that a consistent amount of charcoal is added to each tube. Charcoal may also disturb the equilibrium of the antibody-ligand reaction by stripping off the ligand bound to antibody. For the same reason, charcoal should be left in contact with the reaction mixture for the minimum possible time.

4. Double or Second Antibody Methods.   The introduction of secondary anti-species antisera was a major advance in introducing specificity for the separation mechanism of free from bound form in the immunoassay reaction. First introduced by Utiger et al. (1962) and Morgan and Lazarow (1962), precipitation of the bound complex with secondary antibody directed against the primary antibody is widely used as a separation procedure in RIA systems. The second or double antibody method is rapidly becoming the method of choice for an even wider range of as-says, from those for small molecules to enzyme-labeled immunoassays.

Precipitation reactions occur only at high concentrations of ligand and anti-body. The aim of this separation method is to create a sufficiently large immuno-logical complex or micelle by incorporating the first antibody-specific antigen complex, so that it is possible to separate the bound phase by ordinary centrifuga-tion. In this method, the first or primary antibody becomes the antigen of a second antigen-antibody reaction. Separation by this technique requires a relatively large amount of secondary antibody and, therefore, a correspondingly large amount of the species of IgGs of which the first antibody forms a part must be included. Thus, a second antibody system always involves addition of carrier protein, either whole serum or IgGs from the species in which the first antibody is raised.

To use this separation technique, it is necessary to test not only the concentra-tion of antibody but also the appropriate concentration of carrier IgGs, because the optimal amounts will vary with each antiserum tested. The following factors thus need to be considered.

1. Completeness of precipitation of the bound complex. In the presence of an excess of the first antibody, this should represent nearly 100% of the immunoreactive tracer.

2. The minimum quantity of the "second" antibody required to achieve complete pre-cipitation. Excessive amounts are likely to be both expensive and to lead to the problem of the "prozone" phenomenon, i.e., in the presence of an excess of anti-body, immunoprecipitation may not occur.

3. The "assay blank," or the amount of tracer precipitated by second antibody in the absence of primary antibody. Ideally, this should be less than 5%. A high value can occasionally be due to the presence in the antiserum of antibodies directed to the ligand.

4. A second antibody system should be evaluated in the presence of the test sample (e.g., urine, plasma, serum, etc.) for which the assay is designed. A procedure that appears satisfactory in the presence of diluent buffers may nevertheless, in the pres-ence of biological material, be subject to striking nonspecific effects. This is often reflected in precipitation of the bound complex or an increase in the assay blank, or both.

The double antibody separation can either be performed as a postprecipitation method, which is the most frequently used variant, or the precipitation may be done before the primary reaction (preprecipitation). The latter works if the antibody-binding sites of the primary antibody are not sterically hindered after the precipitate has developed. Otherwise, it will prevent the primary antibody-ligand reaction. This modification can actually be regarded as a solid phase coupled antibody system where the immunoglobulin precipitate constitutes the solid phase. However, this method does not work with all double antibody precipitating antisera, and may have a tendency to decrease the sensitivity of the assay.

An alternative in which the double antibody is coupled to a solid phase (DASP) avoids some of the variability of the precipitation reaction. Coupling of the second antibody to an insoluble matrix such as cellulose (den Hollander and Schuurs 1971) yields a system that is convenient and that does not require the use of carrier immunoglobulins. However, the precipitation and evaluation is time consuming, and therefore, the method is not widely used except in the form of commercially available reagents.

The main advantage of second antibody separations is the very low NSB (~ 2%). However, this method requires the need for a second incubation (1 hr to overnight, depending on the secondary antisera) that can more than double the overall time it takes to perform the assay. This problem is sometimes overcome by the use of accelerators of the precipitate formation, such as ammonium sulfate or PEG.

PEG-assisted secondary antibody precipitation in fact combines the benefits of the second antibody and PEG methods, i.e., specificity and speed of the assay are enhanced. When used at low concentrations of 4%, PEG hastens the precipitation of the cross-linked matrix. Taking performance and convenience into account, this is probably the best of all liquid-phase separation methods.

5. Surface-Coated Solid-Phase Systems.   The liquid- or solution-phase separation systems described above suffer from several disadvantages, including time-consuming preparations, longer assay times, high NSB, and the need for careful washing of the immunoprecipitates and refrigerated centrifuges. In order to overcome these difficulties and to provide a means of automating the separation technique, the solid-phase group of separation methods was developed. At present, they are the most popular separation methods used in EIAs.

In solid-phase assays, the antibody or the receptor is coupled to a nonsoluble phase, either noncovalently by simple physical adsorption or by covalent coupling. This coupling is done in advance so that the antibody remains insoluble from the beginning of the assay. Antibodies coupled to such a solid phase or matrix are often referred to as "immunoadsorbents" or "immunosorbents." Because the antibody is now no longer freely in solution, it becomes much easier to separate it from the supernatant after the primary incubation step. The separation of the free from bound form can be achieved by centrifugation or filtration for particulate solid phases such as agarose, polyacrylamide, or polystyrene beads. Magnetic fields can be used to separate particles that have an iron core. The larger, dispensable forms of solid phases such as tubes, cuvettes, microtiter plates, balls, dipsticks, and adsorbent devices allow easier and efficient separation of the two phases through simple rinse and decantation steps.

In the case of polymer-coated magnetic iron particles ("magnetizable" particles), separation involves placing the assay tubes after the primary incubation in a holder over a strong permanent magnet and decanting the supernatant. The immobilized antibody-bound fraction is retained at the bottom of the tube.

The Serono MAIAclone™ assays use such magnetizable particles. These are liquid-phase immunometric assays in which the capture antibody is conjugated to fluorescein isothiocyanate (FITC). Following the first incubation, which occurs in solution, the generic solid phase (magnetizable particles that have an attached antibody to FITC) is added and, following a second incubation, the particles are sedimented in a magnetic field. This design uniquely combines the fast kinetics of a liquid-phase assay with the low background signal of a solid-phase assay.

The first solid-phase system introduced was the coated tube method of Catt and Tregear (1967). Since then, an enormous range of solid-phase supports have become available for performing immunoassays. These range from particles of dextran and cellulose, continuous surfaces such as polystyrene or polypropylene tubes and microtiter plates, latex beads, and various membranes to more recently polymer-coated magnetic iron particles. Although they were first introduced in the mid-1960s, solid-phase supports gained popularity only in the 1980s. The relatively slow introduction of these techniques was primarily due to difficulties in producing reliable coating/coupling procedures to link the antibodies to the solid phase. This was particularly a major problem with tube and plate supports, and as a result, particulate solid phases were the first to receive widespread acceptance as their large surface area enabled more reproducible reaction conditions (Donini and Donini 1969; Bolton and Hunter 1973; Axon et al. 1967; Chapman et al. 1983).

The binding of antibodies to a solid phase may change its ligand binding characteristics. Therefore, immunosorbent antibody must be tested after coupling of the antibody. Both changes in avidity and significant loss of antibody quantity (titer) are known to occur as effects of coupling to solid materials.

Some solid-phase systems also offer distinct advantages with regard to the coating process, which requires considerable investment of time and money. It is therefore preferable not to have to do this for each individual antiserum. Moreover, this also results in over 90% wastage of the precious antibody. Thus, universal solid systems coated with substances such as streptavidin or protein A can be used for adsorption of the primary antibody. The former reagent can be used in conjunction with biotinylated primary antibodies.

Irrespective of the nature of the solid support being used, the use of solid-phase secondary antibodies offers several advantages over their first antibody counterparts. Such systems also provide a universal reagent for use in a wide range of assays provided the primary antibody is raised in the same animal species. This method also allows efficient use of the primary antibodies and does not affect the kinetics of the primary antigen-antibody reaction as sometimes happens when the primary antibody is attached to the solid phase. Moreover, being an excess reagent, the secondary antibody does not require precise dispensing, unlike a solid-phase primary antibody, whose concentration in the assay system is critical.

Antigen-coated solid phases are also used in heterogeneous EIAs. Antigen-coated solid surfaces are often used in limited reagent systems and are, therefore, most appropriate to the measurement of small molecular weight analytes. However, the recent development of anti-idiotypic antibodies has now enabled the development of excess-reagent systems for such analytes (Barnard and Kohen 1990).

Solid phases as separation systems offer several advantages. Some of these include the following:

1. They can be applied to virtually any binding protein capable of physical or covalent attachment to solid surface.
2. They are highly efficient and produce virtually complete separation of the bound fraction.
3. NSB is very low (< 1%).
4. They give excellent precision if carefully tested.
5. They are not as liable as solution-phase separation systems to nonspecific effects introduced by plasma and serum.
6. The stability of the immunosorbent is considerably enhanced. For example, latex particle coupled antibodies are stable for over a year when stored as a liquid reagent (Thakkar et al. 1991). Similarly, if solid phases are dried and stored desiccated, the antibody is stable almost indefinitely (Voller et al. 1979).

In spite of their several obvious advantages over solution-phase separation systems, solid-phase systems do suffer from certain drawbacks, which at least partially explains why these sophisticated systems are not in universal use. These include the following:

1. The preparation of primary reagent is tedious.
2. The recovery of antibody activity on the solid phase is only 30% or less of that in the original IgG preparation. This is probably due to the fact that many of the molecules attach to the solid phase through their binding sites. Such wastage is tolerable only if the supply of the antiserum is abundant.
3. In the case of antibodies developed for larger molecules, their attachment to solid phase often results in a loss of affinity, and hence, sensitivity in the assay. However, this factor is only critical in assays where extreme sensitivity is required.
4. The coupling process, whether covalent or noncovalent, sometimes results in the partial denaturation of the antibody, thereby changing its affinity constant. This

TABLE 7.2.   Relative Merits of Solid-Phase Separation Techniques

Solid phase	Advantages	Limitations
Beads	Quantity can vary to suit assay, good surface areas, material can vary to suit assay, surface coating of other materials (magnetics), generally applicable, can be automated	Requires separations, multiple steps, large assay volumes
Tubes	Eliminates centrifugation, convenient, allows automation, generally applicable	Large assay volumes, expensive, limited selection of material, slow kinetics
Plates	Small reagent volumes, eliminates centrifugation, convenient, generally applicable	Waste for unused wells, requires special readers, slow kinetics, limited selection of material
Paddles	Flexible format, faster kinetics, larger selection of materials	Higher cost per test, difficult to automate
Membranes	Can be fabricated to suit assay, high binding capacity, faster kinetics, flow-through capability, dot binding format	Potentially more expensive solid phase

tends to increase the usage of precious primary antibody. However, the availability of monoclonal antibodies now renders this a financial rather than a volume consideration.

5. Immobilized antibodies tend to "leak" off during storage and during the assay. This results in loss of precision in the assay.

6. The actual assay procedure may become more complex and the washing of the solid phase to separate the free and bound fractions may require several washing and aspiration steps. This is particularly critical with assay designs that utilize particulate solid phase support systems.

The relative merits and weaknesses of different solid-phase separation techniques are summarized in Table 7.2.

Errors of Separation

A continuing problem with solid-phase separation technologies is choosing between high antibody binding capacity and low NSB. Generally, these two parameters exhibit an inverse relationship. All solid materials have a tendency to adsorb small amounts of the label. In EIAs, because of the signal amplification effect, even very small quantities adsorbed will influence the assay results.

The degree of NSB is affected by the type of diluent, the concentration of the enzyme label used, the volume of the solid-phase material, pH, ionic strength, and the presence of other reagents. The inclusion of buffer salts, chaotropic reagents, proteins, and detergents in the wash solution is well known in most immunoassay development programs.

The physical separation of the bound and free forms is still one of the main sources of assay imprecision in heterogeneous EIAs. When the soluble phase is decanted from the solid phase, a small portion of the fluid may still be left in the tube after decanting. If it constitutes a substantial fraction of the soluble phase, variations in its volume can cause considerable assay variation. This type of error is particularly prominent if the total volume of the assay is small.

Repeated washing of the solid phase to separate the two phases often improves the assay precision, primarily by lowering the NSB of the enzyme label. The physical flushing that introducing a wash buffer produces is only one component of this effect. Repeated washings remove entrapped label, reduce adsorption of label to surfaces, and aid in the removal of the reaction supernatant.

The manner in which the wash buffer is applied depends upon the solid phase used. Particles, beads, microtiter plates, and tubes can be washed actively, whereas membranes will be washed by capillary flow or radial partition. When the wash flow rate is slow, as in the latter case, the use of detergents and proteins in wash fluid becomes even more important in reducing the NSB (Newman and Price 1991).

Washing and separation steps have been automated in a variety of systems. Some examples include microtiter plates for which various commercial plate washers are available, membranes (Abbott IMx®, OPUS-PB®), magnetic particles (Serono-Baker™ SR1®, TOSOH AIA®-600, and the Technicon Immuno-1®), and coated tubes (Boehringer Mannheim ES-600® and the Becton Dickinson Affinity® systems). The automation of these important steps of heterogeneous EIA systems constitutes one of the most important developments in automating immunoassays.

Finally, regardless of theoretical arguments, it should never be assumed that any separation technique is perfect and produces total separation of the free and bound fractions. Also, it should not be assumed that different separation techniques differ only in their efficiency. The fact that two different procedures yield identical results for assay blank and zero standard does not necessarily mean that the composition of these fractions is identical. Similarly, the results obtained with any separation procedure are almost certain to vary with the actual medium used in the assay (e.g., serum, plasma, urine, milk, etc.). Comparison of standards prepared in a diluent buffer with unknowns will generally reveal nonidentity. This is in essence an artifact of the separation technique itself. The only solution to this problem is to ensure that standards or calibrators are prepared in media as nearly as possible identical to that of the sample.

In the following section, important characteristics of the various solid-phase systems commonly used in EIAs are briefly described.

SOLID-PHASE SYSTEMS

Introduction

The solid phase on which the antibodies or antigens are immobilized is one of the most important elements in heterogeneous EIAs. It contributes to the reproducibility, accuracy, precision, and sensitivity of the EIA (Kemeny and Challacombe 1988; Ngo 1991). The simple manipulations that are generally required to separate the free from bound form immobilized covalently or noncovalently are probably the most important reason for the rapid increase in popularity of solid-phase EIA systems.

A wide range of solid phases are commonly used in immunodiagnostic products (Figure 7.2). They are also used in about 70% of all new assays currently being developed (Figure 7.1). In some solid-phase assays, the use of first (De Boever et al. 1983) or second antibody (Zaidi et al. 1988) conjugated to microcellulose particles or "microbeads" acts largely to accelerate precipitation of the immune complexes. However, mainly because they require use of a centrifuge, the use of microbeads has fallen from about 40% of new solid-phase assays in the early and mid-1980s to 10% or lower today (Figure 7.2). Magnetizable particles, which also offer large surface-to-volume ratios, have had continual popularity, being found in about 10% of new assays (Gosling 1990). Antibody-coated tubes were used in about 15% of new solid-phase assays in the early part of the 1980s, but their use has declined with the more widespread use of 96-well microtiter plates and strips of similar wells. The growth in the use of microtiter wells has been steady, increasing from about 15% of new assays in 1980–1982 to about 70% in 1989. Precision-molded plastic balls and plastic sticks are used in about 5% of solid-phase assays. Antibodies immobilized on membranes are now an important factor, particularly in immunoassays formulated for over-the-counter sale or patients' bedside use.

According to Tijssen (1985), an ideal solid phase should have the following desirable characteristics:

1. High capacity for binding immunoreagents, i.e., high surface area-to-volume ratio,
2. Possibility of immobilization of many different immunoreagents,
3. Minimal desorption or "leakage" of the immunosorbent,
4. Negligible denaturation of immobilized molecule, and
5. Orientation of immobilized antibody with binding sites towards the solution and the Fc to the solid phase.

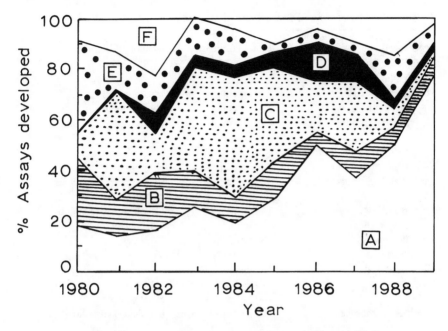

Figure 7.2. Trends over the past decade with respect to the use of different solid-phase separation methods. The general criteria for the inclusion of data were the same as for Figure 7.1. The microtiter wells were in plates (a total of 71 assays over the decade) or in strips (16 assays), with the latter increasing in relative popularity since they first appeared in 1983. The microbeads included microcrystalline cellulose, Sepharose® 4B, etc., coated with primary antibody or second antibody. The other solid phases included anion-exchange column (3), thread (2), capillary tube (2), indium slide (1), and "in situ generated solid phase" (1).

A: Microtiter wells   B: Tube
C: Microbeads   D: Large beads
E: Magnetizable beads   F: Other solid phases
From: Gosling (1990)

Solid-phase matrices commonly employed in heterogeneous EIAs are listed in Table 7.3. Polystyrene and plastics in general are by far the most popular solid phases, because their use makes the coating procedures extremely simple. Membranes, especially nitrocellulose, are replacing plastic in several EIA applications. Particulate solid phases are also very efficient because they may be dispersed throughout the reaction mixture and have a much higher ratio of surface area-to-volume.

The important characteristics of various solid-phase systems are described below.

TABLE 7.3.   Types of Solid-Phase Supports Used in EIA

Type	Protein capacity	Method of binding
Plastic Plates and Tubes		
Polystyrene (PS)	300 ng/cm^2	Hydrophobic
Polyvinyl chloride	300 ng/cm^2	Hydrophobic
Derivatized microtiter plates	100 μg/cm^2	Covalent, hydrophobic, hydrophilic
Beads and Microparticles		
PS beads	300 ng/cm^2	Hydrophobic
Derivatized PS beads	300 μg/cm^2	Covalent, hydrophobic, hydrophilic
Microparticles[a]	10 mg/ml	Covalent, hydrophobic
Protein A beads[b]	20 mg/ml	Noncovalent
Activated beads[c]	10 mg/ml	Covalent
Membranes		
Nitrocellulose	100 μg/cm^2	Hydrophobic
PVDF[d]	100 μg/cm^2	Hydrophobic
Nylon	100 μg/cm^2	Hydrophobic
Modified membranes	100 μg/cm^2	Covalent, hydrophobic
Paper		
Diazotized paper	> 10 mg/cm^2	Covalent

[a]Less than 1 μm in diameter and remain in colloidal suspension throughout the assay.
[b]For adsorption of antibodies only through Fc region.
[c]Include agarose, cross-linked agarose, mixed agarose/polyacrylamide, and polyacrylamide.
[d]Polyvinylene difluoride.

Plastics

Plastic was one of the first solid-phase systems used in immunoassays. The separation of free from bound form on plastic is achieved simply by decanting, aspirating, or retrieving the solution. This simplicity has a direct positive effect on the reproducibility and sensitivity of the assay. Because they can be molded into many shapes, plastics are a convenient solid support for many immunoassays. Polystyrene, polypropylene, and polyvinyl chloride (PVC) are the most common plastics used in the diagnostics industry. Because PVC can be cut easily, it is an excellent matrix for RIAs in which quantitation is done by counting individual wells. However, because it is not translucent, EIAs that will be quantitated by a plate reader should be performed in polystyrene and polypropylene and not in PVC plates. A general description of the chemistry and manufacturing aspects of these and other types of plastics commercially available can be found in an annual publication, Modern Plastics Encyclopedia (McGraw-Hill, New York).

The forms of plastic supports for EIAs are varied. Some of these include tubes, microtiter plates, beads, discs, plastic-coated dipsticks, matchsticks, paddles, and even cocktail stirrers. Many other sizes and shapes can also be used as a solid phase plastic support in EIA systems. The commonly used forms are described below.

1. Tubes. Plastic tubes were first introduced by Catt and Tregear (1967). Tubes often give the best choice of assay surface. They offer flexibility in number of assays and in work volume, combined with acceptable handling possibilities and equally good use in ELISA and IRMA/RIA. Commercially, several competitive and immunometric EIAs have been developed that rely on antibody- or antigen-coated plastic tubes. They have also been automated by a number of manufacturers.

The choice of plastic is critical. Because it is easier to handle and is less breakable, polypropylene is generally preferred over polystyrene. However, there is little or no difference in the coating characteristics of these two plastics. The size of the tubes used in commercial EIA kits is typically 10–12 $\times$ 70–75 mm. They may be conical, round, or flat bottomed; however, to provide increased surface area and better mixing during the assay, star tubes are generally preferred. These make the assays work better and faster. The surface-to-volume ratio can also be increased by rotating the tubes almost horizontally in circular test tube racks. Upon slow rotation, a relatively large surface is covered. Tubes generally require larger volumes than microtiter plates. Although 0.5–1-ml volumes are commonly used, up to 3–4 ml can be used.

Although they were one of the earliest solid-phase systems exploited, coated tubes are also one of the lowest capacity solid phases used in EIAs. Moreover, they have several other drawbacks. Although they are well suited for screening methods, tubes lack adequate precision for quantitative EIAs, primarily because of the variability in the amount of immunosorbent immobilized during the coating stage. Not only do tubes from different batches of the same manufacturers vary, but individual tubes of the same batch do so as well. The immobilized reagent also tends to "leak" off during storage and during the assay.

Uniform coating of the tubes with the immunoreagent is therefore absolutely critical to ensure precision in the final assay and to avoid very high rejection rates of batches of coated tubes. Especially for competitive assays, it is essential that all of the tubes are coated with an identical amount of antibody. This is generally achieved by carefully controlling all of the manufacturing conditions, including the processes used to manufacture these tubes.

Another major drawback associated with plastic tubes is the small surface area for antibody binding as compared to other solid-phase supports such as membranes and coated particles. In addition, the ligand-antibody reaction takes place only at the interface between the solid phase and liquid, thereby slowing down the rate of the assay. However, assay kinetics can be greatly improved by agitation of the tubes during the incubation steps.

These drawbacks notwithstanding, the low binding capacity of coated plastic tubes can produce extremely low NSBs; values of less than 0.01% are not uncommon. Moreover, precoating and storing of tubes gives an optimal choice of protecting reaction surfaces provided airtight packs containing desiccant are used as packaging materials. Furthermore, the exact number of precoated tubes can be taken from storage without manipulating the reaction surfaces that are not going to be used.

Analogous to the use of coated tubes is the use of "dipsticks" and other such similar devices. Although not widely used, they are best suited for qualitative or semi-quantitative EIAs, and offer extremely simple assay protocols. Disposable polystyrene cuvettes are also used in some cases because they can be directly used with a common spectrophotometer. The volumes of such containers are quite large.

2. Microtiter Plates.   Microtiter plates are probably the most popular solid phase in use at the present time. Their widespread application in the diagnostics industry required the development of automated plate washing systems as well as suitable plate-based detection systems. It was the delays in the development of these instruments that held back their introduction as long as it did (Newman and Price 1991). A major reason for their popularity is that less labeling of the reagents is required, thereby reducing a very tedious and labor intensive process. Microtiter plates in general are not very well suited for RIAs. They are most commonly used with nonisotopic labeling systems such as enzyme, chemiluminescent, and fluorescence-based immunoassays.

Microtiter plates are available in a range of plastics treated in a variety of manners. Polystyrene and PVC are the most commonly used plastics for this purpose. Although PVC plates may adsorb antibodies more efficiently, they also tend to give high background levels and release little of the adsorbed immunoreactant. The most commonly used format is the 96-well (12×8 wells of 0.3 ml) microtiter plate form, where the wells provide convenient reaction chambers for immunoassays. However, strips as well as individual wells are also available commercially. The wells may have flat, U-shaped, or V-shaped bottoms. The choice between these shapes depends on the method used for quantitation of the EIA. With a read-through photometer, flat-bottomed wells give less variation, whereas for visual inspection, U-shaped wells are desirable (Tijssen 1985).

Microtiter plates suffer from similar drawbacks as those associated with coated plastic tubes. In addition, drift across the plate due to pipetting delays and temperature gradients across the plate due to their thermal insulating qualities are major problems. Assay kinetics can be greatly improved by incubating the plate at elevated temperatures (37°C) and using continuous vibration. In a conventional air incubator, the wells at the corners warm up more quickly than ones in the center; thus, special plate incubators have been developed that provide consistent warming of all of the wells. Similarly, as their volume capacity is only about one-tenth that of the coated plastic tube, they also require strong signal generation systems to provide sufficient signals.

In spite of their theoretical limitations and difficulties associated with their manufacture, immunoassays based on microtiter plates as well as coated tubes have been very successful commercially. These assays provided early indications that the success of an assay system could be driven more by convenience than by superior technical performance. Because of their compact size, microtiter plates are the most convenient formats for manually performed immunoassays.

3. Beads.   Until the advent of fully automated analyzers, bead assays were commonly used for immunometric sandwich assays. They are much larger than microparticles, so that just one bead is used for each test. The surface area offered by the bead is similar to that of a coated tube. They are also easier to manufacture than tubes and can be coated in bulk rather than individually.

Membranes

Polymeric, microporous membranes are also used as a solid support for filtration-based immunoassays of a wide variety of analytes. They overcome many of the problems inherent in the solid phase immunoassays as they combine the qualities of a solid substrate with a range of expanded capabilities. The flow-through capability of membranes offers a distinct advantage to improve assay kinetics. Large fluid volumes can be actively moved past the membrane surface, improving the reaction kinetics between immobilized and soluble reactants. Thus, assay times can be significantly reduced to meet the demands of the market.

Microporous membranes are generally described as polymer membranes with a defined and measurable (rated) porosity in the range of 0.05–12 μm. A microporous membrane is generally a thin-film membrane (100–200 μm thick), composed of a polymeric lattice that is largely void volume (80% void volume). This provides a flat sheet material with a large surface area-to-volume ratio, which in the case of some polymer materials, binds proteins at a high density (Harvey 1991). Membranes are also available sealed to the bottom of wells in microtiter plates. With certain types of polymers, this large surface area can be chemically activated to provide a mechanism for covalent attachment of proteins.

Commercially, a wide variety of membranes is available, including nitrocellulose, cellulose acetate, regenerated cellulose, nylon, and polyvinylidene fluoride (PVDF). Glass fibers can also be used as a membrane support. The protein binding capacity differs with the type of polymeric material, with nitrocellulose and nylon providing the highest binding capacity. Nitrocellulose membranes also provide lower background signals than the nylon membranes. Cellulose acetate and regenerated cellulose membranes are generally insufficient to support a rapid immunoassay. The use of the hydrophobic PVDF membrane in immunoassays is generally limited, primarily because the membrane must first be wetted with a nonaqueous solvent. This characteristic renders hydrophobic PVDF membranes unusable for membrane-based immunoassay configurations.

Among the different types, nitrocellulose microporous membranes have found wide applications in qualitative and semiquantitative immunoassay systems. Nitrocellulose is prepared by the direct nitration of cellulose, usually to 10–15% nitrogen by weight. Nitrocellulose membranes have a high protein binding capacity of high affinity and offer a surface that is easily blocked. The binding of proteins to nitrocellulose is extremely fast, and therefore preparation of the membrane is very rapid. In addition, these membranes are available in a variety of different pore sizes, yielding different flow properties and protein binding capacities. The membranes also can be cast with fabric supports to significantly increase the tensile strength for subsequent device manufacture. Nitrocellulose membranes can be cast on nonporous plastic supports, creating a form of the material that plays a different role in some immunoassay systems. Such examples include lateral flow devices and dipstick configurations.

Although compatible with isotopic detection, the majority of membrane-based immunoassays have been developed for use with nonisotopic signals. Of these two types, EIA and particle capture agglutination assays have been most important. With the former, the use of substrates that yield insoluble colored products predominates, although enzyme-linked chemiluminescent assays have been developed more recently. Nitrocellulose membranes are particularly useful in assays where only the presence, and not the quantity, of an immunoreactant is to be established, when only very small amounts of samples (e.g., < 1 µl) are available, and when ionic detergent-solubilized antigens are to be tested. Nitrocellulose membranes also offer great potential for providing a simple multianalyte approach by immobilizing several antibody "spots" on a single strip of membrane (Pappas 1988; Ekins and Chu 1991).

Paper

As material purity is a primary consideration in the production of papers for analytical and diagnostic use, only pulps of the highest quality are used for their production. These are high α-cellulose (that fraction of a pulp stock remaining undissolved after treatment of the pulp by 17.5% NaOH at 20°C for a total elapsed time of 45 min) pulps, which are obtained primarily from cotton linters and secondarily from sulfite-processed wood chips.

Diazotized paper is widely used for dot blot immunoassays in which the amount of protein bound to the sheet must be particularly high. The linkage to the paper is covalent through free amino groups on the antigen or antibody. There are a number of commercially available sources for diazotized paper. The most stable of the derivatized papers is aminophenylthioether cellulose (APT paper). Diazotized paper is prepared by treating the APT paper with an acidic nitrite solution, converting it to diazophenylthioether (DPT) paper. The coupling is then done by adding the protein solution to the paper.

This most inexpensive form of solid phase has the advantage that large amounts of coated solid phase may be prepared. Paper also offers several additional advantages in the diagnostic devices as a support medium. The bulk of the applied test solution is rapidly wicked away from the point of application/site of reaction and the products that are formed there. This capillary/filtering action assists in the creation of sharply defined reaction zones and the removal of potentially obscuring soluble secondary reaction products and/or other colored materials contained in the originally applied sample solution. Furthermore, the smooth, white background of the paper itself provides an ideal contrast for any colored reaction products formed thereon. Paper thus provides a multiplicity of functional benefits as a reagent support and dispersing media and as a sample application, test component immobilizing, concentrating, and filter media.

Glass

Antigens or antibodies can also be immobilized onto glass surfaces by heating, fixation with formaldehyde, or coupling with glutaraldehyde to aminoalkylsilyl glass surfaces. However, glass is not widely used in EIA systems.

Particulate Solid Phases

Particulate solid phases have originally been used for the separation of radiolabeled ligand-antibody complex from free labeled antibody in immunometric assays. Since then, their applications have been extended to both hapten and protein assays and with all detection systems developed for heterogeneous EIAs.

Particulate solid phases for immunoassays offer several advantages. These include:

1. The speed of the assay compared with microtiter-based assay increases considerably. A comparison of particle-bound antibodies versus an equal amount of antibody on the wall of the microtiter wells demonstrates that the particles captured all the antigen in 5–10 min, whereas a microtiter well required 24 hr (Nustad et al. 1991).

2. Batchwise preparation of solid-phase antibody allows easy monitoring and quality control.

3. They provide freedom to modify surface characteristics in order to reduce interference and allow antibody to be immobilized with a variety of methods.

4. There is no limitation on the amount of solid-phase antibody that can be used in an assay and no antibody variation between samples.

5. Magnetizable particles are applicable to rapid isolation of particulate antigens.

6. The combined use of defined particle-size mixtures, each coated with its own uniquely specific capture antibody and flow cytometry, allows development of multianalyte assays. These also allow for the inclusion of an interference assay in the package.

7. There is no limitation on sample size.

Polymer particles also have some drawbacks. They are not readily adapted to automated washing procedures. Furthermore, they tend to sediment during the incubation period, which necessitates continuous shaking during the incubation step.

Particulate solid phases can be broadly divided into two types: nonmagnetic and magnetic. Their properties are briefly described below.

1. Nonmagnetic Particles.   The nonmagnetic particulate solid supports include glass, latex, Sepharose®, Sephadex®, Sephacryl®, and nylon particles and beads. The choice of a particular support depends on the relative coupling capacities of the different plastics and the size and density of the particles.

The carbohydrate polymers, viz., Sephadex® and cellulose, have higher antibody binding capacities than polystyrene, nylon, or glass particles. However, the absolute binding capacity of the particles depends upon the surface area available, i.e., on the size of the individual particles.

Among the various nonmagnetic particulate solid phases, latex is the most common. Latex particles were first used by Singer and Plotz (1956) in an agglutination test for rheumatoid factor. Since then, they have been widely used for several agglutination tests in clinical diagnostics including pregnancy detection kits, in serodiagnostics for bacterial typing and identification, and for measuring serum blood levels of several antibiotics. In contrast, their use in EIAs has been fairly limited. Latex particles have a similar density to water, so they stay in suspension during the incubation. Hence, the kinetics of the assay are efficient, requiring shorter incubation times.

Most of the currently available latex particles are produced by emulsion polymerization. This process is best suited for producing submicron (< 1 μm) particles, but can be extended for particles up to 3 μm. A suspension polymerization process is used to make particles of up to 10-μm size. Very large particles (> 100 μm) are produced by the suspension process. However, it yields particles of a rather broad size distribution as compared to the emulsion process.

There are essentially two types of latex polymers: homopolymers, produced by the copolymerization of a single type of monomer and represented by polystyrene and polyvinyltoluene; and copolymers, produced by the polymerization of more than one type of monomer, and represented by styrene-divinylbenzene, styrene-butadiene, vinyltoluene-*t*-butylstyrene, and styrene-vinylcarboxylic acid.

The four basic ingredients of the emulsion polymerization process are a monomer; water; an emulsifying (surface-active) agent such as potassium laurate, sodium dodecyl sulfate, or sodium dihexylsulfosuccinate; and a free-radical-initiating agent such as potassium persulfate or hydrogen peroxide.. The polymerization process involves a particle-nuclear initiation phase and a subsequent particle growth phase.

Commercially available latex suspensions are primarily composed of polymer particles (30%) and water (> 69%), with small amounts of surfactants (0.1–0.5%) and inorganic salts (0.2%). Surfactants dissolved in the aqueous phase and adsorbed on the particles assist in the stabilization of the particles. Their complete removal from the suspension makes the latex particles less stable and more prone to flocculation. Latex particles have an inherent negative surface charge provided by covalently bound sulfate groups on their surfaces. Carboxylate-modified and amide-modified latex particles contain active carboxylic and carboxyamide hydrophilic groups, respectively, on their surfaces. Latex particles produced in water look like milk or white paint, and may occasionally develop an iridescent or metallic sheen. Dyed latex particles are also available commercially.

Separation of particulate solid-phase systems is usually by centrifugation and decantation. In microparticle capture enzyme immunoassay (MEIA), the latex particles are captured by a filter consisting of a glass-fiber matrix. Particles may also be captured by membrane filtration. The advantage of MEIA and membrane-based systems is that the particles may be easily washed, thereby considerably reducing the background signal.

2. Magnetizable Particles.   A natural progression from latex particles led to the inclusion of iron oxide in cellulose or latex particles to form paramagnetic microparticles. These are commonly used as a solid phase with the advantage of a separation that does not require centrifugation. Separation is achieved by the application of a magnetic field, which draws the particles to the side or base of the tube. The supernatant is then removed by aspiration or decantation (Wild and Davies 1994).

For use in EIAs, the magnetizable particles should fulfill the following requirements.

1. The incorporation by the matrix of high amounts of magnetic material,
2. The possibility of easily activating and subsequently coupling proteins to the magnetized matrix,
3. A relatively high mechanical stability to prevent the beads from fragmenting and thereby avoiding interference from the magnetic material in the assay,
4. Ease of manufacture, and
5. Homogeneity with regard to size and amount of incorporated magnetic material.

The use of magnetic supports for use in immunoassays dates back to the early 1970s. The advantages of a magnetizable particle are a high surface area, rapid analyte capture, and properties that lead to efficient separation and washing. Paramagnetic ferrous oxide-based magnetizable particles are the most commonly used in commercial applications. The ferrous oxide is incorporated into a cellulose matrix to which the antibody is coupled to provide a stable reagent with low NSB. This was later extended to develop a magnetizable charcoal reagent by cotrapping charcoal and ferrous oxide in a polyacrylamide gel. This approach considerably enhanced the use of charcoal as a separation matrix for hapten assays. It not only eliminated the centrifugation step, but also minimized adverse effects on the primary ligand-antibody reaction (Al-Dujaili et al. 1979).

Ferrous oxide has thus far proved far superior to other magnetic components, primarily because of its small particle size (10–20 nm) and good magnetic response. The size is, however, dramatically increased during entrapment in the cellulose, the final particle size being determined by milling (1–3 μm). Chromium dioxide has also been employed as a particulate solid phase as it has less residual magnetism than iron oxide, and hence, is easier to resuspend during the important washing process. Chromium dioxide also results in lower NSB.

Not all magnetic separation systems are based on suspensions of microparticles. The solid-phase AIA®-PACK range of assays manufactured by TOSOH consists of twelve 1.5-mm polymer beads coated with ferrite. An external magnetic field keeps the beads in motion during the incubation (Wild and Davies 1994).

The first magnetizable particle assays used either batch processing in magnetic racks (e.g., Corning), or automation in continuous flow systems. More recent automation has included unit dose systems such as the Serono-Baker™ SR1®, and the TOSOH AIA®-600. Magnetizable particles have also been used in assays as secondary antibody separation systems and coupled to primary antibodies, for use in assays for haptens and proteins using a range of detection systems from isotopes, enzymes, and fluorophores to chemiluminescence (Newman and Price 1991).

IMMOBILIZATION OF IMMUNOREAGENTS

General Considerations

Immunoreagents, either antibodies or antigens, can be immobilized onto solid phases in one of three different ways.

1. Adsorption to predominantly hydrophobic surfaces
2. Covalent attachment to activated surface groups, and
3. Noncovalent, electrostatic and hydrophilic bonds between a molecule immobilized by either of the above two methods, which in turn is coupled to a ligand, e.g., an antibody-antigen or biotin-avidin bond.

The extent to which these methods are chemically discrete is not absolute (Butler 1991). In any event, the solid phase should not be considered a passive component in the process of adsorption or coupling of antibody or antigen. The physical coupling of an antibody restricts its movement. The degree to which this influences the reaction kinetics depends upon the nature of the solid-phase surface and on the surface area of coupled antibody in relation to the volume and concentration of the other immunoreactants.

Compared to solution-phase assays, the ligand-antibody reactions in solid-phase asays are much slower. In such assays, the forward rate constant k_a is reduced while the dissociation constant k_d is increased, thereby increasing the equilibrium association constant (Newman and Price 1991). However, because the binding reactions are generally irreversible, as generally is the case with high affinity antibodies, the surface ligand-antibody reaction can be considered multivalent. This significantly increases the chances of reassociation. The overall reaction rates, however, can become limited by the rate of diffusion of the solution-phase components as the surface concentrations become depleted.

A number of studies have been conducted, especially with polystyrene and membranes, to understand the processes involved in the immobilization of immunoreagents onto the solid phases. Indeed most solid phases used in the EIAs are mildly to strongly hydrophobic, thus providing the potential for secondary hydrophobic interaction even when the primary interaction is covalent. In fact, studies on covalent immobilization of proteins to microparticles have shown that, although the initial bonding is > 80% covalent, the effective bonds after incubation for 16 hr can be > 80% noncovalent (Butler 1991). Direct covalent coupling of immunoreagents to hydrophobic surfaces might therefore be considered a means of facilitating secondary hydrophobic interactions. The immobilization of solid-phase immunoreagents is thus chemically heterogeneous and varies with the specific surface, ligand chemistry, and conditions of immobilization.

Similar solid phases manufactured by different companies also tend to exhibit different behaviors regarding protein immobilization. Kenny and Dunsmoor (1983) evaluated 11 different polystyrene microtiter plates from four different companies and found essentially two types: one that adsorbs albumin poorly, and another that adsorbs it well. IgG was adsorbed well on both plate types. Plates that bind albumin well are best suited for mixtures of antigens; however, background signal tends to be higher. A 100-fold excess of a nonspecific protein during adsorption essentially prevents the detection of the antigen.

Plastic plates exhibit a significant variability, not only among the various lots, but also among the wells of the same plate. Irrespective of the origin of the plates, the coefficient of variation of absorbance can range from 5% for the wells of one plate to 30% of the other for the same lot. The "edge effect" (i.e., wells of the perimeters adsorbing more protein than those in the interior of the plate) is particularly notorious when one works with the microtiter plates (Chessum and Denmark 1978; Kricka et al. 1980). Because of this reason, as many as 40% of the available wells may not be used by the researchers.

The edge effect has been attributed to differences in surface characteristics of the plastic (Burt et al. 1979) and to thermal characteristics (molding temperature, cooling) being different from those in the interior (Denmark and Chessum 1978). Thermal gradients generated during incubation may thus play a critical role in producing the edge effect. Polystyrene is a poor conductor of heat, and therefore, a thermal gradient may exist between the outer and inner wells during an incubation period of 30 min or longer, with initial and final temperatures of 20°C and 37°C, using a routine laboratory incubator. The use of a forced air incubator may eliminate this edge effect, giving a higher assay precision and reproducibility. Warming both the plate and the solution to the incubation temperature prior to the addition of the solution also seems a simple alternative (Tijssen 1985).

Furthermore, each protein also appears to have its own unique affinity for the hydrophobic polymer surface (Figure 7.3). Thus, within the linear binding region, the ratio of bound:free protein is constant, but the percentage adsorbed is protein specific (Butler 1991). A similar difference in affinity was observed by several researchers (Brash and Lyman 1969; Lee et al. 1974; Tijssen 1985). In this regard, high molecular weight biomolecules appear to adsorb best, probably because of multisite attachment to the solid phase.

Passive adsorption thus takes place predominantly by means of hydrophobic interactions between the solid-phase and the liquid-phase immunoreagents. In order to adsorb one molecule onto the solid phase, other molecules must be removed. If this process leads to a gain in entropy, a firm attachment can be achieved. The presence of polar groups in the solid phase also increases the binding capacity of water-soluble molecules such as antibodies (Rasmussen 1990). The binding strength in hydrophobic interaction is dependent on the number of binding sites (interactions) between the molecules and the solid phase. Larger molecules with potentially many binding sites, as mentioned earlier, will therefore adsorb better and with higher stability than smaller molecules with low affinity for hydrophobic interaction.

Although immobilization of antibodies and antigens on a plastic solid phase by hydrophobic interactions has proven to work extremely well in an increasing number of EIAs, the need for a more specific coupling as well as the need for coupling a wide range of molecules is quite obvious. For example, ELISAs of small peptides are sometimes difficult to perform because of the poor adherence by these peptide antigens to the plastic surfaces of commercially available, inexpensive ELISA plates. With small peptides, the hydrophobic interactions are frequently too weak to permit detection of the final antigen-antibody complex (Dagenais et al. 1994). In such cases, covalent attachment of the solid-phase immunoreagent may permit a higher concentration of reagent to be immobilized than by simple adsorption. Furthermore, a specific binding with potential for specific orientation

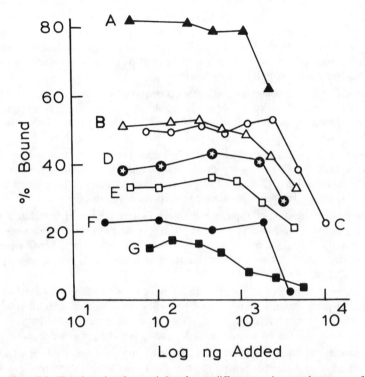

Figure 7.3. The adsorption characteristics of seven different proteins on polystyrene surface. Within the linear binding region, the ratio of bound:free protein is constant, but the percentage adsorbed is protein specific, because of each protein's own unique affinity for the hydrophobic polymer surface.

A: Bovine IgM   B: Bovine SIgA
C: Ovalbumin   D: Bovine IgG1
E: Bovine IgG2   F: Bovine serum albumin
G: α-Lactalbumin
From: Cantarero et al. (1980)

of the molecules can only be established by covalent coupling where only one binding site is needed to give a stable coupling with high binding strength.

    Covalent coupling of immunoreagents to solid phases offers some advantages. These include the following:

1. Studies have shown that one can attach 10–40% more protein covalently than by adsorption (Douglas and Monteith 1994; Bangs and Meza 1995).

2. Covalent coupling may provide more precise control of the coating level when the desired protein coverage is low.

3. Covalent coupling binds protein more securely. This is beneficial in production of tests or assays that are so sensitive that they would be influenced by minute quantities of IgG that might leach off the solid phase over time.

4. The covalent bond is thermally more stable. For example, after 1 hr at 56°C, 99.7% of covalently linked IgG remained bound, compared with only 70% of adsorbed IgG (Bangs and Meza 1995). This property could be essential with solid phases, such as microparticles, that are to be used in polymerase chain reaction or other applications requiring thermocycling.

5. Covalent coupling conserves costly reagent because it does not require the large excess of protein necessary for the adsorption process. Because excess reagents are not used, the multiple layer phenomenon present with passive adsorption is also less of a problem.

6. Hydrophilic molecules must be covalently linked to solid phase. Unless bound to the surface, they will surely desorb when the equilibrium is disturbed by removal of unbound soluble molecules from solution. The smaller antibody fragments, such as $F(ab')_2$, Fab, or Fv portions, do not normally adsorb well.

7. Covalent introduction of a spacer arm allows a secure but flexible attachment of many different molecules to the solid phase. Covalent attachment of hetero- or homo-bifunctional and trifunctional crosslinkers (available from Pierce Chemical Co., Rockford, IL) also facilitates coupling of ligands with unusual available chemical groups.

8. Directional binding (e.g., periodate oxidation of vicinal hydroxyls on the carbohydrate portion at the Fc portion of IgG and binding of the oxidized IgG to hydrazide-activated matrices) ensures that binding sites are directed outward to the solution and remain accessible. Thus, binding is no longer a random event.

9. Covalent attachment at relatively few sites may overcome the "Gulliver effect," whereby large, well-adsorbing protein molecules become tightly adsorbed over so wide an area or at so many contact points that they become distorted or denatured.

10. With some solid phases, covalent coupling may be the only viable alternative. For example, monoclonal antibodies with pI around 4.0 cannot be adsorbed on latex particles, which tend to flocculate in suspension at this pH. Thus, covalent coupling of such monoclonal antibodies to latex particles may be easier than adsorbing them.

11. Similarly, proteins do not adsorb well to hydrophilic silica microparticles; they must therefore be linked covalently to any of several coupling groups.

12. Covalent coupling also permits the use of high concentrations of surfactants (up to 1% Tween®) in the assay design to eliminate nonspecific binding. Such a high concentration of surfactant could easily dislodge the adsorbed but not the covalent coupled protein. Moreover, it reduces matrix interferences.

13. Covalent coupling allows for small peptides, nucleic acids, and other "nonadsorbable" molecules to be bound to a solid phase. Also, molecules that are present in too low a concentration to establish stable immobilization by adsorption may be firmly bound by covalent coupling.

14. Biomolecules that need detergents for solubilization cannot be adsorbed by passive methods. Thus, covalent coupling is preferred in such instances.

The concern over the stability of noncovalently adsorbed immunoreagents onto solid phases has therefore stimulated a great deal of research aimed at improving the bond stability between the immobilized protein and the solid phase. In this regard, particular attention has been given to modify polystyrene surfaces for higher binding capacity and/or improved binding stability.

Several companies at present market a variety of solid phases containing reactive groups to covalently couple immunoreagents. For example, Nunc Inc., Denmark, has introduced microtiter plates that are coated with a secondary amine (methyl amine). Thus, any ligand carrying a carboxylic acid group can be linked covalently to the plate using a condensing reagent such as a water-soluble carbodiimide. Costar (Cambridge, MA) also markets covalent plates and strips that are prepared by modifying polystyrene surfaces by a post molding treatment to yield either a carboxylated or aminated surface. Biomedical Products Ltd., Rockaway, NJ, recently introduced microtiter plates with a hydrophilic surface of reactive aldehyde groups that enable the binding of amino group-containing ligands. Covalent immobilization of antibodies through its amino groups is generally not preferred, as these groups are involved in antigenic binding reactions. Non-amino acid groups such as carbohydrate chains on the proteins are often not involved in biological functions; thus, they can be used to covalently couple the antibodies onto the solid phase. BioProbe International (Tustin, CA) recently introduced microtiter plates (Avidplate-HZ™) with hydrophilic hydrazide surface groups capable of binding to the Fc portion of oxidized antibodies. Unlike other microtiter plates designed for covalent binding, no expensive and unstable cross-linking reagents are required for this purpose. Furthermore, any immobilized antibodies have their antigen-binding sites facing away from the solid phase, thus allowing maximal antigen binding.

Covalent coupling can also be performed with solid phases other than polystyrene microtiter plates. Surface-modified polymer microparticles are available commercially. These are often made by copolymerizing styrene with a small amount (< 5%) of a functional monomer, such as acrylic acid. This yields microparticles covered with carboxylic acid groups. Other monomers are used for this purpose with different surface chemistries. For example, native silanol groups on the surface of silica microparticles are readily reacted with aqueous or solvent-based silane coupling agents to yield preactivated silica particles with a large variety of surface functional groups. Examples include chloromethyl, carboxyl, and amino groups (Bangs and Meza 1995). Thus, oligonucleotides can be covalently bound to surface-modified silica via the 5'-amino end, while lipids can be bound via the ω-carboxyl group on the fatty acid chain and propylamine surface groups on the silica (Haggin 1994).

Chemically modified surfaces often exhibit greater protein binding capacity. For example, bromoacetyl-functionalized polystyrene beads exhibit up to a 10-fold greater capacity for protein binding than unmodified polystyrene (Peterman et al. 1988). Moreover, no detectable dissociation occurs, unlike that observed with simple adsorption. The bromoacetyl coating is produced by chlorosulfonation of the aromatic ring of the polystyrene surface, followed by sulfonamide formation using excess di- or tri-amine and, finally, reaction of the amine groups with bromoacetyl bromide.

Proteins or other biological materials can be attached to the metal surface by first adding a linking coordinating polymer, such as polyimine, polythiol, or poly(iminoethene)*N*-dithiocarboxylate. After chemisorption of these polymers to the metal surface, the amino or thiol groups that are accessible can be used to link proteins or other ligands either directly or through bifunctional reagents. Using this method, the active ester derivatives of biotin and rabbit IgG were immobilized with the aid of bifunctional cross-linkers such as SMCC or glutaraldehyde (Chadwick and Hudson 1989).

Conformational Effects

Immobilized reactants often behave quite differently than soluble reactants. Because most proteins tend to adopt a conformation in which hydrophilic groups tend to be on the surface and hydrophobic groups buried within its structure, it seems inevitable that hydrophobic binding to polymer solid-phase surfaces will cause conformational changes in the adsorbed protein.

In one of the earliest studies, Bull (1956) observed that adsorption of albumin on glass resulted in a conformational change of the protein. His findings were later confirmed by several researchers using thermodynamic analyses (Nyilas et al. 1974; Lyklema and Norde 1973; Kennel 1982; Soderquist and Walton 1980; Dierks et al. 1986; Kochwa et al. 1967; Michaeli et al. 1980).

The various conformational changes and the changes in epitope or antibody binding site availability and concentration that can occur as a result of immobilization onto the solid phase are shown in Figure 7.4. In an effort to achieve thermodynamic stability, molecules incubated with plastic surface often orient their hydrophobic region toward the adsorptive surface (Butler 1991). Achieving energetically the most favorable conformation on the surface may thus alter or destroy epitopes that are normally expressed by these molecules in solution (Figure 7.4D). Molecules that undergo little conformational change may nevertheless express some of their epitopes cryptically after adsorption (Figure 7.4B). Although covalent attachment may allow a higher concentration of reagent to be immobilized than adsorption, it may also induce conformational changes, block epitopes, or chemically alter them (Figure 7.4A).

When adsorbed at low concentrations, proteins tend to undergo the most conformational change. For example, IgG adsorbed on polystyrene latex at low concentration behaved immunogenically like heat-denatured material (Kochwa et al. 1967). A progressive loss of antigenicity thus is observed when correspondingly lower concentrations of antibody are adsorbed onto the plastic surface (Figure 7.4E). Such effects can be overcome if albumin or other nonspecific protein is coadsorbed (Lyklema and Norde 1973).

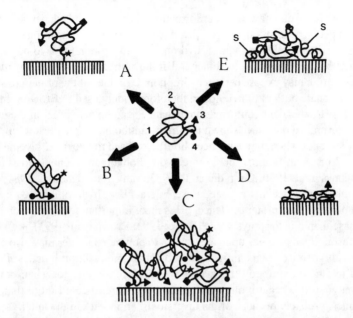

Figure 7.4. The effects on a soluble protein of immobilization on a solid phase. The hypothetical soluble globular protein at the center of the diagram expresses four functional groups, antigenic epitopes, enzyme active sites, etc., indicated by different symbols 1 through 4. Five different examples of conformational or steric effects resulting from immobilization on a solid phase (shaded vertical bars) are illustrated on the peripheral diagrams A through E.

(A): Covalent attachment resulting in loss of functional group 2, and, if the hypothetical protein were an antigen, possible steric hindrance of epitope 3.

(B): Immobilization by adsorption without conformational change but with functional groups 3 and 4 becoming cryptic and nonfunctional.

(C): Adsorptive immobilization, which proceeds through the process of aggregation resulting in the steric hindrance of certain functional groups but an increase in the local concentration of others.

(D): Adsorptive immobilization resulting in conformational change and loss of all but one functional group.

(E): Immobilization by adsorption in the presence of a stabilizing protein (S), which prevents the formation of secondary surface interaction and thus reduces conformational deformation.

From: Butler (1991)

At the other end of the spectrum is concentration dependency of protein conformation. Increasing the concentration of protein generally increases its tendency for aggregation. This in turn increases the tendency toward insolubility. According to Butler (1991), insolubility or aggregation might be a prelude to adsorption such that adsorbed proteins occur as aggregated clusters (Figure 7.4C). Because aggregation is typically concentration dependent, molecules adsorbed at different concentrations might be expected to be more or less aggregated, and hence, express different conformations.

Although it is well known that conformation of proteins is pH dependent, the influence of pH on adsorption onto solid surface still remains controversial. Because proteins are least soluble at their pI, hydrophobic adsorption tends to be maximal near the pI of the proteins. At the other end, if they are exposed to alkaline pH, partial denaturation may expose certain hydrophobic residues of the protein to solvent phase. Carbonate buffers of pH 9.6 are therefore widely used in ELISAs. In some instances, pH-dependent conformational changes can be beneficial in the coating process. Several studies have shown that many antibodies show greater binding to polymer surfaces when they have been briefly exposed to a low pH glycine buffer prior to adsorption to render them more hydrophobic than when stored at neutral pH (Ishikawa et al. 1980; Conradie et al. 1983; Vanderbranden et al. 1981). Blotting membranes, especially nitrocellulose, interact both hydrophilically as well as hydrophobically with proteins. Hence, pH is more important for adsorption on this solid phase than it is for polystyrene (Butler 1991; Brown et al. 1991).

Conformational changes upon adsorption to polymer surfaces may also change the antibody affinity for its antigen, thereby changing reaction kinetics. It is not unusual to find an antibody that works well in solution but not when bound to solid phase. In this regard, monoclonal antibodies are affected more than polyclonal ones, probably because of the diversity of antibodies in the latter.

Given the fact that not all solid surfaces are created equal, and because different immunoreagents behave quite differently, much remains to be done to understand the various processes involved in their immobilization onto solid phases. A better understanding of this important aspect of solid-phase EIAs will undoubtedly lead to easier optimization of the coating processes.

REFERENCES

AL-DUJAILI, E. A. S.; FORREST, G. C.; EDWARDS, C. R. W.; and LANDON, J. 1979. Evaluation and application of magnetizable charcoal for separation in radioimmunoassays. *Clin. Chem.* 25:1402–1405.

AXON, R.; PORATH, J.; and ERNBACK, S. 1967. Chemical coupling of peptides and proteins to polysaccharides by means of cyanogen bromide. *Nature* (London) 214:1302–1304.

BANGS, L. B., and MEZA, M. B. 1995. Microspheres, Part 2: Ligand attachment and test formulation. *Med. Device Diagnostic Ind.* 17(4):20–26.

BARNARD, G., and KOHEN, F. 1990. Idiometric assay: Noncompetitive immunoassay for small molecules typified by the measurement of estradiol in serum. *Clin. Chem.* 36:1945–1950.

BOLTON, A. E., and HUNTER, W. M. 1973. The use of antisera covalently coupled to agarose, cellulose, and sephadex in radioimmunoassays for proteins and haptens. *Biochim. Biophys. Acta* 329:318–330.

BRASH, J. L., and LYMAN, D. J. 1969. Adsorption of plasma proteins in solution to uncharged hydrophobic polymer surfaces. *J. Biomed. Mater. Res.* 3:175–189.

BROWN, W. R.; DIERKS, S. E.; BUTLER, J. E.; and GERSHONI, J. M. 1991. Immunoblotting: Membrane filters as the solid phase for immunoassays. In *Immunochemistry of Solid-Phase Immunoassay,* ed. J. E. Butler, pp. 151–172, CRC Press, Boca Raton, FL.

BULL, H. B. 1956. Adsorption of bovine serum albumin on glass. *Biochim. Biophys. Acta* 19:464–471.

BURT, S. M.; CARTER, T. J. N.; and KRICKA, L. J. 1979. Thermal characteristics of microtiter plates used in immunological assays. *J. Immunol. Meth.* 31:231–242.

BUTLER, J. E. 1991. Perspectives, configurations and principles. In *Immunochemistry of Solid-Phase Immunoassay,* ed. J. E. Butler, pp. 3–26, CRC Press, Boca Raton, FL.

CANTARERO, L. A.; BUTLER, J. E.; and OSBORNE, J. W. 1980. The adsorptive characteristics of proteins for polystyrene and their significance in solid phase immunoassays. *Anal. Biochem.* 105:375–382.

CATT, K., and TREGEAR, G. W. 1967. Solid phase immunoassay in antibody-coated tubes. *Science* 158:1570–1572.

CHADWICK, A. T., and HUDSON, M. J. 1989. Bonding to metal surfaces. Eur. Pat. Application 1989, Pub. No. 0 3339 821.

CHAPMAN, R. S.; SUTHERLAND, R. M.; and RATCLIFFE, J. G. 1983. Application of 1,1'-carbonyldiimidazole as a rapid practical method for the production of solid phase immunoassay reagents. In *Immunoassays for Clinical Chemistry,* 2d ed., eds. W. M. Hunter and J. E. T. Corrie, pp. 178–190, Churchill-Livingstone, Edinburgh.

CHARD, T. 1978. *An Introduction to Radioimmunoassay and Related Techniques.* North-Holland, Amsterdam.

CHESSUM, B. S., and DENMARK, J .R. 1978. Inconstant ELISA. *Lancet* 1:161.

CONRADIE, J. E.; GOUENDER, M.; and VISSER, L. 1983. ELISA solid phase: partial denaturation of coating antibody yields a more efficient solid phase. *J. Immunol. Meth.* 59:289–299.

DAGENAIS, P.; DESPREZ, B.; ALBERT, J.; and ESCHER, E. 1994. Direct covalent attachment of small peptide antigens to enzyme-linked immunosorbent assay plates using radiation and carbodiimide activation. *Anal. Biochem.* 222:149–155.

DE BOEVER, J.; KOHEN, F.; and VANDEKERCHKHOVE, D. 1983. Solid phase chemiluminescence immunoassay for plasma estradiol-17β during gonadotropin therapy compared with two radioimmunoassays. *Clin. Chem.* 29:2068–2072.

DEN HOLLANDER, F. C., and SCHUURS, A. H. 1971. Discussion. In *Radioimmunoassay Methods,* eds. K. E. Kirkham and W. M. Hunter, pp. 419–425, Churchill-Livingstone, Edinburgh.

DENMARK, J. R., and CHESSUM, B. S. 1978. Standardization of enzyme-linked immunosorbent assay (ELISA) and the detection of Toxoplasma antibody. *Med. Lab. Sci.* 35:227–232.

DIERKS, S. E.; BUTLER, J. E.; and RICHARDSON, H. B. 1986. Altered recognition of surface-adsorbed compared to antigen-bound antibodies in the ELISA. *Mol. Immunol.* 23:403–411.

DONINI, S., and DONINI, P. 1969. Radioimmunoassay employing polymerized antisera. *Acta Endocrinol.* (Suppl). 142:25–28.

DOUGLAS, A. S., and MONTEITH, C. A. 1994. Improvements to immunoassays by use of covalent binding assay plates. *Clin. Chem.* 40:1833–1837.

EKINS, R. P. 1960. The estimation of thyroxine in human plasma by an electrophoretic technique. *Clin. Chim. Acta* 5:453–459.

EKINS, R. P., and CHU, F. W. 1991. Multianalyte microspot immunoassay—Microanalytical "compact disk" of the future. *Clin. Chem.* 37:1955–1967.

GOSLING, J. P. A decade of development in immunoassay methodology. *Clin. Chem.* 36:1406–1427.

HAGGIN, J. 1994. New applications touted for immobilized artificial membranes. *Chem. Eng. News* 72:34–35.

HALES, C. N., and RANDLE, P. J. 1962. Immunoassay of insulin by isotope dilution. *Biochem. J.* 84:79.

HARVEY, M. A. 1991. *Optimization of Nitrocellulose Membrane-Based Immunoassays.* Schleicher and Schuell, Keene, NH.

HERBERT, V.; LAU, K. S.; GOTTLIEB, C. W.; and BLEICHER, S. J. 1965. Coated charcoal immunoassay of insulin. *J. Clin. Endocrinol. Metab.* 25:1375–1384.

HUNTER, W. M., and GREENWOOD, F. C. 1962. Radioimmunoelectrophoretic assay for human growth hormone. *Biochem. J.* 85:39–40.

ISHIKAWA, E.; HAMAGUCHI, Y.; IMAGAWA, M.; INADA, M.; IMURA, H.; NAKAZAWA, N.; and OGAWA, H. 1980. An improved preparation of antibody-coated polystyrene beads for sandwich enzyme immunoassays. *J. Immunoassay* 1:385–398.

KEMENY, D. J. and CHALLACOMBE, S. J. 1988. Microtiter plates and other solid phase supports. In *ELISA and Other Solid Phase Immunoassays: Theoretical and Practical Aspects,* eds. D. J. Kemeny and S. J. Challacombe, pp. 31–56, John Wiley and Sons, Chichester, England.

KENNEL, S. 1982. Binding of monoclonal antibody to protein antigen in fluid phase or bound to solid supports. *J. Immunol. Meth.* 55:1–12.

KENNY, G. E., and DUNSMOOR, C. L. 1983. Principles, problems and strategies in the use of antigenic mixtures for the enzyme-linked immunosorbent assay. *J. Clin. Microbiol.* 17:655–665.

KOCHWA, S.; BROWNELL, M.; ROSENFIELD, R. E.; and WASSERMAN, L. R. 1967. Adsorption of proteins by polystyrene particles. I. Molecular unfolding and acquired immunogenicity of IgG. *J. Immunol.* 99:981–986.

KRICKA, L. J.; CARTER, T. J. N.; BURT, S. M.; KENNEDY, J. H.; HOLDER, R. L.; HALLIDAY, M. I.; TELFORD, M. E.; and WISDOM, G. B. 1980. Variability in the adsorption properties of microtiter plates used as solid supports in enzyme immunoassays. *Clin. Chem.* 26:741–744.

LEE, R. G.; ADAMSON, C.; and KIM, S.W. 1974. Competitive adsorption of plasma proteins onto polymer surfaces. *Thrombosis Res.* 4:485–490.

LYKLEMA, J., and NORDE, W. 1973. Biopolymer adsorption with special reference to the serum albumin-polystyrene latex system. *Croatica Chem.* Acta 45:67–84.

MICHAELI, I.; ABSOLON, D. R.; and VAN OSS, C. J. 1980. Diffusion of adsorbed protein within the plane of adsorption. *J. Coll. Interface Sci.* 77:586–587.

Modern Plastics Encyclopedia. McGraw-Hill, New York.

MORGAN, C. R., and LAZAROW, A. 1962. Immunoassay of insulin using a two antibody system. *Proc. Soc. Expl. Biol. Med.* 110:29–32.

MORRIS, B. A. 1985. Principles of immunoassay. In *Immunoassays in Food Analysis,* eds. B. A. Morris and M. N. Clifford, pp. 21–52, Elsevier Applied Science Publishers, London.

NEWMAN, D. J., and PRICE, C. P. 1991. Separation techniques. In *Principles and Practice of Immunoassay,* eds. C. P. Price and D. J. Newman, pp. 78–95, Stockton Press, New York.

NGO, T. T. 1991. Immunoassay. *Current Opinion in Biotechnol.* 2:102–109.

NUSTAD, K.; PAUS, E.; and BORMER, O. 1991. Monosized polymer particles in immuno-assays. Applications and immunochemistry. In *Immunochemistry of Solid-Phase Immunoassay,* ed. J. E. Butler, pp. 243–250, CRC Press, Boca Raton, FL.

NYILAS, E.; CHIU, T. -H.; and HERZLINGER, G. A. 1974. Thermodynamics of native protein/foreign surface interactions. I. Colorimetry of the human γ-globulin/glass system. *Trans. Am. Soc. Artif. Intern. Organs* 20:480–490.

PAPPAS, M. G. 1988. Dot enzyme-linked immunosorbent assays. In *Complementary Immunoassays,* ed. W. P. Collins, pp. 113–134, John Wiley and Sons, Chichester, England

PETERMAN, J. H.; TARCHA, P. J.; CHU, V. P.; and BUTLER, J. E. 1988. The immuno-chemistry of sandwich-ELISAs. IV: The antigen capture capacity of antibody covalently attached to bromoacetyl surface-functionalized polystyrene. *J. Immunol. Meth.* 111:271–275.

RASMUSSEN, S. E. 1990. Covalent immobilization of biomolecules onto polystyrene microwells for use in biospecific assays. *Ann. Biol. Clin.* 48:647–650.

SINGER, J. M., and PLOTZ, C. M. 1956. The latex fixation test. I. Application to the sero-logic diagnosis of rheumatoid arthritis. *Am. J. Med.* 21:888–892.

SODERQUIST, M. E., and WALTON, A. G. 1980. Structural changes in proteins adsorbed on polymer surfaces. *J. Coll. Interface Sci.* 75:386–397.

THAKKAR, H.; DAVEY, C. L.; MEDCALF, E. A.; SKINGLE, L.; CRAIG, A. R.; NEWMAN, D. J.; and PRICE, C. P. 1991. Stabilization of turbidimetric immunoassay by covalent coupling of antibody to latex particles. *Clin. Chem.* 37:1248–1251.

TIJSSEN, P. 1985. *Practice and Theory of Enzyme Immunoassays.* Elsevier, Amsterdam.

UTIGER, R. D.; PARKER, M. L.; and DAUGHADAY, W. H. 1962. Studies on human growth hormone. I. A radioimmunoassay for human growth hormone. *J. Clin. Invest.* 41:254–261.

VANDERBRANDEN, M.; DECOEN, J. L.; JENNER, R.; KANAREK, L.; and RUYSCHAERT, J. M. 1981. Interaction of γ-immunoglobulins with lipid mono- or bilayers and lipo-somes. Existence of two conformations of γ-immunoglobulins of different hydropho-bicities. *Mol. Immunol.* 18:621–632.

VOLLER, A.; BIDWELL, D. E.; and BARTLETT, A. 1979. *The Enzyme-Linked Immunosorbent Assay (ELISA).* Dynatech Europe, England.

WILD, D. 1994. *The Immunoassay Handbook.* Stockton Press, New York.

WILD, D., and DAVIES, C. 1994. Components. In *The Immunoassay Handbook,* ed. D. Wild, pp. 49–82, Stockton Press, New York.

YALOW, R. S., and BERSON, S. A. 1960. Immunoassay of endogenous plasma insulin in man. *J. Clin. Invest.* 39:1157–1175.

ZAIDI, M.; GIRGIS, S. I.; and MACINTYRE, I. 1988. Development and performance of a highly sensitive carboxyl-terminal-specific radioimmunoassay of calcitonin gene-related peptide. *Clin. Chem.* 34:655–660.

Part II

Product Development

8

Immunoassay Classification and Commercial Technologies

INTRODUCTION

Immunoassays encompass techniques for the detection and quantification of antigens or antibodies, and are one of the most powerful of all immunochemical techniques. Although the term "immunoassay" generally refers to a quantitative method, in a much broader sense it also includes characterizing methods for analyzing the immunological properties of analytes.

The nomenclature of immunoassay systems is at times quite confusing. This is at least in part due to several variations that exist in the design of particular immunoassays. Generally, all assay names contain the word "immuno" combined with another word indicating the type of label used, along with the word "assay." For example, radioimmunoassays (RIAs) describe systems in which the detection label is a radioisotope. Radioreceptor assays are analogous systems except that the antibodies are replaced with specific receptors.

In contrast, in nonisotopic immunoassays, none of the reactants is labeled with a radioisotope. In all such systems, a variety of markers or labels, individually or in combination, are used to follow and measure the assay reactions. Examples of this type include enzyme immunoassays (or enzymoimmunoassay, EIA), fluorescent immunoassay (or fluoroimmunoassay), and chemiluminescent immunoassay (or chemiluminoimmunoassay).

When the immunoassay involves the use of reagents in stoichiometric excess, the nomenclature involves the words "immuno" and "metric." In general, the term "immunometric assay" is used to describe reagent excess assays. Thus, according to the label used, such an assay may be designated as an immunoradiometric assay (IRMA), immunoenzymometric assay, or immunofluorometric assay.

The term "enzyme-linked immunosorbent assay" (ELISA), first coined by Eng-vall and Perlmann (1971), is generally used for reagent-excess enzyme immuno-assays of specific antibodies or antigens. When the ELISA is applied to antigens, it is the equivalent of the IRMA. However, in the diagnostics industry, the term ELISA is often used interchangeably with enzyme immunoassay (or enzymo-immunoassay) and immunoenzymometric assay.

In this chapter, the more commonly used classification systems for describing immunoassays, the various assay designs and formats, and important commercial technologies built around enzyme immunoassays are described.

CLASSIFICATION

Immunoassays are performed in several different ways, and, therefore, can be classified on the basis of many different criteria. Some of these include the type of analysis (quantitative or qualitative), test sample (antigen or antibody), assay system (labeled versus nonlabeled, competitive versus noncompetitive, separation free versus separation required, and end-point detection either visibly or by in-strumentation), and assay conditions (liquid versus solid phase, equilibrium versus nonequilibrium, manual or automated).

One of the earliest classification systems divided immunoassays as nonlabeled or labeled reagent assays. The nonlabeled immunoassays, because of a lack of sig-nal amplification system, have limited sensitivity, because large antigen:antibody complexes must be formed for their detection. These include immunoprecipitin and agglutination methods and the corresponding light-scattering techniques (e.g., nephelometry and turbidimetry) for the detection of antigen:antibody com-plexes either by equilibrium or by kinetic analyses. These methods with sensitiv-ity in the range of micromoles per liter are generally used for analyzing proteins.

Fundamental Assay Designs

As described earlier in Chapter 3, the basic principle of an immunoassay is a reversible reaction between an antigen (Ag) and its antibody (Ab) as follows:

$$Ag + Ab \underset{k_2}{\overset{k_1}{\rightleftharpoons}} Ag : Ab$$

The antigen combines with the antibody to form the Ag:Ab complex at rate constant k_1. At equilibrium, the complex dissociates with a rate constant k_2 to form free Ag and Ab. The term antigen in the following discussion is used inter-changeably with analyte or ligand.

Based on this principle, immunoassays may be broadly divided into two classes (Ekins 1981,1985; Nakamura et al. 1986) as follows:

1. Type I: reagent observed (reagent excess or excess antibody):

    analyte + antibody → analyte:antibody complex + residual antibody
2. Type II: analyte observed (limited antibody or reagent/excess analyte):

    analyte + antibody → analyte:antibody complex + residual analyte

1. Type I Immunoassays. The type I assays (synonymous with two-site immunometric reagent excess assays, labeled antibody assays, immunometric assays, and noncompetitive immunoassays) rely on labeling of the antibody reagent and utilize a stoichiometric excess of reagent antibody over analyte. These assays utilize direct measurement of antibody binding sites occupied by analyte. The sample is first incubated with the "capture" antibody, which reacts with the first epitope of the analyte and binds it to the solid phase. The solid phase is then washed to remove unreacted components and further incubated with a labeled detecting antibody that binds to the antibody-analyte complex. Unreacted excess detecting antibody is then removed by washing. The enzyme activity in the solid phase is directly proportional to the concentration of the analyte.

In type I assays, if both capture and detecting antibody reactions proceeded to completion, then the dose-response curve would approximate a straight line. Many assays approach this ideal, particularly at low concentrations, and some have sufficient linearity to enable single-point calibration of the standard curve (Davies 1994). Type I assay designs have the obvious advantage of increasing the specificity of the assay. This design only became practicable with the advent of monoclonal antibodies with single-epitope specificity. Several commercially available two-site immunometric assays have a single-step incubation with binding of both capture and detecting antibody proceeding simultaneously.

Type I assays are also more sensitive than type II assays, with the theoretical sensitivity approaching detection of one analyte molecule. Because the antibody is introduced in excess, type I assays are not significantly influenced by environmental substances such as salt and urea, which may modulate the kinetics of antigen-antibody reactions.

Immunometric assays are also often described as "sandwich" assays. Type I assays with a radiolabeled antibody are called IRMAs. Analogous methods with enzyme labels on the analyte and antibody are called enzyme immunoassays and immunoenzymometric assays, respectively. Immunometric assays that have antibody or antigen coated onto a solid phase are also known as enzyme-linked immunosorbent assays (ELISA).

2. Type II Immunoassays. The type II (reagent or antibody limited/excess analyte) assays are based on labeling of the analyte and a limited antibody concentration. If the equilibrium constant K_{eq} and antibody concentration are known, then it is possible to derive the bound/free (B/F) ratio, and hence, the percentage bound for any given concentration of antigen (see Chapter 3 for details). Conversely, if the percentage bound is measured, then the antigen concentration of an unknown solution can be estimated. In fact, this principle formed the basis for the first type of immunoassay described by Yalow and Berson (1959) for the assay of insulin in human serum.

Type II assays require a method of separating the bound antigen from the free, and a means of determining the relative quantities of antigen in each. Commercially, a variety of separation and detection systems have been employed in type II assay designs. In practice, neither K_{eq} nor the antibody concentration is known with sufficient certainty to enable accurate predictions to be made of the concentration of unknown antigen; therefore, samples of known antigen concentration are included as standards or calibrators. A calibration curve is then drawn of percentage of activity in the bound fraction/total activity against the antigen concentration in the standards. The concentration of antigen in unknown samples may then be interpolated from the calibration curve. In type II assays, the specific activity of the bound labeled analyte fraction is inversely proportional to the concentration of the free analyte in the sample.

Type II assays are commonly but incorrectly referred to as "competitive" assays. In type II assays, true competition does not occur because the antibody binding sites are never fully saturated. According to Ekins (1981; 1991), the term "competitive" as commonly applied to immunoassay methodology encompasses two different concepts. The first is that labeled and unlabeled analyte molecules "compete" for a limited number of antibody binding sites; the second, that the amount of (labeled) antibody employed in the assay system is "small." The fundamental reasons underlying the existence of two alternative immunoassay strategies and the essential differences between them, however, are obscured by these concepts. This has resulted in the existing confusion regarding the fundamental distinctions between the "competitive" and "noncompetitive" immunoassay designs.

Based on the principle of "antibody occupancy," Ekins has further discussed the controversy surrounding this issue in several elegant articles (Ekins 1980; 1981a,b, 1985, 1991). All immunoassays essentially depend upon measurement of the "fractional occupancy" by the analyte of antibody receptor binding sites following reaction of analyte with a "sensor" or "capture" antibody.

The fractional occupancy of antibody may be determined in one of two basically different ways (Ekins 1991) (Figure 8.1) as follows:

1. The occupied sites may be determined "indirectly," i.e., by measurement of "unoccupied" sites from which the fraction of occupied sites is inferred by subtraction, and
2. The occupied sites are measured "directly."

Immunoassay methods that rely on measurement of residual, unoccupied antibody binding sites optimally require capture-antibody concentrations tending towards zero as the analyte concentration itself approaches zero. Under such cir-

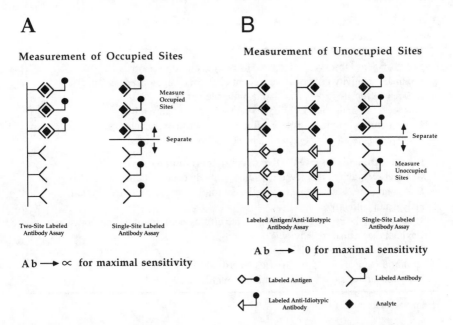

Figure 8.1. "Noncompetitive" assays (A) rely on direct measurement of antibody binding sites occupied by analyte; "competitive" assays (B) rely on measurement of unoccupied sites, from which the occupied sites are estimated essentially by subtraction. Labeled antigen (Figure B, left) or labeled anti-idiotypic antibody methods (Figure B, center) rely on measurement of sites unoccupied by analyte, and are therefore of "competitive" design. Single-site labeled antibody methods (Figure B, right) may be either of noncompetitive or competitive design, depending on whether labeled antibody bound or unbound to analyte is measured following separation of the two fractions. From: Ekins (1991)

cumstances, fractional occupancy is maximal, and its value is measured with maximal precision. Such techniques may be termed "competitive." Conversely, those techniques in which occupied sites are "directly" measured often permit the use of relatively high capture antibody concentrations, and may thus be termed "noncompetitive."

Hence, Ekins (1981) has termed such type II assay designs as "reagent limited," primarily and logically to distinguish them from type I reagent excess assays. This terminology, however, has not gained wide acceptance, and the term "competitive assay" has remained in common usage. The maximum theoretical sensitivity of type II assays is on the order of 10^{-14} mol/l, or approximately 10^7 molecules/ml.

Type II assays where the antigen is labeled, however, have several disadvantages (Davies 1994). The sensitivity of such assays is primarily governed by the equilibrium constant of the antibody, and thus may not fully exploit the potential

TABLE 8.1.   Characteristics of Type I "Reagent Observed" and Type II "Analyte Observed" Immunoassays[a]

Type I "Reagent Observed" (Excess Antibody) Immunoassays
- Maximal sensitivity is attained as amount of antibody approaches infinity
- Theoretical sensitivity of assay is one molecule of analyte
- Cross-reactive antigens will be equipotent with excess antibody
- Antigen-antibody reaction is less influenced by substances such as salt and urea
- Assay time is relatively rapid with labeled antibody procedures

Type II "Analyte Observed" (Excess Analyte) Immunoassays
- Maximal sensitivity is achieved when the antibody concentration approaches zero
- This saturation assay is regulated by the equilibrium constant of the reaction between analyte and antibody
- Sensitivity of the assay is dependent on the affinity constant of the antibody
- Maximal theoretical sensitivity is $1 \times 10^{-14} M$
- Cross-reactive antigen will demonstrate a relative potency dependent on the rate of the equilibrium constants of the analyte and cross-reactive antigen
- Assay reaction is slow, because equilibrium must be reached

---

[a]From Nakamura et al. (1986).

that immunoanalytical techniques offer with alternative designs. Labeling the analyte sometimes also reduces or abolishes the recognition by antibody because a critical epitope may be affected or hidden. In other instances, labeling may actually enhance recognition when it is directed to the same position (i.e., bridge recognition, see Chapter 5 for detailed discussion) that the hapten was originally conjugated to on the immunogen.

The characteristics of type I and type II assays are summarized in Table 8.1.

Classification Based on Analytic Methodology

Based on their analytic methodology, immunoassays can be formatted into two basic systems: heterogeneous (separation-required) and homogeneous (separation-free assays). In the heterogeneous systems, because the activity of the enzyme label is not affected by the antigen:antibody reaction, it must be separated into antibody-bound and free enzyme fractions. The enzyme activity of either of these two fractions can then be measured. In the homogeneous systems, the enzyme activity of the assay solution is measured without a prior physical separation of the antibody-bound enzyme label from the free, primarily because the activity of the bound enzyme label is significantly different from the free (unbound) one. The various heterogeneous and homogeneous immunoassays can be further characterized as either competitive or noncompetitive (immunoenzymometric) assays depending on whether the unlabeled antigen and the antigen linked to an enzyme or attached to a solid phase compete for a limited number of antibody binding sites, or whether the antigen or antibody to be measured is allowed to react alone with an excess of immune reactant.

Furthermore, immunoassays can be either heterologous or homologous assays. In heterologous assays, the haptens used as antigen to produce the antibody are different from the haptens used for labeling the enzyme. In homologous immunoassays, the haptens are similar in both cases. Enzyme immunoassays represent the most rapidly growing nonisotopic methods in the diagnostics industry. Some of their advantages and disadvantages are listed in Table 8.2.

The different types of homogeneous and heterogeneous immunoassay systems can be further classified into six groups (Gosling 1990). According to this classification, groups 1 through 4 and 6 are mainly assays of antigens/haptens, and group 5 assays are for specific antibodies. These groups are briefly described below.

1. Group 1 Immunoassays: Competitive Assays of Antigens or Haptens with Labeled Analyte.   Immunoassays of antigens or haptens in this group primarily use labeled analyte with a limited concentration of antibody, as is used in classical RIAs. The label may differ substantially from the analyte being analyzed. It may be formed by tagging the analyte (or its derivative) with radioiodine, a fluorescent or luminescent compound, or an enzyme. The detection limit of the analyte may be lowered, i.e., the assay sensitivity improved, by promoting disequilib-

---

TABLE 8.2.   Advantages and Disadvantages of Enzyme Immunoassays (EIAs)[a]

Advantages
- Sensitive assays can be developed by the amplification effect of enzymes
- Reagents are relatively cheap and can have a long shelf life
- Multiple simultaneous assays can be developed
- A wide variety of assay configurations can be developed
- Equipment can be inexpensive and is widely available
- No radiation hazards occur during labeling or disposal of wastes
- Rapid, simple EIA adaptable to automation can be developed
- Homogeneous EIA can be developed for haptens and proteins

Disadvantages
- Measurement of enzyme activity can be more complex than measurement of the activity of some types of radioisotopes
- Enzyme activity can be affected by plasma constituents
- Many of homogeneous assays at the present time have the sensitivity of 10^{-9} M and are not as sensitive as RIA
- Homogeneous EIAs for large protein molecules have been developed but require complex immunochemical reagents

---

[a]From Nakamura et al. (1986).

rium and adding the labeled analyte to the system some time after the sample analyte (unlabeled) and antibody are mixed together.

On separation of the bound and free fractions, the bound fraction is often measured. A plot of the concentration of bound label against the concentration of analyte gives an inverse relationship. This is because less labeled analyte will be bound to the antibody in the presence of a high concentration of the free analyte in the sample. The detection limits of these reagent-labeled immunoassays may be greatly improved by the use of high-specific-activity label (e.g., ^{125}I rather than ^{3}H for RIAs), but the smallest amount of analyte detectable is ultimately limited by the affinity of the antibody used. However, very low concentrations of haptens are generally determined with group 1 immunoassays, the most sensitive of which can detect < 1 fmol (10^{-15} M) of testosterone (Howard et al. 1989).

2. Group 2 Immunoassays: Competitive Assays for Antigens and Haptens with Labeled Antibody (Reagent Limited).   This approach is useful for antigens or haptens that may affect the label properties on conjugation. Therefore, labeled antibody is used in this approach. In this type of assay, it is essential that the solid-phase immobilized analyte also be present in a constant, limited amount. The partition of the label between the immobilized analyte and the free analyte depends on the concentration of analyte in the test sample, the bound label often being determined after incubation. Similar to group 1 immunoassays, the standard curves are inverse, and the sensitivity is dependent largely on antibody affinity. This approach is often used for acridinium ester chemiluminescent assays (Sturgess et al. 1986) and for Eu^{3+} or Eu^{3+}-chelate fluorescent assays of steroids and other haptens (Helsingius et al. 1986).

3. Group 3 Immunoassays: Antigen-Antibody Immune Complex Assays. Group 3 assays include precipitation, nephelometric, and turbidimetric immunoassays as well as particle agglutination and particle-counting immunoassays. The endpoints in these assays involve the direct detection of immune complexes. Group 3 assay systems are generally characterized by the lack of any labeled reagent. The precipitation of immune complexes may occur in a gel (e.g., immunodiffusion and immunoelectrophoretic methods), or their kinetics may be monitored by nephelometric and turbidimetric techniques. With specialized automatic nephelometers or centrifugal analyzers acting as turbidimeters, a wide range of serum and urinary proteins can be determined at concentrations as low as 10^{-4} g/l (about 10^{-8} mol/l). The major limitations of this type of assay are their limited ranges and the requirement for controls and checks to avoid underestimations caused by antigen (analyte) excess (Gosling 1990).

In particle-labeled group 3 immunoassays, the antibody or the antigen/hapten may be coupled to polystyrene latex particles, inorganic colloidal particles, or erythrocytes to give an agglutination response.

4. Group 4 Immunoassays: Sandwich Assays.   Group 4 immunoassays utilize all reagents in excess. These include the classical two-site "sandwich" assays such as immunochemiluminometric assays, immunoradiometric assays (IRMA), immunofluorometric assays (IFMA), and most ELISAs. Unlike competitive assays, they are not largely dependent on antibody affinity. If problems such as high nonspecific binding of the label, degradation of assay specificity, and interference from heterophilic antibodies (all from the use of excess reagents) are minimized, then very low detection limits can be achieved by maximizing the signal-to-noise ratio of the label. The two-site sandwich assays, in which the antigen is sandwiched between two antibodies and the combined selectivity determines the specificity, are inherently more specific than the single-site assays (Seth et al. 1989). The most sensitive of such assays are capable of detecting < 1 amol (10^{-18} *M)* of analyte (Ishikawa et al. 1989; Johannsson et al. 1986). However, the analyte must have two antigenic sites that can be recognized simultaneously by the two antibodies. Therefore, such assays cannot be used for simple steroids, small peptides, and most drugs.

Several variations of the basic two-site sandwich assay for antigens have been used. These include the use of alternative labeling substances, antibody fragments to decrease nonspecific binding, secondary label reagents, and anchor ligands (Gosling 1990).

Some variations of group 4 reagent-excess, one-site immunoenzymometric assays for haptens use incubation of analyte with a calculated excess of labeled antibody. The unoccupied labeled antibody is then removed by means of excess immobilized antigen before the label associated with the analyte is determined (Grenier et al. 1987; Freytag et al. 1984). Although similar to the labeled-antibody competitive assays of group 2, the use of excess label and the subsequent determination of the analyte-label complex completely alter the character of the assay. These assays are capable of detecting about 100 amol (10^{-16} *M)* of analyte whereas about 400 amol (4×10^{-16} M) can be detected by the most sensitive competitive immunoassays for hapten (Freytag 1985; Munro and Lasley 1988).

5. Group 5 Immunoassays: Noncompetitive Assays for Specific Antibodies. These assays are used for quantifying specific antibodies. Usually, diluted test serum is added to excess antigen immobilized on a solid phase. The amount of specific antibody that binds to the antigen is then quantified by the use of labeled antibodies that specifically bind to the constant regions of the immunoglobulin class of interest. The sensitivity of antibody quantification by such "antibody capture" can often be improved by using a complex sequential double-solid-phase assay system to reduce interference from nonspecific antibodies (Gosling 1990; Deshpande 1994).

Conversely, "antigen-capture" assays can also be developed by using a reverse protocol. Antigen capture assays are useful for assay of specific antibodies of the minor immunoglobulin classes. In addition, relevant immunoglobulin complexes containing intact antigen may be detected if labeled antigen is not used.

6. Group 6 Immunoassays: Separation-Free Assays Using Labeled Reagents. This class is characterized by separation-free homogeneous assays in which the label signal is modulated by antigen-antibody binding reactions. Characterized by speed and simplicity, group 6 immunoassays are widely used to monitor concentrations of therapeutic drugs and drugs of abuse in blood and urine when detection limits of less than 10 μmol/l are not required (Gosling 1990). The activity of the label is either decreased or increased by the binding reaction. In an ideal homogeneous assay, antibody binding modulates 100% of the signal from the label. In practice, this is very difficult to achieve. Hence, homogeneous immunoassays are generally much less sensitive than separation-required heterogeneous immunoassays.

In the following sections, various enzymatic heterogeneous and homogeneous immunoassay formats are described.

HETEROGENEOUS ENZYME IMMUNOASSAYS

Heterogeneous enzyme immunoassays can be formatted in many different ways. Their sensitivity is primarily determined by (1) antibody affinity or binding constant, (2) type of label and detection system employed for the label, (3) type of format used, and (4) manipulation of reagents in a given format to achieve desired sensitivity. Sensitivity ranging from 1 μmol (10^{-6} M) to 1 amol (10^{-18} M) has been reported in the literature.

Heterogeneous, separation-required enzyme immunoassays were simultaneously reported by van Weeman and Schuurs (1971) and by Engvall and Perlmann (1971). These assays were analogous to the well-known radioimmunoassay of Yalow and Berson (1959). Because of the large size and other properties of the enzyme label relative to the antigen or antibody, the systems used for the separation of free and bound labeled molecules are mostly solid phase. Antibody-coated polystyrene tubes, microtiter plates, cellulose, and magnetic polyacrylamide agarose particles can be used for the separation of bound and free enzyme labels. A separation can also be effected by precipitation of the immune complex formed with polyethylene glycol or a second antibody and by using a precipitated complex of first and second antibody (Oellerich 1984). These separation systems were described in the previous chapter.

The principle of a direct competitive assay (group 1) is shown in Figure 8.2A. Enzyme-labeled ligand (E-L) and unlabeled antigens or ligand (L) compete for the binding sites of a limited amount of solid-phase immobilized antibody (AB). The saturation of the antibody occurs simultaneously, provided all reactants are

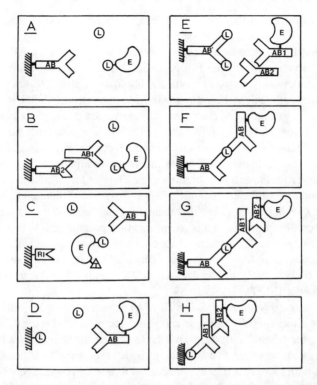

Figure 8.2. Principles of heterogeneous enzyme immunoassays (ELISA). The abbreviations used are as follows: L = ligand; E-L = enzyme-labeled ligand; L-E-T = tagged enzyme-ligand conjugate; E-AB, E-AB2 = enzyme-labeled antibodies; AB', AB, AB1 = antibodies against ligand under test (A through G) or antibody to be determined (H); AB2 = antibody directed against the immunoglobulin of the species that was immunized with the antigen to be determined (A through G) or against the antibody to be detected (H); RI = immobilized receptor. A, C, and E represent different variants of competitive enzyme immunoassays; B is a competitive sandwich assay; D and F are two variants of sandwich antigen assays; G represents an immunoenzymometric assay; and H represents a sandwich antibody assay.
Adapted from Oellerich (1984)

incubated together. If the antigen concentrations in test samples are very low, sequential saturation (antigen is added first and incubated, followed by E-L addition and incubation) can be used to enhance the assay sensitivity. After separation of the unbound label, the antibody-bound label is then assayed for enzymatic activity, which is inversely related to the concentration of the test ligand. The sensitivity and specificity of the assays depend on the binding constant and specificity of the antibodies used. Competitive heterogeneous EIAs can detect picogram quantities of hormones and drugs (Bosh et al. 1978; Deshpande 1994; Hefle 1995; Pestka et al 1995).

In a modification of the above procedure, a second antibody bound to a solid phase can be used for the separation step (Figure 8.2B). E-L, AB1, and L (sample) are incubated together and added to the immobilized second antibody, AB2. The AB2 binds AB1, which is bound with E-L depending on the L concentration in the sample. This variation is known as the "double-antibody solid-phase" (DASP) technique.

In a novel approach to heterogeneous EIAs, the enzyme label is covalently linked not only to a ligand of the type to be determined but also to a tag molecule (L-E-T) (Figure 8.2C) (Ngo and Lenhoff 1981). In this assay, a fixed concentration of L-E-T, antibody (AB), and insolubilized receptor (RI) is used. The amount of free L-E-T available to bind RI is dependent on the amount of L, as both L and L-E-T compete for the antibody binding sites. Thus, at lower concentrations of L, more antibody is available to complex L-E-T. If the L-E-T conjugate is bound to the antibody, its tag is masked so that it can no longer bind to the RI. The amount of enzyme conjugate bound to this receptor is thus proportional to that of the free ligand to be analyzed. If this assay is used with biotin as tag and avidin as the insoluble receptor, analytes at nanomolar levels can be measured in about 1 hr (Ngo and Lenhoff 1981). This procedure is called "antibody-masking enzyme-tag immunoassay" (AMETIA).

The immunoenzymometric assays (IEMA) (group 4, group 5) work with ligands having either single or multiple antigenic determinants (Figure 8.2D). An excess of enzyme-labeled antibodies (AB-E) to a given antigen (L) is added to the test sample. The remaining free labeled-antibody is separated in a subsequent step by binding with an excess of antigen coupled to a solid phase. The immobilized antigen associated with the bound enzyme-labeled antibodies is then separated from the soluble antigen-enzyme-labeled antibody complexes by a brief centrifugation. The enzyme activity of the soluble fraction increases linearly with the amount of ligand in the test sample.

Gnemmi et al. (1978) described a variation of the above procedure for the multivalent antigens (Figure 8.2E). Two antibodies are produced that bind to different sites on the multivalent antigen. A fixed amount of one antibody (AB1') is attached to solid phase and saturated with the antigen. The other antibody (AB1) and an enzyme-labeled anti-IgG antibody, AB2 (a second antibody that is immunologically reactive to the AB1), are mixed together to form an enzymic immunocomplex. In a typical assay, a sample is incubated with the preformed enzymic immunocomplex, and the mixture is then added to the solid phase. If binding sites on AB1 are free (low concentration of antigen in the sample), then it will bind to the antigen on the AB1', forming a sandwich. In case of a high concentration of antigen in the sample, most of the binding sites on AB1 will be occupied, thereby disabling the sandwich formation. After a wash step, enzyme activity in the immobilized phase is determined. Enzyme activity is inversely proportional to the concentration of antigen in the sample. This assay is referred to as a competitive sandwich assay.

The sandwich assays (Figure 8.2F) are applicable only to bivalent or multivalent antigens. In these procedures, the solid-phase immobilized antibody is incubated with the test antigens. After a separation step, the immobilized antibody-antigen complex is incubated with an excess of enzyme-labeled second antibody that binds to one of the remaining antigenic sites. After a washing step, the enzymatic activity of the bound fraction is measured. The activity is directly proportional to the concentration of the standard or analyte antigen.

The above procedure may be modified into an indirect method by using an unlabeled second antibody (Figure 8.2G). The immunocomplex is then reacted with an excess of enzyme-labeled third antibody specific for the IgG of the species of the second antibody. Such an indirect labeling can also be applied to immunoenzymometric tests and inhibition enzyme immunoassays (Oellerich 1984). Indirect labeling has the advantage that the same enzyme-antibody conjugate can be used in assays for various different antigens, thereby circumventing the often difficult direct labeling of the antibody.

In yet another modification of the sandwich antibody assay (Figure 8.2H), the antibody to be determined (AB1) is reacted with an antigen (L) bound to the solid phase. It is then detected by a second enzyme-labeled generic antibody (AB2-E) to the first antibody (AB1) in the sample. The technique is useful for the determination of antibody in blood in response to an infection such as with *Salmonella dublin* in dairy cattle.

Castro and Monji (1985) reported a novel separation technique for heterogeneous EIAs. The assay uses enzyme-labeled ligands and an affinity gel that is capable of binding the enzyme portion of the enzyme-labeled ligand conjugates when they are free (i.e., not bound by antibodies). The antibody-bound, enzyme-labeled ligand conjugate, however, does not bind to the affinity gel (Figure 8.3), presumably because of a steric hindrance created by the antibody bound to the la-

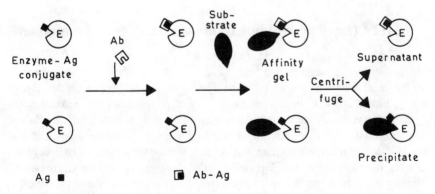

Figure 8.3. Schematic representation of steric hindrance enzyme immunoassay (SHEIA)
Adapted from Castro and Monji (1985)

beled ligand. For this reason, the assay is called a "steric hindrance enzyme immunoassay" (SHEIA). The enzymic activity in the supernatant is inversely related to the analyte concentration in the test sample.

Heterogeneous immunoassays are generally more sensitive because the unbound reactants are removed as part of the assay protocol. They are, therefore, also more sensitive than the separation-free, homogeneous immunoassays and generally have lower detection limits approaching the attomole $(10^{-18}\ M)$ level. The major disadvantages include the separation step itself, increased incubation times, and increased difficulty in terms of automation.

HOMOGENEOUS ENZYME IMMUNOASSAYS

Homogeneous EIAs are based on antibody-mediated changes in enzyme activity. The assays are performed by simply mixing the sample with reagents and measuring the resulting enzyme activity, which correlates with the concentration of the analyte.

The basic principle of homogeneous EIA involves a competitive binding mechanism and the differences between different homogeneous assay techniques lies in the way in which the reactants are labeled, the molecular recognition properties of antibodies, and the detection label. The antibody-hapten interaction is used to either inhibit or enhance the enzymatic activity. These assays require no separation step, and hence, simplified assay protocols can be developed. In addition, all homogeneous EIAs are based on measurement of enzyme activity in the kinetic mode of detection to eliminate endogenous sample interference and to reduce background noise.

Homogeneous EIA formats commonly used in the diagnostics industry are briefly described below.

EMIT® (Enzyme Multiplied Immunoassay Technique) Assays

These assays belong to the group 6 assays described earlier. EMIT® assays developed by Syva Company, San Jose, California, still comprise probably the most widely used homogeneous EIA system. The basic principle on which the EMIT® assays function is that the enzymatic activity of conjugates between marker enzyme and the analyte can be modulated by antibodies directed against the analyte. Modern EMIT® assays use drug-labeled glucose-6-phosphate dehydrogenase (G6PDH). The assay is based on competition for antibody binding sites between drug in the sample and G6PDH-drug conjugate. Enzyme activity decreases on binding to antibody, so the drug concentration in the sample can be measured in terms of enzyme activity (Figure 8.4A). Active enzyme converts NAD to NADH, resulting in an absorbance change that is measured spectrophotometrically.

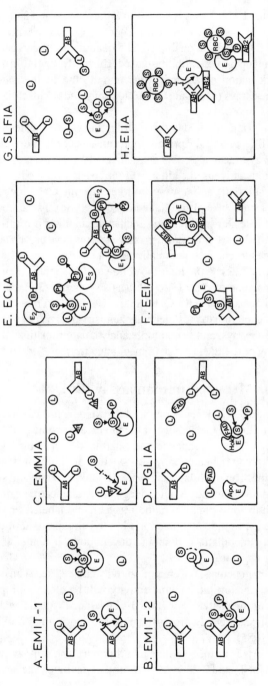

Figure 8.4. Principles of homogeneous enzyme immunoassays. All reactants are present in one reaction medium. The abbreviations used are: L = ligand; E-L, E1-L = enzyme-labeled ligands; M-L = ligand-substituted enzyme modulator; FAD-L = ligand covalently linked to flavin adenine dinucleotide; S-L = ligand-labeled substrate; P-L = ligand-labeled product; AB = limited amount of antibody; AB1 = antibody to be detected; AB2 = succinylated antibody; AB1-E, AB2-E = enzyme-labeled antibodies; E2-B-AB = antibody coimmobilized with an enzyme (E2) on fine beads (B); E, E1, E2, E3 = enzymes; ApoE = apoenzyme; HoloE = holoenzyme; S = substrate; RBC = erythrocyte; P, P1, P2, Q = products of enzymic reactions; i = inhibition of enzymes. The abbreviations for the types of homogeneous immunoassays are as follows: EMIT = enzyme multiplied immunoassay technique; "marker enzyme." The abbreviations for the types of homogeneous immunoassays are as follows: EMIT = enzyme multiplied immunoassay technique; EMMIA = enzyme-modulator-mediated immunoassay; PGLIA = prosthetic group-labeled immunoassay; ECIA = enzyme-channeling immunoassay; SLFIA = substrate-labeled fluorescence immunoassay; EEIA = enzyme enhancement immunoassay; and EIIA = enzyme inhibition immunoassay. Adapted from Oellerich (1984)

245

In the two variations of the EMIT® assays shown in Figures 8.4A and 8.4B, the activity of the marker enzyme is either inhibited (A) or increased (B) when the conjugate is bound to the antibody. Both qualitative and quantitative assays can be developed depending on the method of calibration used. Currently, EMIT® assays are widely used for the detection of abused drugs, monitoring of therapeutic drugs, and thyroid hormone assays (Meenan et al. 1990; Levy and Jaklitsch 1993; Khanna 1991).

EMIT® assays have been primarily used to detect drugs and small analytes such as thyroxin. The use of EMIT® assays for protein molecules has been more difficult because of the large size of the analyte. When relatively large molecules of an enzyme are coupled to protein antigens, the binding of the specific antibody to the protein antigen may occur at a point so far away from the enzyme that a modulation of the enzyme activity cannot take place (Nakamura et al. 1986). The antibody binding to a protein may cause steric inhibition of substrate binding if a macromolecular substrate is used in the reaction. EMIT® assay systems have proven quite useful in clinical laboratories to enhance the quality of patient care. Some of the specific advantages of EMIT® assays include the following: (1) minimal or no sample preparation, (2) small sample size (3 to 50 μl is sufficient), (3) excellent correlation with other methods such as HPLC, RIA, and GC-MS, (4) rapid assay time (< 1 min), (5) ease of adaptation to most general chemistry analyzers, (6) similar assay protocols for most assays, and (7) multiple analyte determinations from the same sample. Sensitivity of the EMIT® assays may range from 10^{-6} to 10^{-8} M.

Enzyme-Modulator-Mediated Immunoassay (EMMIA)

The EMMIA assay (Figure 8.4C) depends on the ability of a ligand-labeled enzyme modulator (L-M) to influence the activity of the indicator enzyme (E) (Ngo and Lenhoff 1980; Oellerich 1984; Deshpande and Sharma 1993). The enzyme modulator may be an antibody to indicator enzyme as well as an inhibitor or activator of the indicator enzyme. The modulator is covalently linked to a ligand similar to the analyte such that the modulator is able to regulate the indicator enzyme, depending on the concentration of the analyte in the assay. Ligand-labeled enzyme modulator and ligand from the sample compete for a limited amount of antibody. If the ligand-labeled enzyme modulator is bound to the antibody, it cannot affect the activity of the indicator enzyme.

Modulators that increase or decrease the enzyme activity can be used. Thus, for EMMIA with a positive modulator, the enzyme activity will be directly proportional to the concentration of the analyte. For EMMIA developed with a negative modulator, the activity will be inversely proportional to the concentration of the analyte (Ngo and Lenhoff 1985).

EMMIAs have been reported for theophylline and thyroxin. The sensitivity of the assay is 2.56×10^{-9} M for T_4 and 2.22×10^{-6} M for theophylline. The advantage of the EMMIA is that no enzyme conjugate preparation is required, thereby avoiding all the problems associated with enzyme conjugate preparation. Modulator-ligand conjugates are more stable than the enzyme conjugate. In addition, modulator-ligand conjugates can be mass-produced by a microorganism genetically engineered by recombinant DNA technology (Ngo and Lenhoff 1980).

Prosthetic Group-Labeled Immunoassay (PGLIA)

The principles of the PGLIA (also known as ARIS for "apoenzyme reactivation immunoassay") assay are shown in Figure 8.4D). Prosthetic groups such as flavine adenine dinucleotide (FAD), which are necessary for the activity of glucose oxidase, can be used as labels in these homogeneous assays. The ligand-FAD (L-FAD) and the free ligand (L) from the sample compete for a limited amount of antibody. If the ligand-FAD conjugate is bound by the antibody, it can no longer combine with the apoenzyme (ApoE) to form an enzymatically active holoenzyme (HoloE). Thus, the observed enzyme activity is directly related to the concentration of the analyte (L).

Glucose oxidase is an excellent enzyme for PGLIAs for the following four reasons: (1) FAD can be dissociated from the intact enzyme to leave a stable apoenzyme, (2) the glucose oxidase apoenzyme can be easily converted to the holoenzyme, (3) the enzyme has a relatively high turnover, and (4) the reaction product H_2O_2 can be determined with high sensitivity by varying methods.

The PGLIA principle can be used for the determination of haptens and antigens (Oellerich 1984). PGLIAs have been used to assay IgG, theophylline and phenytoin (Nakamura et al. 1986).

Enzyme-Channeling Immunoassay (ECIA)

The ECIA (Figure 8.4E) represents a rather complex assay system. The channeling immunoassay employs two enzymes related in such a way that the product of one serves as the substrate for the other. Two basic configurations are possible: either the first or the second enzyme serves as an immunochemical label. Hexokinase (HK):glucose-6-phosphate dehydrogenase (G6PDH), and glucose oxidase (GOD):horseradish peroxidase (HRP) have been used as enzyme pairs for the channeling assays (Deshpande and Sharma 1993).

In the ECIA procedure, the ligand (L) is covalently labeled with an enzyme (E_1). This conjugate competes with the ligand in the sample for the binding sites of a limited amount of antibody, which is coimmobilized with the second enzyme (E_2) on a solid phase such as agarose beads (B). The overall enhancement in the rate of formation of the reaction product (P2) from the substrate (S) depends on the ligand (L) concentration. In the presence of a high ligand concentration, there would be less ligand-enzyme conjugate (E_1-L) bound to the antibody, and consequently less substrate would be converted to the product (P2). Because the formation of the final product P2 from P1 occurs at the solid phase, the P1 formed by unbound conjugate (in solution) can be converted to the product Q by a further soluble, scavenger enzyme (E_3) in order to reduce background reactions.

The porosity of agarose is not found to be required; therefore, conventional microtiter well surfaces are sufficient to support the channeling reactions. Steps for a human IgG (hIgG) microtiter assay using the glucose oxidase:horseradish peroxidase pair are as follows: (1) coat microtiter well with glucose oxidase and hIgG, (2) incubate wells with varying amounts of hIgG, horseradish peroxidase-labeled anti-hIgG, and catalase as a scavenger enzyme, (3) add ABTS [2,2'-azino-di-(3-ethylbenzothiazoline-6-sulfonate)] and glucose directly to the assay mixtures. No separation and washing steps are required. Sensitivity for IgG approached 10 ng/well $(4 \times 10^{-14} \ M)$. Extremely sensitive assays can be developed by combining latex coagglutination with the enzyme-channeling method; these allow detection at the 100 pg/ml level (Ullman et al. 1983).

Chromatography paper can also be used as a solid surface to develop channeling assays. The method is called "enzyme immunochromatography." An assay developed by Syva Company using paper is called the AccuLevel® test. Briefly, the test has three main components: a test cassette that contains chromatographic paper coated with antibody to a drug being assayed; reagent 1, which contains horseradish peroxidase-labeled drug and glucose oxidase; and a developing reagent that contains glucose and 4-chloro-1-naphthol (reagent 2). The test cassette is placed in a mixture of the patient's blood and reagent 1. As the solution migrates by capillary action up the chromatographic paper, drug from the sample and the peroxidase-labeled drug bind to the immobilized antibody sites. The number of occupied antibody sites depends on the concentration of drug components in the solution. Because the concentration of the labeled drug is always the same, the drug concentration in the patient sample determines how far the two components travel up the chromatographic paper. When the cassette is immersed in reagent 2, an insoluble blue product forms on the portion of the chromatographic paper to which enzyme-labeled drug is bound. The height of the color bar is proportional to the amount of drug in the patient sample (Ullman et al. 1983).

ECIAs can also be used without immobilization of any of the enzymes. These assays operate on similar principles to the immunometric assays, using a large excess of labeled antibody to saturate the antigenic sites of the analyte. The solid phase is created during the course of the assay with production of a colloidal immune aggregate (Nakamura et al. 1986; Ullman et al. 1983; Deshpande and Sharma 1993).

Enzyme Enhancement Immunoassay (EEIA)

The homogeneous EEIA (Figure 8.4F) used for polyvalent ligands and antibodies avoids the need for a labeled antigen. Enzyme-labeled antibody (E-AB1) and succinylated antibody (AB2) form an immune complex with a polyvalent antigenic analyte (L) present in the sample. The product (P1) formed by the free enzyme conjugate outside of this immune complex remains soluble, but the enzyme within this negatively charged microenvironment of the immune complex converts the substrate (S) to a product (P2) that forms a second light-scattering phase. The concentration of the second-phase product (P2) is proportional to the amount of sample antigen (L) and can be determined by turbidimetry (Oellerich 1984). EEIAs have been used to assay IgG and C-reactive protein as well as specific antibodies (Gibbons et al., 1980). These assays, however, require sophisticated and complex reagents, although the procedure is one of great simplicity and is adaptable to automation.

Substrate-Labeled Fluorescence Immunoassay (SLFIA)

Enzyme substrates have been used as immunoassay labels. However, as the hapten and hapten-labeled substrates are involved in competition, only a small amount of hapten-labeled substrate can be used. This prevents the normal amplification of the signal usually associated with enzymic catalysis. In order to achieve adequate sensitivity at micromolar concentrations, it is customary to use a fluorescence product (Gould and Marks 1988). This approach, which uses the enzyme β-galactosidase, is termed "substrate-labeled fluorescence immunoassay" (SLFIA, Figure 8.4G). The common substrate employed is an umbelliferyl galactoside derivative of the hapten that does not fluoresce until hydrolyzed by the enzyme.

In SLFIA, a ligand (L) labeled with a fluorogenic substrate (S) is used. This conjugate is nonfluorescent and competes with the analyte (L) for a limited amount of antibody (AB). When the substrate-ligand conjugate (S-L) is bound by the antibody, it cannot be converted by an enzyme to a fluorescent product (P-L). The amount of conjugate available for reaction with the enzyme and the resulting fluorescence intensity are proportional to the concentration of the analyte.

SLFIAs have been described for the determination of several drugs, haptens, and antigens (Nakamura et al. 1986; Oellerich 1984). The disadvantage of the system is the limited sensitivity of 10^{-9} to 10^{-10} M concentration of the analyte in biological samples of sera. The SLFIA procedure has been adapted to automation, and the assays can be performed rapidly (within 30 min). The enzymatic reaction can be monitored by either kinetic or single-point measurements of fluorescence. The latter are made possible with a large dilution factor of the serum samples so that the background fluorescence is negligible (Nakamura et al. 1986).

Enzyme Inhibition Immunoassay (EIIA)

Human IgG (AB1) was determined by EIIA. In this procedure (Figure 8.4H), rabbit antihuman IgG (AB2) was labeled with phospholipase C enzyme (E) (Wei and Riebe 1977). The enzymatic activity of the conjugate is inhibited by human IgG (AB1) present in the sample. The substrates (S) used are phospholipids, which are components of the erythrocyte membranes and therefore may be viewed as being immobilized. When sample IgG (AB1) forms a complex with the enzyme-antiIgG conjugate, it sterically prevents an interaction between the enzyme label (E) and the substrate (S).

Radial-Partition Immunoassay (RPIA)

The radial-partition immunoassay (RPIA) combines solid-phase immunological techniques with radial chromatography to produce a novel technology for the rapid and convenient detection and quantitation of both low and high molecular weight analytes in biological fluids (Evans et al. 1991; Giegel 1985).

In RPIA, the antibody is first immobilized on a small section of a filter paper that is inserted into a plastic tab holder for transport. The filter paper provides a very high surface area-to-volume ratio, approximately 2000 times greater than a coated tube. Moreover, because of the smaller diffusional distances, the antigen is in close contact with the antibody on the solid phase throughout the reaction. A variety of labels, including enzymes, fluorophores, chromophores, and radioisotopes, can be used with this technology.

RPIA can be performed in three different formats. In the "sequential" assay, sample containing the analyte is added to the dry tabs. After about 2 min, an excess of enzyme-labeled analyte is then added to saturate the remaining antibody sites on the tab. Following a brief incubation period, substrate for the enzyme is added to the tab in such a way as to effect a radial chromatographic wash of excess, unbound labeled analyte. This step is common to all three types of assay and is an essential feature of this technology (Giegel et al. 1982). The center portion of the tab containing the antibody-bound fraction is monitored by front-surface fluorescence to determine the rate of enzyme activity. This rate is then converted to concentration units by reading from an appropriate calibration curve. Sequential assays are useful for the direct determination of low-concentration analytes such as digoxin in serum (Giegel 1985).

In the "competitive" RPIA, the sample and enzyme label are premixed prior to addition to the antibody tab. The remaining procedure is similar to that in the sequential assay. Competitive assays are useful for high-concentration analytes such as most therapeutic drugs. In addition, the competitive mode is also useful for an assay of thyroxin, in which it is essential to displace thyroxin from its binding proteins prior to measurement.

In the "immunometric" or "sandwich" RPIA, the antibody is labeled with an enzyme. These assays are performed in a manner similar to the sequential assays. Sandwich assays are useful for large, polyvalent antigens such as ferritin, TSH, or hepatitis virus (Giegel 1985; Evans et al. 1991). This technology has potential applications in the detection of cancer antigens as well as in serological and micro-biological testing.

RPIA has been commercialized in the Stratus® Immunoassay System. It is a versatile immunoassay technology that has both the sensitivity and flexibility of heterogeneous immunoassay techniques and the ease of use of a homogeneous technique. It can be combined with many different detection systems and assay formats to produce immunological methodologies for the quantitation of a wide range of analytes in a variety of assay specimens.

Combined Enzyme Donor Immunoassay (CEDIA®)

More recently, a novel strategy has been adopted to develop a new homoge-neous EIA system called "combined enzyme donor IA" (CEDIA®) (Khanna et al. 1989; Khanna and Worthy 1993). For this system, recombinant DNA technology was exploited to produce new strains of *E. coli* that can synthesize large inactive fragments of β-D-galactosidase (enzyme acceptors, EA) and small inactive frag-ments of the same enzyme (enzyme donors, ED). These fragments spontaneously recombine to form a fully active enzyme.

The basic assay methodology employs these EA and ED enzyme fragments. The analyte ligand to be measured is covalently attached to a unique amino acid site on ED so that the complementation activity of the resulting EA-ED-ligand conjugate remains essentially unchanged (Khanna and Worthy 1993). In the pres-ence of specific antiligand antibody, the complementation activity is greatly re-duced. When sample is added, the free analyte competes with limiting antibody binding sites, allowing the ED-ligand conjugate to react with EA to form active enzyme. Thus, the presence of more analyte in the sample results in higher levels of free ED-ligand conjugate, and consequently more enzyme activity is observed. This antibody modulation of enzyme complementation forms the basis of CEDIA® technology (Figure 8.5).

Most of the CEDIA® assays employ two reagents. The first reagent consists of anti-hapten antibody and EA; the second reagent contains ED-hapten conjugate and sub-strate. Addition of the patient's sample to the first reagent permits the binding of the analyte in the sample with antiligand antibody; EA does not participate in any reaction during this step. On addition of the second reagent, the remaining antiligand antibody binds to the ED-hapten conjugate, reducing the amount of complemented enzyme formed. The amount of enzyme is measured by the formation of product by kinetic rate analysis. The concentration of analyte in the sample is determined by comparison of observed enzyme rates from a calibration curve. CEDIA® products contain two or three calibrators, depending on the instrument employed. The enzyme rates observed with these calibrators can be used to construct a linear standard curve by least-squares regression analysis or simple arithmetic calculation commonly used for enzyme de-termination on clinical chemistry analyzers (Khanna and Worthy 1993).

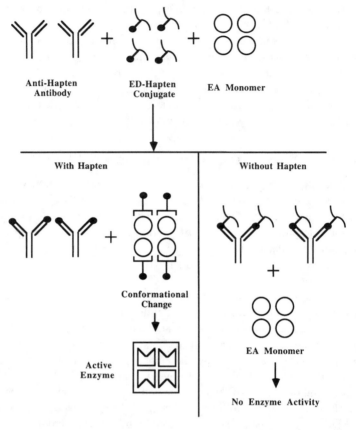

Figure 8.5. Schematic representation of CEDIA®. Antibody modulation of complementation is the basis of the CEDIA® immunoassay technology.
Adapted from Khanna and Worthy (1993)

CEDIA® sensitivity can be increased by employing a three-reagent assay format. The first reagent contains only antibody; the second reagent, ED-hapten conjugate along with substrate; and the third reagent, EA. In this assay, the immunoassay is performed in a sequential manner so that there is no competition between antibody and EA for binding to ED-hapten conjugate. The combination of all these approaches has been utilized to develop homogeneous CEDIA® assays for triiodothyronine, where high affinity antibody availability is limited because of the presence of endogenous thyroid hormones.

The CEDIA® assays are totally homogeneous. They require no sample pretreatment, and the assay protocols are simple to use on chemistry analyzers present in most clinical laboratories. Potential applications of CEDIA® technology make it a highly versatile, convenient, and state-of-the-art immunoassay system. Furthermore, because of its relative insensitivity to sample interference, it can tolerate larger sample volumes that ultimately result in increased assay sensitivity. The demonstrated sensitivity achievable with CEDIA® suggests further applications in clinical diagnostics, including the analysis of vitamins, hormones, drugs, and cancer markers.

The technology, however, does have some limitations. CEDIA® assays are sensitive to protease, which can digest and gradually inactivate the ED label (Khanna and Worthy 1993; Deshpande and Sharma 1993). Use of protease inhibitors or a mixture of random peptides can minimize this effect. The liquid reagent stability of enzyme fragments is also low over a long period of time. Enzyme acceptors have been stabilized by the addition of antioxidants and metal scavengers. Stabilization with high concentration of detergents such as SDS, however, interferes with enzyme complementation and antibody binding with both hapten and enzyme donor conjugate.

Enzyme Membrane Immunoassay (EMIA)

After Bangham and Horne's (1964) discovery of artificial membranes, the first enzyme membrane immunoassay (EMIA) techniques were described by Kinsky et al. (1969). The artificial membranes consist of several layers of phospholipid-based lipid bilayers within which aqueous spaces are entrapped. These model membranes are called "multilamellar liposomes" or simply "liposomes" (Vistnes 1988). Only amphiphatic antigens with well-spaced hydrophilic and hydrophobic regions are suitable for EMIAs. The hydrophobic tail is required to anchor the antigen on lipid bilayers, while the hydrophilic portion serves as the antibody-binding site. Nonamphiphatic haptens need to be covalently bound to an appropriate carrier.

Liposome-entrapped EMIAs can be broadly divided into two categories depending on whether or not complement proteins are required for the lysis of liposomal membranes (Deshpande and Sharma 1993). For homogeneous type complement-dependent systems (Figure 8.6), liposomal membranes are prepared in the presence of a membrane-impermeable, water-soluble marker. The marker should ideally be a stable, preferably charged compound and exhibit low permeability through the membrane. Glucose, dichromatic ions, fluorescent dyes, ATP, spin labels, radioisotopes, and enzymes have been used as markers in EMIAs. The detection system accordingly varies with the type of marker used. After washing of the nonentrapped marker, the liposomal membranes are sensitized with the amphiphatic antigens. In the presence of specific antibody and complement (provided in the form of human or animal sera), the resulting membrane lysis releases the entrapped marker, which is then measured. The marker signal is inversely proportional to the concentration of free antigen or analyte in the sample.

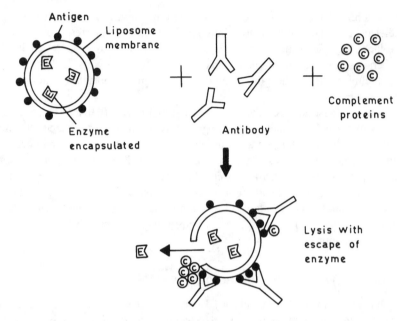

Figure 8.6. Homogeneous enzyme membrane immunoassay (EMIA). The enzyme alkaline phosphatase is encapsulated within the liposomes, and antigen is incorporated into the lipid bilayer of the liposomes. The specific antibody in the presence of complement will cause lysis of liposomes with release of enzyme. The enzyme activity is inversely proportional to free antigen or analyte in the sample
Adapted from Nakamura et al. (1986)

Commercially, EMIA systems based on the use of immunoreactive liposomes have been developed and marketed by Collaborative Research Inc. (Lexington, MA) for thyroxin and theophylline.

In the complement-independent system, the antigen is labeled with cytolysin, capable of lysing the liposomal membranes. The lytic activity of this conjugate is lost on binding with the antibody, presumably because of steric hindrance, which prevents the cytolysin from contacting the membrane. In addition, the antigen-cytolysin conjugate also competes with the free antigen from the test sample for antibody binding sites. Thus, in the absence or low levels of free analyte in the sample, more antibody will be available to bind the antigen-cytolysin conjugate with a concomitant loss of lytic activity. In contrast, higher levels of analyte in the test samples will leave more antigen-cytolysin conjugate free to lyse the membrane, thereby releasing greater amounts of the entrapped marker. Thus, for an enzyme marker, its activity is directly proportional to the levels of analyte in the test sample. Using such an approach, Litchfield et al. (1984) reported a sensitive and rapid assay for digoxin based on liposome-entrapped alkaline phosphatase marker enzyme and ouabain-melittin conjugate. Ouabain is a digoxin analog, and melittin is a 26-amino acid lytic peptide from bee venom.

Two other variations of EMIA technique include the development of the spin membrane immunoassay (SMIA), which utilizes nitroxide radical as a spin label marker. The detection system in this case is an electron spin resonance (ESR) spectrometer (Vistnes 1988). Yasuda et al. (1988) have described a second EMIA technique that utilizes fluorescent dye-trapped liposomes. This technique, called the liposome immune lysis assay (LILA), suffers from similar drawbacks associated with the EMIA techniques.

EMIAs offer sensitive, relatively rapid, and specific as well as general methods. However, because of several intrinsic problems, large scale applications for serological measurements have not yet been successful. A major problem is related to the unavailability of well-defined homogeneous vesicles with long term stability after washing of the external marker (Vistnes 1988; Deshpande and Sharma 1993). Nonspecific leakage of marker from the liposomes is yet another serious drawback of EMIA systems. Similarly, in complement-dependent EMIAs, the presence of lipoproteins and other immunoglobulins in the serum source also influences the results. Thus, in spite of their widespread use in research laboratories, commercial applications of EMIA technology must await further advances in liposomal technology.

Other competitive procedures in which coenzymes are used to label the ligand may also be classified as homogeneous enzyme immunoassays. If the ligand-coenzyme conjugate is bound to an antibody, its cycling in a suitable enzymic cycle is proportionately reduced. This inhibition is reversed by unconjugated ligands in competitive binding reactions. So far, however, this assay principle has found very few applications in clinical diagnostic reagents.

Homogeneous immunoassays are advantageous because they are easy to automate, generally rapid (i.e., incubation times are short), and simple to perform. They also offer the advantages of extended shelf life and ease of adaptability to a variety of available instruments in clinical laboratories. However, there are three factors that cause the major limitations in assay sensitivity observed with most homogeneous immunoassays (Nakamura et al. 1986; Deshpande and Sharma 1993). The first relates to the sensitivity of enzyme detection, which in turn is limited by the turnover rate of the enzyme, the choice of available substrates, and the extinction coefficient of the resulting product of the enzyme action. The second factor relates to the nonselectivity of enzyme derivatization, with multiple ligands resulting in poor assay sensitivity. The third factor relates to the significant matrix effect shown by most of the enzymes that are currently used in homogeneous EIAs. As there is no separation step involved in homogeneous EIAs, the background signal coupled with sample matrix effects also limits the inherent sensitivity. Nonetheless, because of their simplicity, homogeneous EIAs are the most preferred systems in clinical diagnostics.

A summary of commercial homogeneous immunoassays is presented in Table 8.3.

TABLE 8.3. Summary of Commercial Homogeneous Enzyme Immunoassay Systems[a]

Parameter	EMIT	SLFIA (ARIS)	PGLIA	Inhibitor-based	CEDIA
Assay principle	Modulation of hapten-labeled enzyme activity by antibody binding	Antibody induced steric hindrance for hapten-labeled substrate binding to an enzyme	Antibody effect on hapten-labeled prosthetic group reactive to apoenzyme	Antibody induced steric hindrance for hapten-labeled inhibitor binding to an enzyme	Antibody induced modulation of hapten-labeled recombinant enzyme fragment for enzyme complementation
Enzymes commonly used	G6PDH, MDH, Lysozyme, β-galactosidase (special substrate)	β-Galactosidase	Glucose oxidase	Cholinesterase	β-Galactosidase
pH Optimum	G6PDH 7–8 MDH 8.5–9.5	6–8	6–8	7–9	6–8
Choice of substrates	Limited to NAD/malate; G6PDH NAD/G6P	Umbelliferone galactoside	HRP enzyme substrate coupling system	Limited substrates	A variety including umbelliferone galactoside, ONPG, CPRG
Detection system	MDH, G6PDH chromogenic or fluorogenic; lysozyme light scattering	Fluorogenic	Chromogenic	Chromogenic or fluorogenic	Chromogenic or fluorogenic
Signal enhancement	NADH/Meldola blue	None	Better HRP substrates	Unknown	Better enzyme substrates

TABLE 8.3. (continued)

Parameter	EMIT	SLFIA (ARIS)	PGLIA	Inhibitor-based	CEDIA
Serum matrix effects	Sensitive about 2%	Sensitive above 1%	Sensitive above 5%	Unknown	Insensitive up to 12%
Assay calibration	6-point nonlinear curve	6-point nonlinear curve	6-point nonlinear curve	6-point nonlinear curve	2-point linear curve
Detection limit (sample concentration)	10^{-9} M (haptens), 10^{-5} M (proteins)	10^{-6} M (haptens), > 10^{-5} M (proteins)	10^{-7} to 10^{-9} M (haptens)	10^{-7} to 10^{-8} M (haptens)	10^{-11} M (haptens)
Applicability	Haptens (MDH, G6PDH), proteins (β-galactosidase with macromolecular enzyme substrate)	Haptens, proteins	Haptens	Haptens	Haptens, proteins
Commercial viability	Very successful	Limited to specialized fluorescent detection based systems	Limited application	Very limited application	Wide applicability

[a]From Khanna (1991).

COMMERCIAL TECHNOLOGIES

At present, several different immunoassay product technologies are available commercially. The agglutination type immunoassays, based on antigen-coated latex beads or erythrocytes, are still used in several laboratories and blood banks to test for the presence of anibodies. Precipitin assays such as radial immunodiffusion and immunoelectrophoresis, although still used for certain applications, are confined to specialty centers and tend to be low volume assays. The nephelometric and turbidimetric assays are primarily used for the detection of macromolecular proteins at high concentrations on clinical chemistry analyzers.

RIAs still continue to be used in clinical diagnostics because of their obvious advantages of precision and sensitivity at low analyte concentrations and a wider dynamic range (Wild 1994). Moreover, because of the low cost of the reagents, their proven reliability and the ability to run kits from more than one supplier using the same equipment, many laboratories still continue to use RIAs and IRMAs.

Since their introduction in the early 1970s, EIAs first proved to be most successful in the immunometric (sandwich) formats in the fields of infectious diseases and oncology. However, early EIAs were more susceptible to interferences than RIAs due to effects on the enzyme-mediated signal generation steps. The introduction of homogeneous EIAs proved to be one of the greatest breakthroughs in immunoassay technology.

At around the same time, increased emphasis was placed on automation of immunoassays. Abbott's TDx® and Baxter's Stratus® proved to be the most popular automated systems. Further advances led to the introduction of dual technology batch analyzers (e.g., Abbott's IMx® system); automated, multianalyte batch analyzers; random access analyzers with bulk reagent packs; and unitized random access systems. The market of physicians' office automated systems, however, did not prove to be a big success, primarily because of increased regulatory requirements and higher costs. However, currently several nonautomated, single-use tests are available for physicians' offices that do not require any processing equipment. For home use, several pregnancy (serum or urinary hCG) tests are now available that have simple protocols and a colorimetric end point. In recent years, similar tests are also made available for cholesterol testing.

In this section, some of the important commercial technologies based on EIA principles are described briefly.

Syva ETS® Plus System (EMIT®)

The ETS® Plus System and EMIT® assays are manufactured by Syva Company, San Jose, California. The ETS® is a fully automated instrument and reagent system designed specifically for the primary screening of drugs of abuse (d.a.u.)

in urine, the qualitative analysis of various analytes in serum, and that of ethyl alcohol in urine, serum, and plasma (Syva Co. 1994). It provides full paneling, random access (i.e., samples may be loaded at any time with the capability to perform several tests on a given sample), and batch capability. The ETS® system is designed to run 13 EMIT® d.a.u.™ assays, 3 EMIT® tox serum assays, and the EMIT® ETS® Plus Ethyl Alcohol Assay. The last test uses alcohol dehydrogenase as the enzyme; the rest are classical EMIT® assays using G6PDH label.

More than 2000 ETS® and ETS® Plus Systems are currently in use worldwide in a variety of drug testing environments, including clinical laboratories, hospital emergency rooms, drug rehabilitation clinics, industrial health offices, and criminal justice facilities (Centofanti 1994).

Typically, patient samples are placed in the analyzer in sample cups, a work list is programmed, and testing is completed by the analyzer at a throughput of 50–80 tests per hour. Calibration is automated and is based on a qualitative cutoff level. The analyzer combines specified volumes of sample and reagents in a thermally controlled, disposable reaction cuvette. After a 15 second incubation, it takes a photometric reading of the reaction and calculates the absorbance rate over a 30-second interval. The results are interpreted as positive or negative by comparing them to the stored cutoff calibration value.

The reagents, calibrators, and controls for the EMIT® d.a.u. and tox assays are provided lyophilized and require reconstitution before use. Once reconstituted, the reagents are stable for up to 12 weeks.

Syva MicroTrak® EIA

The MicroTrak® EIA product line is targeted for the antigen and/or antibody detection of *Chlamydia,* HIV, hepatitis A and B viruses, *Toxoplasma,* rubella, and cytomegalovirus in microtiter plate formats. Syva offers two instrument systems for this purpose: a semiautomated MicroTrak® EIA system consisting of a modular 96-well washer and a microplate reader, and a fully automated, random-batch analyzer system, MicroTrak® XL, that can run any combination of up to 8 assays on any of up to 95 specimens, with a maximum of 192 tests per batch (Houts 1994). The reagents can also be used on a variety of instruments designed for testing microtiter plates.

MicroTrak® EIAs are heterogeneous assays that use peroxidase as the enzyme label and TMB and hydrogen peroxide as substrate. Both competitive and sandwich assay formats are used with either antibody or antigen immobilized on the surface of the microwell. The reagents are supplied as liquids, either in ready-to-use form or as concentrates, and have long shelf lives. Typically, two control samples are included in each run. The cutoff between reactive and nonreactive samples is usually based upon absorbance values for these controls. The tests provide qualitative results, the antibody tests generally being specific for either IgG or IgM.

Syva Vista® Immunoassay System

The Vista® Immunoassay System is a benchtop, fully automated, multitest analyzer that offers random test selection. It is designed to measure a wide range of special chemical analytes including those for thyroid function, fertility, anemia, cancer markers, hepatitis, and HIV tests in medium to large hospitals and laboratories (Li 1994). The instrument is manufactured by Hitachi and the reagents provided by Syva. The Vista® system is a heterogeneous enzyme fluorescent immunoassay system utilizing chromium dioxide magnetic particles as solid phase and alkaline phosphatase as the enzyme label. 4-Methylumbelliferone phosphate in diethanolamine buffer is used as the fluorescent substrate.

The Vista® Immunoassay System is designed to eliminate the need for multiple batch operation by offering random access for up to 19 different onboard assays. The user enters a sample number and can order single tests, panels, or profiles for each patient sample. Up to 50 samples at one time can be loaded onto the sample carousel and up to 19 different reagents can be onboard the instrument for any one run. Cartridges can be removed and replaced without affecting reagent inventory management. Thus, any of up to 15 different methods can be selected on any sample without sample splitting (Li 1992, 1994).

The Vista® Immunoassay System utilizes a heterogeneous competitive antibody-binding assay format for small molecule assays in which the magnetic particles are coated with streptavidin and specific antibodies are biotinylated. For large molecule assays, a heterogeneous antibody sandwich format is used. A typical assay protocol involves totally automated sample and reagent addition, mixing, incubation, washing, and fluorescence measurement. For quantitative assays, five calibrators need to be run once, in duplicate, with each new reagent lot. The Vista® Immunoassay System features extended curve storage capability to ensure accurate results without frequent recalibration.

Abbott IMx®

The Abbott IMx® is a fully automated, benchtop, batch analyzer that incorporates two immunoassay technologies: a novel technology known as microparticle capture enzyme immunoassay (MEIA) for high molecular weight, low concentration analytes; and a fluorescence polarization immunoassay (FPIA) for low molecular weight analytes (Chou 1992; Groff 1994). Currently, it is the most widely used immunoassay system with over 22,000 analyzers in use worldwide in early 1993. Reagents are available for a wide range of assays, including thyroid, anemia, cardiac, reproductive endocrinology, oncology, HIV, infectious disease, and allergy analytes.

The FPIA assays are homogeneous and competitive. The sample, fluorescein-labeled analyte, and antibody are incubated in the reaction cell for 3 min prior to signal detection. No separation is required.

The MEIA technology is based on two key features: increased kinetics through the use of very small particles (0.47 μm in diameter) as the solid phase, and efficient separation of bound from unbound material by capture of the microparticles in a glass-fiber matrix.

The MEIA assays are heterogeneous, requiring a separation of the bound and unbound fractions during the procedure. The bound fraction is immobilized on uniform latex microparticles, which bind to a glass-fiber matrix. Unbound material is washed through the glass-fiber matrix into an absorbent, opaque blotter below, using prewarmed buffer.

The MEIA can be used in either a immunometric sandwich or a competitive ligand format. In the sandwich format, the latex microparticles are covalently coated with capture antibody. When incubated with the unknown sample, they form the capture complex. An enzyme-labeled antibody conjugate is then incubated with the particles to form the second half of the sandwich complex (Brown 1987). In the competitive format, antibody-coated microparticles are incubated with sample containing the analyte and an enzyme-labeled analyte conjugate. Removal of unbound conjugate after equilibrium is accomplished with the glass-fiber surface. Substrate is then added and the resulting fluorescence signal is measured. The enzyme/substrate system used in both MEIA assay formats is alkaline phosphatase/4-methylumbelliferyl phosphate (4-MUP). The analyzer measures the rate of fluorescence development.

The instrument provides three calibration modes: CAL, a full calibration of six calibrators run in duplicate; mode 1, a one-point adjustment of the stored calibration that is required for some MEIA assays; and mode 2, where the stored calibration curve is used.

Abbott IMx® provides very high precision for all assays, thus eliminating the need to run these assays in duplicate or triplicate. It is designed to be used by medium-size hospital laboratories. The throughput is much too low for high-volume users. The IMx® system is also not a random access instrument, and it becomes very ineffective for profile analyses. To run multianalyte tests, the reagent pack has to be changed in between each analyte test run. Moreover, there is no positive specimen identification for this instrument. The IMx® Select™, which is a recent instrument enhancement, allows up to three analytes to be processed on a single carousel run.

Despite its disadvantages, the IMx® system continues to provide a very attractive alternative for immunoassays. It is easy to operate and maintain, and provides stat analyses around the clock. It is a true walkaway system that reduces hands-on time and improves productivity in clinical laboratory settings.

Baxter Stratus® Immunoassay Systems

First introduced in 1983, the Stratus® methods use RPIA (described earlier in this chapter), which is a combination of paper chromatography and EIA techniques. The introductory test menu of the Stratus® was not extensive, consisting only of assays for eight therapeutic drugs and thyroxine. Since then, the Stratus® II system was introduced in 1989, followed by its more advanced version, Stratus® IIntellect® in 1992.

Both qualitative and quantitative assays are available on Stratus® systems. Each of these three fully automated analyzers has a capacity of 30 samples per run that can be processed in less than 40 min. Reagents are available for a wide range of therapeutic drugs as well as for thyroid, anemia, cardiac, reproductive endocrinology, and allergy analytes. Certain tumor marker assays can also be run on these analyzers through an alliance with Hybritech, Inc., San Diego, California (McLellan and Plaut 1994; Kahn and Bermes 1992).

Each Stratus® assay uses filter paper containing embedded glass fibers as the solid phase. Antibodies to the analyte are immobilized on the glass fibers and the filter paper squares are mounted in plastic casings, referred to as tabs. The tabs for the folate and vitamin B_{12} assays are exceptions. The vitamin B_{12} tabs have immobilized goat antimouse IgG, while the folate tabs have folate-binding protein immobilized on the glass fibers. When the sample antigen is applied to the tab by the analyzer's sample dispenser, the antigen binds to the immobilized antibody. The enzyme-labeled reagent conjugate can be added simultaneously with the sample, as in competitive assays, or in separate steps (sequential and sandwich assays).

If the conjugate is added separately, it is dispensed two min after the sample has been added. Three min later the next reagent, the substrate wash solution, is placed on the tab. The tabs are incubated during the entire process at a constant temperature of 38°C. The unbound enzyme-labeled antigen is radially eluted (or partitioned) to the periphery of the reaction tab, forming a second "ring" of reactivity around the reaction zone in the center of the tab. The final stage of the assay process is the measurement of the rate at which alkaline phosphatase converts a fluorogenic substrate to product. During a 20-second period, as many as 100 readings are taken at the fixed read station (McLellan and Plaut 1994).

The analyzer is completely automated. For most assays, the first result is reported 8 min after the sample is added to the tab; subsequent results are reported at 1-min intervals. A run of 30 samples is thus completed in less than 40 min. The reagent sets for the quantitative assays have six individual levels of serum calibrators that range from one that is free of the analyte of interest to one elevated well above the upper limit of the analyte's reference interval or therapeutic range. The calibration curves are stable for at least 14 days.

The Stratus® Immunoassay System is likely to continue to be one of the most widely used automated immunoassay analyzers. The Stratus® II and its advanced version offer the clinical laboratory a unique combination of broad test menu and numerous optional advantages that are difficult to obtain with other types of immunoassay equipment.

Boehringer Mannheim ES-300 Immunoassay System

Since its entry into the immunodiagnostics market in the early 1970s, Boehringer Mannheim introduced in 1972 its first immunoassay analyzer, ES-11, a photometer that had the capability to do curve calculations. In the next few years, the ES-11 was upgraded to ES-22, and then to a semiautomated version, the ES-33. In 1986, the company introduced the ES-600, a totally automated, multichannel immunoassay instrument (Sagona et al. 1992).

In 1990, the ES-300 immunoassay system was launched worldwide. It is a fully automated immunoassay system that is capable of performing up to 12 different tests within one run, including multiple tests on a single sample. The ES-300 system was developed by Boehringer Mannheim in Germany and the analyzer manufactured by Hamilton in Switzerland. Currently, there are about 2500 ES-300 analyzers in use (Bush 1994).

The ES-300 system runs the Enzymun-Test® range of assays. These are heterogeneous ELISAs with antibody-coated tubes as the solid phase and a colorimetric signal generation system comprised of horseradish peroxidase and ABTS. Some assays use streptavidin-coated tubes and biotinylated capture antibodies.

The assay formats are either competitive, one- and two-step sandwich, or modified competitive. Each assay has five to seven standards for a full curve calibration that is stable for up to two weeks. Generally, a one-point recalibration is suggested with each run using one of the designated standards.

The assay reagents for the ES-300 system cover a wide range of applications, including thyroid, tumor markers, reproductive endocrinology, and infectious disease analytes. The only limitation of this system is that it is not a full random access system. Thus, new patient samples cannot be added to the current run once it has been started.

DPC Milenia®

The Milenia® product range, manufactured in the United Kingdom by Euro/ DPC Ltd., utilizes kinetic EIAs based on ligand-coated microtiter wells, a ligand-labeled antibody, anti-ligand separation, and a horseradish peroxidase label.

The Milenia® system includes assays that utilize both competitive and immunometric sandwich immunoassay methods. Competitive assays are used for small analytes such as steroids and drugs of abuse, whereas either a sequential or a nonsequential sandwich method is used for larger molecules such as tumor markers or allergen-specific IgEs (Milenia® AlaStat) (Hand 1994a).

The Milenia® system features microplates that are universal within one analyte panel. For example, with the tumor marker panel, both prostate-specific antigen (PSA) and prostatic acid phosphatase (PAP) assays can be performed at the same time on the same plate.

The basis of the universal microplate is the ligand/antiligand separation system. Microplates are coated with a ligand and antibodies are conjugated with the same ligand. Immune complexes are unreactive with the microplate until the addition of a multivalent antiligand after the primary liquid-phase immunological reaction. This antiligand is reactive with the ligand-labeled antibodies and ligand-coated plates, and binds the antibodies to the well of the plate to allow separation of bound and free fractions by washing the microplate (Hand 1994a).

The Milenia® system measures the rate of color development rather than a single endpoint reading of total absorbance. This allows for expanded dynamic ranges without the loss of low-end sensitivity for Milenia® assays. The system uses serum- or urine-based calibrators supplied with the assay kits. Qualitative assays use positive and negative reference preparations.

The DPC Milenia® system has been applied to a variety of molecules of both high and low molecular weights, including steroid hormones, thyroid hormones, drugs of abuse, tumor markers, peptides, and compounds for allergy testing.

DPC Immulite®

The DPC Immulite®, manufactured in the United States by DPC, is a benchtop chemiluminescent system that allows total automation of immunoassays with random access capabilities (Hand 1994b). It uses a proprietary assay tube; thorough and efficient washing of an integral antibody-coated bead is provided by spinning the tube on its vertical axis. The Immulite® assays use bound alkaline phosphatase label, which is quantified by activation of the sensitive chemiluminescent reaction of adamantyl dioxetane phosphate to the unstable adamantyl dioxetane anion.

Competitive assays for haptens such as T_4 and cortisol, and sandwich assays for large molecules such as TSH and hCG have been developed for the Immulite® system.

Kodak Amerlite® Modular System

The Amerlite® product range is manufactured by Kodak Clinical Diagnostics Ltd. in the United Kingdom. The Amerlite® Modular System is a complete immunoassay system based on chemiluminescent principles. All of the individual steps are (or can be) automated; however, the operator must occasionally and briefly intervene (Wild 1994). The products use competitive and immunometric formats for quantitative measurement of both antigens and antibodies. The infectious disease product range is primarily qualitative and uses antiglobulin and antibody class capture assay formats.

Amerlite® assays use horseradish peroxidase as a label. The signal generating system contains a peracid salt, luminol and a *para*-substituted phenol as an enhancer. The peracid salt generates hydrogen peroxide in solution, which oxidizes the luminol. The reaction is catalyzed by the enzyme label. In a series of reactions, the oxidized form of luminol breaks down with the generation of light. The enhancer increases the level of light produced by at least 1000 times and prolongs its emission. The light is read in the analyzer between 2 and 20 min after the addition of the signal reagent (Hand 1994b; Faix 1992).

The Amerlite® product range includes assays for thyroid, reproductive endocrinology, pregnancy, oncology, and adrenal and infectious disease analytes.

Hybritech ICON® and Tandem® ICON® QSR®

The Hybritech ICON® (immunoenzymatic device, Hybritech) and Tandem® ICON® QSR® (quantitative stat result) devices provide a quick and reliable method for testing individual samples for a variety of conditions, including pregnancy, and acute myocardial infarction, as well as for a number of infectious disease analytes (Payne et al. 1994). ICON has also been adapted by IDEXX Corp., Maine, for a number of veterinary applications.

The ICON® is a single-test device, providing a qualitative result without the need for instrumentation. The Tandem® ICON® QSR® is an internally referenced device that, when used with a portable reflectometer (ICON® reader reflectometer, Hybritech), provides a quantitative result.

Several assay formats are available on the ICON® device. The most common is the dual monoclonal antibody immunometric sandwich assay. The capture antibody is immobilized onto polystyrene latex particles that are dispersed onto a nylon membrane. Analyte present in the sample is bound to the solid-phase antibodies as the sample passes through the membrane and a porous polyethylene disc into an absorbent pad of cellulose acetate. Enzyme-labeled secondary antibody is then added, binds to any analyte bound to the solid phase, and is allowed to drain through the membrane. Wash solution is then added to remove excess unbound conjugate. Finally, the substrate/dye solution is added to the cylinder membrane to generate color in the presence of bound conjugate (Payne et al. 1994).

Alkaline phosphatase is used as the enzyme label. The color, if present, can be directly visualized by the user, providing a qualitative result, or compared to the color density of two calibrator zones by a portable reflectometer, thereby providing a quantitative result.

Qualitative ICON® assays have been developed for several analytes. In each of these assays, a negative control zone using latex particles coated with an irrelevant antibody has been incorporated into the assay system (Achord et al. 1991). This negative control zone serves as an indicator of nonspecific binding (NSB) or for incorrect procedure performance such as a failure to remove unbound conjugate because of incorrect washing. If the antibody is chosen correctly, any NSB present should be shown in parallel on both the negative control and test zones.

A positive control or reference/calibrator zone can be prepared by coating latex with an anti-alkaline phosphatase antibody. The amount of this antibody-coated latex can be adjusted so that the color in the positive reference/calibrator zone can be used as a calibrator or as an indicator of low performance in the assay. Usually, monoclonal antibodies are chosen over immobilized antigen, as they mimic the reagent stability characteristics of antibodies on the test zone.

ICON® assays are designed to be simple and quick to perform. However, they are not necessarily as precise or sensitive as the most robust immunoassays that require bulky and expensive equipment to carry out.

Kodak SureCell®

Initially developed in conjunction with Cetus Corporation (Emeryville, California) and marketed by Eastman Kodak Company (Rochester, New York), the Kodak SureCell® test kits provide quick and easy protocols for testing for pregnancy (serum or urinary hCG), *Streptococcus* A (strep A), *Chlamydia,* and herpes simplex virus (type 1 and 2).

Two slightly different technologies based on membrane filtration are employed in the SureCell® test kits (Wild 1994). The tests for strep A and hCG are solid-phase immunometric (sandwich) assays employing capture antibody immobilized onto a solid phase. The latter consists of polymer particles dried onto the surface of a nylon membrane. In the *Chlamydia* kit, the antigen is adsorbed directly onto a modified nylon membrane. In each of these assays, a nonspecific conjugate or IgG of the same subclass is used as a negative control, while the positive control contains preadsorbed analyte.

The SureCell® assays are qualitative, and provide a positive or negative result.

Microgenics CEDIA®

The CEDIA® immunoassay line consists of reagent kits for performing homogeneous EIAs on photometric clinical chemistry analyzers. The kits utilize Microgenics' patented, genetically engineered β-galactosidase fragment technology described earlier. CEDIA® tests primarily measure low molecular mass analytes such as hormones, therapeutic drugs, and vitamins. The technology, however, is amenable for the detection of low concentration proteins such as ferritin (Coty 1994).

Each CEDIA® kit contains lyophilized reagents, their respective reconstitution buffers, and calibrators. Assays for small analytes are designed to give a linear calibration curve. Tests are calibrated with two points in duplicate or three single-point calibrators depending on instrument software. Controls and unknown samples are quantified from the calibration curve by linear interpolation or regression analysis. Out-of-range samples can be diluted with low calibrator and retested. CEDIA® tests do not give a high dose "hook" effect.

CEDIA® assays can be used on a wide variety of clinical chemistry analyzers, thus eliminating the costs of acquisition and maintenance of specialized instruments. Most tests are fully automated and require no special sample handling or pretreatment (Coty 1994). CEDIA® assays are also compatible with random access operation of many analyzers, and thus allow integration of special chemistry testing with routine profiles. This eliminates labor costs due to sample "splitting," and errors arising during sample identification and the combination of data from separate workstations.

CEDIA® assays are primarily designed for use on automated clinical chemistry analyzers. They are not suited for manual application to a spectrophotometer.

PB Diagnostics OPUS®, OPUS® PLUS, and OPUS® MAGNUM

The OPUS® instruments are continuous-access, automated, benchtop immunoassay analyzers. The system performs sample handling, test incubation at constant temperature, signal detection, and conversion of signal into a test result. The reagents are unitized, with each test module containing all the reagents required for a single test. The OPUS® PLUS is an advanced version that can also perform dilutions and sample pretreatments. The OPUS® MAGNUM is a high-volume analyzer that features both primary tube sampling with positive sample ID and 360 assay onboard stage. All instruments use the same reagents (Crowley and Bauduin 1994).

OPUS® analyzer operation is totally dry; there are no liquid reagents or tubings to change. There is no sample or reagent carryover, because disposable pipette tips are used for each pipetting step.

OPUS® reagents are packaged in two types of test modules: one for multilayer dry-film assays, which are competitive in nature, and the other for fluorogenic ELISA assays, which are immunometric. The ELISA assay module contains wells for two liquid reagents, the conjugate and the substrate. The capture antibody is immobilized on a fibrous matrix.

In OPUS® assays, separation of the unbound analyte is by washing. Unbound fractions are washed along the fibrous matrix by the substrate away from the signal detection area. The substrate therefore doubles as a wash reagent. In the ELISA assays, alkaline phosphatase is used as the label and 4-MUP as the substrate.

OPUS® assays are calibrated using assay-specific sets of calibrators. Most quantitative assays require six calibrators, run in duplicate (Crowley and Bauduin 1994; Lehrer et al. 1992). Depending on the analyte, the curves are retained in the instrument memory until replaced by a new calibration, and are stable for 2–8 weeks.

Serono SR1® Immunoassay System

The Serono SR1® is a random access, cartridge-based, benchtop, fully automated immunoassay system designed to provide technologists with walkaway immunoassay technology with minimal operator intervention. The basic reagent system utilizes a bar-coded, self-contained cartridge that includes all of the antibody reagents and calibrators necessary to carry out the assay following the addition of patient sample to the cartridge.

Serozyme assays use both immunometric assays (both coated tubes and magnetizable particle-based) for large protein molecules and competitive EIAs for small analytes such as thyroid hormones and ovarian steroids. The system utilizes a magnetic solid-phase separation technology and a common reaction signal for all assays based on the color development of phenolphthalein. The SR1® instrument itself is a dual-channel photometer that can accommodate up to 80 tests at a time on a continuous basis.

Serozyme assays for haptens are calibrated using pure antigen, and in the case of proteins, either against the international reference preparation or reference sera for antibody tests. In routine use, the assays are run with a calibration curve from which the levels of the sample and controls are measured (Palmer and Bacarese-Hamilton 1994).

The Serozyme products are manual assays. Semiautomation, however, can be achieved using a pipetting system (Serobot®) to pipette and dispense reagents.

TOSOH AIA®-600 and AIA®-1200

The AIA®-600, AIA®-1200, and AIA®-1200DX are fully automated immunoassay analyzers that use a common reagent preparation. These systems are manufactured and distributed by TOSOH Corporation, Tokyo, Japan. The assay specific reagents are prepackaged in foil-sealed, unitized test cups that are provided in trays (Loebel 1994). In addition to the analyte-specific test cups, there are three common reagents: diluent, wash, and substrate. The first two are made up from concentrates. The substrate, which requires reconstitution, is stable for up to one week at 2–8°C.

The AIA®-600 is a benchtop, random access analyzer with a smaller capacity and throughput, whereas the AIA®-1200 and AIA®-1200DX floor-standing models can be used in batch or random access modes. The latter two have high sample capacity and throughput.

The assays are either immunometric (large molecular weight analytes such as thyrotropin, carcinoembryonic antigen, α-fetoprotein, etc.) or competitive binding assays for small analytes such as thyroxine and digoxin. Antibody-coated magnetic beads are used as the solid phase. The assays utilize alkaline phosphatase as the enzyme label and 4-MUP as the fluorogenic substrate. The system is calibrated by using two to six calibrators periodically.

The major advantages of AIA® analyzers are their ability to improve the turnaround time and the technical time required to perform the testing.

Eclipse ICA®

The Eclipse ICA® analyzer is a benchtop, random access, automated centrifugal immunoassay and chemistry system. It was designed for the small hospital, the large group practice, and satellite laboratories. The system provides a cost-effective and time-effective alternative to sending assay samples to reference laboratories. It can accommodate 16 tests per operation with a run time of approximately 17 min. Currently, assays are available for cholesterol, triglycerides, glucose, blood urea nitrogen (BUN), digoxin, and TSH.

The basis of the Eclipse technology is found within the cassette, a disposable device that includes all of the reagents and compartmentalization necessary to perform an assay (Schueler et al. 1993). Precision pipetting, mixing, incubation, washing, and sample measurements take place within this cassette. For the assays developed, there are no sample pretreatment steps. Once patient samples are obtained and added to the cassettes, and the cassettes are placed in available rotor positions, the Eclipse system will generate immunoassay and clinical chemistry results in 17 min or less. For the tests mentioned above, the system currently provides a level of sensitivity and performance equivalent to those of existing technologies.

The Eclipse ICA® cholesterol test combines coupled enzymatic activities (cholesterol esterase and oxidase) with the peroxidase/4-aminoantipyrine detection system for the measurement of total cholesterol. This is, however, not an immunoassay. Similar enzymatic chemistry is also used for the BUN assay. The TSH assay is a solid-phase, two-site or sandwich enzyme immunoassay, while the digoxin assay is a competitive solid-phase enzyme immunoassay. Both use a fluorometric detection system for signal generation.

The Eclipse ICA® is one of the few systems capable of performing a large battery of tests, both chemistries and immunoassays. System performance is equivalent to that of the majority of the automated analyzers currently available in the diagnostics market. The instrument is compact and easy to use, and provides fast turnaround on results, thus making the system ideal for the alternate-site laboratory or the physician's office. With the automated tracking of quality controls, compliance with governmental regulations is easy.

REFERENCES

ACHORD, D.; PAYNE, G.; SAEWERT, M.;and HARVEY, S. 1991. Immunoconcentration. In *Principles and Practice of Immunoassay,* eds. C. P. Price and D. J. Newman, pp. 584–609, Stockton Press, New York.

BANGHAM, A. D., and HORNE, R. W. 1964. Negative staining of phospholipids and their structural modification by surface-active agents as observed in the electron microscope. *J. Mol. Biol.* 8:660–668.

Bosh, A. M. G.; Van Hell, H.; Brands, J.; and Schuurs, A. H. W. M. 1978. Specificity, sensitivity and reproducibility of enzyme immunoassays. In *Enzyme-Labeled Immunoassay of Hormones and Drugs,* ed. S. B. Pal, pp. 175–187, Walter de Gruyter, Berlin.

Brown, W. E. 1987. Microparticle-capture membranes: Application to TestPack hCG-urine. *Clin. Chem.* 33:1567–1568.

Bush, T. 1994. Boehringer Mannheim ES 300 immunoassay system. In *The Immunoassay Handbook,* ed. D. Wild, pp. 155–160, Stockton Press, New York.

Castro, A., and Monji, N. 1985. Steric hindrance enzyme immunoassay (SHEIA). In *Enzyme Mediated Immunoassay,* ed. T. T. Ngo and H. M. Lenhoff, pp. 291–298, Plenum Press, New York.

Centofanti, J. 1994. Syva ETS® Plus system (EMIT®). In *The Immunoassay Handbook,* ed. D. Wild, pp. 216–219, Stockton Press, New York.

Chou, P. P. 1992. IMx system. In *Immunoassay Automation. A Practical Guide,* ed. D. W. Chan, pp. 203–220, Academic Press, San Diego, CA.

Coty, W. A. 1994. Microgenics CEDIA®. In *The Immunoassay Handbook,* ed. D. Wild, pp. 193–196, Stockton Press, New York.

Crowley, H. J., and Bauduin, M. A. 1994. PB Diagnostics OPUS®, OPUS® PLUS and OPUS® MAGNUM. In *The Immunoassay Handbook,* ed. D. Wild, pp. 197–203, Stockton Press, New York.

Davies, C. 1994. Principles. In *The Immunoassay Handbook,* ed. D. Wild, pp. 3–48, Stockton Press, New York.

Deshpande, S. S. 1994. Immunodiagnostics in agricultural, food and environmental quality control. *Food Technol.* 48(6):136–141.

Deshpande, S. S., and Sharma, B. P. 1993. Diagnostics: Immunoassays, nucleic acid probes, and biosensors. Two decades of development, current status and future projections in clinical, environmental and agricultural applications. In *Diagnostics in the Year 2000,* eds. P. Singh, B. P. Sharma, and P. Tyle, pp. 459–525, Van Nostrand Reinhold, New York.

Ekins, R. P. 1980. More sensitive immunoassays. *Nature* (London) 284:14–15.

Ekins, R. P. 1981a. Toward immunoassays of greater sensitivity, specificity and speed: An overview. In *Monoclonal Antibodies and Developments in Immunoassays,* eds. A. Albertini and R. P. Ekins, pp. 3–4, Elsevier/North-Holland Biomedical Press, Amsterdam.

Ekins, R. P. 1981b. Merits and disadvantages of different labels and methods of immunoassay. In *Immunoassays for the 80s,* eds., A. Voller, A. Bartlett, and D. Bidwell, pp. 5–16, University Park Press, Baltimore, MD.

Ekins, R. P. 1985. Current concepts and future developments. In *Alternative Immunoassays,* ed. W. P. Collins, John Wiley, New York.

Ekins, R. P. 1991. Competitive, noncompetitive, and multianalyte microspot immunoassays. In *Immunochemistry of Solid-Phase Immunoassay,* ed. J. E. Butler, pp. 105–138, CRC Press, Boca Raton, FL.

Engvall, E., and Perlmann, P. 1971. Enzyme-linked immunosorbent assay (ELISA). Quantitative assay for immunoglobulin G. *Immunochemistry* 8:871–874.

EVANS, S.; KIRCHICK, H.; and GOODNOW, T. 1991. Radial partition immunoassay. In *Principles and Practice of Immunoassay,* eds. C. P. Price and D. J. Newman, pp. 610–643, Stockton Press, New York.

FAIX, J. D. 1992. Amerlite immunoassay system. In *Immunoassay Automation. A Practical Guide,* ed. D. W. Chan, pp. 117–128, Academic Press, San Diego, CA.

FREYTAG, J. W. 1985. Affinity column-mediated immunoenzymometric assays. In *Enzyme Mediated Immunoassay,* eds. T. T. Ngo and H. M. Lenhoff, pp. 177–189, Plenum Press, New York.

FREYTAG, J. W.; LAU, H. P.; and WADSLEY, J. J. 1984. Affinity column-mediated immunoenzymometric assay. Influence of affinity column ligand and valency of antibody-enzyme conjugates. *Clin. Chem.* 30:1494–1498.

GIBBONS, I.; SKOLD, C.; ROWLEY, G. L.; and ULLMAN, E. F. 1980. Homogeneous enzyme immunoassay for proteins employing b-galactosidase. *Anal. Biochem.* 102:167–170.

GIEGEL, J. L. 1985. Radial partition enzyme immunoassay. In *Enzyme Mediated Immunoassay,* eds. T. T. Ngo and H. M. Lenhoff, pp. 343–362, Plenum Press, New York.

GIEGEL, J. L.; BROTHERTON, M. M.; CRONIN, P.; D'AQUINO, M.; EVANS, S.; HELLER, Z. H.; KNIGHT, W. S.; KRISHNAN, K.; and SHEIMAN, M. 1982. Radial partition immunoassay. *Clin. Chem.* 28:1894–1898.

GNEMMI, E.; O'SULLIVAN, M. J.; CHIEREGATTI, G.; SIMMONS, M.; SIMMONDS, A. D.; BRIDGES, J. W.; and MARKS, V. M. 1978. A sensitive immunoenzymometric assay (IEMA) to quantitate hormones and drugs. In *Enzyme Labelled Immunoassay of Hormones and Drugs,* ed. S. B. Pal, pp. 29–41, Walter de Gruyter, Berlin.

GOSLING, J. P. 1990. A decade of development in immunoassay methodology. *Clin. Chem.* 36:1408–1427.

GOULD, B. J., and MARKS, V. 1988. Recent developments in enzyme immunoassays. In *Nonisotopic Immunoassay,* ed. T. T. Ngo, pp. 3–26, Plenum Press, New York.

GRENIER, F. C.; GRANADOS, E. N.; SCHICK, B. C.; KOLACZKOWSKI, L.; and PRY, T. A. 1987. Enhanced sensitivity immunoassay for the TDx analyzer. *Clin. Chem.* 33:1570.

GROFF, J. 1994. Abbott IMx®. In *The Immunoassay Handbook,* ed. D. Wild, pp. 143–148, Stockton Press, New York.

HAND, C. 1994a. DPC Milenia®. In *The Immunoassay Handbook,* ed. D. Wild, pp. 165–169, Stockton Press, New York.

HAND, C. 1994b. DPC Immulite®. In *The Immunoassay Handbook,* ed. D. Wild, pp. 170–174, Stockton Press, New York.

HEFLE, S. L. 1995. Immunoassay fundamentals. *Food Technol.* 49(2):102–107.

HELSINGIUS, P.; HEMMILA, I.; and LOVGREN, T. 1986. Solid phase immunoassay of digoxin by measuring time-resolved fluorescence. *Clin. Chem.* 32:1767–1769.

HOUTS, T. 1994. Syva Microtrak® EIA. In *The Immunoassay Handbook,* ed. D. Wild, pp. 222–223, Stockton Press, New York.

HOWARD, K.; KANE, M.; MADDEN, A.; GOSLING, J. P.; and FOTTRELL, P. F. 1989. Direct solid phase enzymoimmunoassay of testosterone in saliva. *Clin. Chem.* 35:2044–2047.

ISHIKAWA, E.; HASHIDA, S.; TANAKA, K.; and KOHNO, T. 1989. Methodological advances in enzymology. Development and application of ultrasensitive enzyme immunoassays for antigens and antibodies. *Clin. Chim. Acta* 185:223–230.

JOHANNSSON, A.; ELLIS, D. H.; BATES, D. L.; PLUMB, A. M.; and STANLEY, C. J. 1986. Enzyme amplification for immunoassays. Detection limit of one hundredth of an atto-mole. *J. Immunol. Meth.* 87:7–11.

KAHN, S. E., and BERMES, E. W. 1992. Stratus II immunoassay system. In *Immunoassay Automation. A Practical Guide,* ed. D. W. Chan, pp. 293–316, Academic Press, San Diego, CA.

KHANNA, P. L. 1991. Homogeneous enzyme immunoassay. In *Principles and Practice of Immunoassay,* eds. C. P. Price and D. J. Newman, pp. 326–364, Stockton Press, New York.

KHANNA, P. L.; DWORSCHACK, R. T.; MANNING, W. B.; and HARRIS, J. D. 1989. A new homogeneous enzyme immunoassay using recombinant enzyme fragments. *Clin. Chim. Acta* 185:231–240.

KHANNA, P. L., and WORTHY, T. E. 1993. CEDIA®: A recombinant-based homogeneous enzyme immunoassay. In*Diagnostics in the Year 2000,* eds. P. Singh, B. P. Sharma, and P. Tyle, pp. 11–20, Van Nostrand Reinhold, New York.

KINSKY, S. C.; HAXBY, J. A.; ZOPF, D. A.; ALVING, C. R.; and KINSKY, C. B. 1969. Complement-dependent damage to liposomes prepared from pure lipids and Forssman hapten. *Biochemistry* 8:4149–4158.

LEHRER, M.; MILLER, L.; and NATALE, J. 1992. The OPUS system. In *Immunoassay Automation. A Practical Guide,* ed. D. W. Chan, pp. 245–267, Academic Press, San Diego, CA.

LEVY, M. J., and JAKLITSCH, A. P. 1993. Homogeneous enzyme immunoassays: EMIT®. In *Diagnostics in the Year 2000,* eds. P. Singh, B. P. Sharma, and P. Tyle, pp. 11–20, Van Nostrand Reinhold, New York.

LI, T. 1992. The Vista immunoassay system. In *Immunoassay Automation. A Practical Guide,* ed. D. W. Chan, pp. 343–350, Academic Press, San Diego, CA.

LI, T. 1994. Syva Vista® immunoassay system. In *The Immunoassay Handbook,* ed. D. Wild, pp. 224–225, Stockton Press, New York.

LITCHFIELD, W. J., FREYTAG, J. W.; and ADAMICH, M. 1984. Highly sensitive immuno-assays based on use of liposomes without complement. Clin. Chem. 30:1441–1445.

LOEBEL, J. E. 1994. TOSOH AIA®-600 and AIA®-1200. In *The Immunoassay Handbook,* ed. D. Wild, pp. 226–232, Stockton Press, New York.

McLELLAN, W. N., and PLAUT, D. 1994. Baxter Stratus® immunoassay systems. In *The Immunoassay Handbook,* ed. D. Wild, pp. 149–154, Stockton Press, New York.

MEENAN, G. M.; BARLOTTA, S.; and LEHRER, M. 1990. Urinary tricyclic antidepressant screening: comparison of results obtained with Abbott FPIA reagents and Syva EMIT reagents. *J. Anal. Toxicol.* 14:273–276.

MUNRO, C. J., and LASLEY, B. L. 1988. Non-radiometric methods for immunoassay of steroid hormones. In *Non-Radiometric Assays: Technology and Application in Polypep-tide and Steroid Hormone Detection,* eds. B. D. Albertson and F. P. Haseltine, pp. 289–329, Alan R. Liss, New York.

NAKAMURA, R. M.; VOLLER, A.; and BIDWELL, D. E. 1986. Enzyme immunoassays: heterogeneous and homogeneous systems. In *Handbook of Experimental Immunology, Vol. 1, Immunochemistry,* 4th ed., ed. D. M. Weir, pp. 27.1–27.20, Blackwell Scientific, Oxford.

NGO, T. T., and LENHOFF, H. M. 1980. Enzyme modulators as tools for the development of homogeneous enzyme immunoassays. *FEBS Lett.* 116:285–288.

NGO, T. T., and LENHOFF, H. M. 1981. New approach to heterogeneous enzyme immunoassays using tagged enzyme-ligand conjugates. *Biochem. Biophys. Res. Commun.* 99:495–503.

NGO, T. T., and LENHOFF, H. M. 1985. *Enzyme Mediated Immunoassay.* Plenum Press, New York.

OELLERICH, M. 1984. Enzyme immunoassay: A review. *J. Clin. Chem. Clin. Biochem.* 22:895–904.

PALMER, R. F., and BACARESE-HAMILTON, T. 1994. Serono Serozyme. In *The Immunoassay Handbook,* ed. D. Wild, pp. 209–213, Stockton Press, New York.

PAYNE, G. P.; SAEWERT, M.; and HARVEY, S. 1994. Hybritech ICON® and Tandem® ICON® QSR®. In *The Immunoassay Handbook,* ed. D. Wild, pp. 1175–178, Stockton Press, New York.

PESTKA, J. J.; ABOUZIED, M. N.; and SUTIKNO. 1995. Immunological assays for mycotoxin detection. *Food Technol.* 49(2):120–128.

SAGONA, M. A.; GADSDEN, R. H.; and COLLINSWORTH, W. E. 1992. ES-300 immunoassay system. In *Immunoassay Automation. A Practical Guide,* ed. D. W. Chan, pp. 191–202, Academic Press, San Diego, CA.

SCHUELER, P. A.; ENFIELD, D. L.; FLECK, T. M.; KINGSLEY, J. T.; and MAHONEY, W. C. 1993. The Eclipse ICA®: An immunochemical and clinical chemistry assay system for near-patient testing. In *Diagnostics in the Year 2000,* eds. P. Singh, B. P. Sharma, and P. Tyle, pp. 177–196, Van Nostrand Reinhold, New York.

SETH, J.; HANNING, I.; BACON, R. R. A.; and HUNTER, W. M. 1989. Progress and problems in immunoassays for serum pituitary gonadotropins: evidence from the U.K. external quality assessment schemes (EQAS), 1980-1988. *Clin. Chim. Acta* 186:67–82.

STURGESS, M. L.; WEEKS, I.; MPOKO, C. N.; LAING, I.; and WOODHEAD, J. S. 1986. Chemiluminescent labeled-antibody assay for thyroxin in serum, with magnetic separation of the solid phase. *Clin. Chem.* 32:532–535.

SYVA CO. 1994. *Syva ETS® Plus System Operators Manual.* Syva Co., San Jose, CA.

ULLMAN, F. F.; GIBBONS, I.; and WENG, L. 1983. Homogeneous immunoassays and immunometric assays employing enzyme channeling. In *Diagnostic Immunology: Technology Assessment and Quality Assurance,* eds. J. H. Rippey and R. M. Nakamura, pp. 31–55, College of American Pathologists, Skokie, IL.

VAN WEEMAN, B. K., and SCHUURS, A. W. H. M. 1971. Immunoassay using antigen-enzyme conjugates. *FEBS Lett.* 15:232–236.

VISTNES, A. I. 1988. Spin membrane immunoassay in serology. In *Nonisotopic Immunoassay,* ed. T. T. Ngo, pp. 361–388, Plenum Press, New York.

WEI, R., and RIEBE, S. 1977. Preparation of a phospholipase C-antihuman IgG conjugate, and inhibition of its enzymatic activity by human IgG. *Clin. Chem.* 23:1386–1388.

WILD, D. 1994. Kodak Amerlite® modular system. In *The Immunoassay Handbook,* ed. D. Wild, pp. 181–185, Stockton Press, New York.

YALOW, R. S., and BERSON, S. A. 1959. Assay of plasma insulin in human subjects by immunological methods. *Nature* (London) 184:1648–1649.

YASUDA, T., ISHIMORI, Y., and UMEDA, M. 1988. Immunoassay using fluorescent dye-trapped liposomes. Liposome immune lysis assay (LILA). In *Nonisotopic Immunoassay,* ed. T. T. Ngo, pp. 389–399, Plenum Press, New York.

9

Assay Development, Evaluation, and Validation

INTRODUCTION

Modern immunoassay systems exist in several different formats with diverse ancillary technologies. An ideal immunoassay for a given need depends upon the kind of information desired and the physical environment of the antigen, i.e., the nature of the test sample, e.g., blood, serum, plasma, urine, biopsy samples, cell culture broths, tissue sections, fecal samples, and environmental samples such as soil, water, and food. The wide range of practical applications that have been demonstrated during the last three decades for such a diverse array of test samples clearly indicates the tremendous potential and adaptability of immunoassay techniques in the field of analyte detection and quantitation (Deshpande 1994).

In addition to the nature of the test sample, immunoassay development targeted to a particular need requires careful consideration of several other factors. For example, an assay that is ideal for the laboratory may not be practical for on-site testing in the field. Consideration must also be given to ease of operation and cost of the accompanying instrumentation. Ideally, the practical assay should conform to available equipment. Unfortunately, the price for improving one factor often comes at the expense of other constituents of any assay development program.

The road to an ideal commercial immunoassay is a three-dimensional puzzle. It requires the immunocharacterization of the antigen or the analyte, the development of appropriate antibody reagents, and the development of the assay protocol. All these factors are interlinked. In fact, overlooking one without understanding its influence on other assay parameters may often lead to stumbling blocks in the assay development program. Solving the puzzle of immunoassay development thus involves multiple cycles that systematically balance competing factors. Careful planning is absolutely vital to keep the final objective in sight.

Given the bewildering array of assay formats, protocols, solid phases, choice of labels, and signal detection technologies, it may appear on the surface that assay development is an extremely complex process. Fortunately, irrespective of the methodology chosen, several basic elements are common to most assay development programs. These include, among others, selection of reagents, optimization of assay conditions, and analytic and clinical validation. Commercial immunoassays are usually developed to a specification that includes items relating to analytic and clinical performance, stability of reagents, resistance to operator and environmental errors ("ruggedness"), and cost of manufacture (Micallef and Ahsan 1994). Aspects such as the instrumentation, label, and separation system to be used may also be fixed by practical considerations or by a requirement for methodological uniformity.

Each of the critical factors in an assay development program can be further grouped into several subclasses. Thus, reagent selection includes identification of useful antibodies, label, calibrant, calibrant matrix, buffers, solid phases, and ancillary components such as enzyme substrates. The formulation of reagents (e.g., solution, suspension, frozen, powdered, tablet, or lyophilized), discussed in Chapter 10 at greater length, must also be determined.

Reagent selection, optimization, and validation, however, are not sequential steps and their roles can be overlapping. Selection of reagents, for example, intrinsically requires some optimization of the process in which they are being evaluated. In actual practice, assay development thus tends to be a cyclic process; decisions taken at early stages (e.g., reagent selection or assay conditions) are provisional and may be changed because of results obtained later (Micallef and Ahsan 1994).

In addition, careful attention must be given to several other parameters during the assay development program. The assay must comply with the needs of the sample. The most practical assays are those that minimize the processing required to make the sample ready for assay. Immunoassays should also provide meaningful results. However, the identification of what constitutes meaningful results is often the most difficult part of immunoassay development. The earlier this is defined, the easier and simpler assay development becomes.

The use of appropriate references is yet another essential criterion for the development of a high performance immunoassay. Generally, the importance of this factor is often overlooked or the reference assays compromised because of issues such as the amount of work required and cost. An ideal assay should always have a positive and a negative reference to establish that the assay conforms and performs to a standard behavior. The importance of positive control is quite evident. Since substandard or less-than-ideal results for the positive reference control quickly identify, among others, factors such as faulty or deteriorated reagents, erratic techniques and interfering substances.

Similarly, negative reference controls can greatly aid in minimizing problems related to sources of assay noise, such as cross-reacting substances and nonspecific binding. It is absolutely essential to have a significant signal spread between the positive and the negative controls. Otherwise, information from the test samples will often be meaningless.

Because most modern immunoassays provide high levels of signal amplification, its misuse can lead to assays that are very unstable and subject to unacceptable levels of noise. The ruggedness of an immunoassay, therefore, is an important parameter that must be conscientiously built into the development program.

Finally, simplicity is the key to a successful immunoassay. Immunoassays that are simple to perform are often backed by sophisticated chemistry and technology. Overall balance of the assay development program is the key to simplicity.

In this chapter, basic principles of immunoassay development are described. The properties of an assay will undoubtedly vary according to the principle and format chosen. Each has its advantages and drawbacks for a particular application. It is impossible to select a universal format that is the best for all purposes; the selection must be made in relation to a particular situation and its needs. Because of their extensive use in most commercial assays, the development and optimization process is primarily presented with microtiter plates as an example of the solid-phase system. Nonetheless, the principles and strategies involved are quite similar and therefore also applicable to other solid-phase systems.

SELECTION OF ASSAY FORMAT

Immunoassays can be performed in a wide variety of designs and formats. They can be considered under two main categories (Davies 1994). The first includes the so-called principal assay designs. These are primarily concerned with exploiting the fundamental properties of immunoanalytical techniques. The second category comprises those designs that have built upon these fundamental principles in order to improve the precision of the assay, reduce assay incubation times, deskill the technology, or render the assays amenable to automation.

The three most commonly used assay formats for immunodiagnostic products commercially available include the antibody capture assays, the antigen capture assays, and the two-antibody immunometric sandwich assays. Within each group, the principle and the order of the steps are similar. However, one can change the variable that is being tested. Thus, by changing certain conditions, an assay can be altered to determine either antigen or antibody level. Even though the assay protocols may be similar, these two assays will yield quite different results.

Within each class of immunoassay design, any immunoassay can be performed with four variations (Harlow and Lane 1988). The assay can be carried out in antibody excess, in antigen excess, as an antibody competition, or as an antigen competition. Assays performed in antibody excess or as antigen competitions are used to detect and quantitate antigens, whereas antigen excess or antibody competition assays are used to detect and quantitate antibodies. These four variations within each of the three classes yield a total of 12 possible immunoassay designs. However, not every combination leads to a useful immunoassay. Thus, only six combinations are widely employed.

The types of antibodies available (e.g., polyclonal, affinity purified polyclonal, single monoclonal, or two or more monoclonals) and the availability of a pure or impure antigen are the two major criteria that need to be considered in choosing the correct design for an immunoassay. The possible choices for an assay protocol based on these two variables are listed in Table 9.1. It groups the assays in an order showing their general usefulness. The actual choice of assay format ultimately depends on analyte concentration, precision and sensitivity requirements, the nature of the sample, and the availability of certain immunochemical reagents. In several situations, it is not necessary to have a precise measure of the concentration of a given analyte. Such is the case in the screening of hybridomas or the routine testing of samples for the presence or absence of an analyte as in pregnancy detection kits. In such instances, it is sufficient to record the result as a simple yes or no. Similarly, it may be sufficient to record the result as high, medium, or low. Many routine test results are reported in this latter format. Such semiquantitative methods, however, still need to be carefully developed.

Given the biological nature of the reactants, immunoassays inevitably deteriorate over time. Therefore, it is important to remember that an assay that was initially sensitive and accurate will tend to be more robust than one that was difficult to design in the first place. Similarly, reagent-excess or immunometric assays, in which all the reactants except the sample are added in excess, are always more precise than the reagent-limited "competitive" immunoassays.

The basic characteristics of the three assay formats are briefly described below.

Antibody Capture Assays

Antibody capture assays can be used to quantitate the levels of both antigen and antibody as well as to compare antibody binding sites. Typically, an antigen is immobilized on a solid phase, and the antibody from a diluted serum sample is allowed to bind to the immobilized antigen. The antibody can be labeled directly or can be detected by using a secondary reagent that will specifically recognize the antibody. Such reagents include labeled anti-immunoglobulin antibodies, protein A, protein G, and avidin-biotin systems. The amount of bound antibody determines the strength of the assay signal.

When labeled antibody assays are performed with excess antigen on the solid phase, the presence and level of antibodies in a test specimen can be measured. Because this design is an indirect assay, the antibodies in the test solution do not

TABLE 9.1.   Criteria for Selecting Immunoassay Protocol[a]

Type of antibody available	Type of antigen available	Assay choices[b]
To measure antigen presence or quantity		
Polyclonal antibodies	Pure	1. Antigen capture (antigen competition)
		2. Antibody capture (antigen competition)
	Impure	1. Antibody capture (antibody excess). (Other assays possible but need secondary techniques)
Affinity-purified polyclonal antibodies	Pure	1. Two-antibody sandwich
		2. Antigen capture (antigen competition)
		3. Antibody capture (antigen competition)
	Impure	1. Two-antibody sandwich
		2. Antibody capture (antibody excess)
One monoclonal antibody	Pure	1. Antigen capture (antigen competition)
		2. Antibody capture (antigen competition)
	Impure	1. Antibody capture (antibody excess). (Other assays possible but need secondary techniques)
Two or more monoclonal antibodies	Pure	1. Two-antibody sandwich
		2. Antigen capture (antigen competition)
		3. Antibody capture (antigen competition)
	Impure	1. Two-antibody sandwich
		2. Antibody capture (antibody excess)
Two or more polyclonal antibodies	Pure	1. Antibody capture (antigen excess)
	Impure	Need secondary techniques
To measure antibody presence or quantity[c]		
Affinity-purified polyclonal antibodies	Pure	1. Antibody capture (antigen excess)
	Impure	Need secondary techniques
One monoclonal antibody	Pure	1. Antibody capture (antigen excess)
	Impure	Need secondary techniques
Two or more monoclonal antibodies	Pure	1. Antibody capture (antigen excess)
	Impure	Need secondary techniques

[a]From Harlow and Lane (1988).

[b]The choices within each group are listed in order of suggested preference.

[c]For most assays used to measure antibody levels, the antibody is detected using an anti-immunoglobulin antibody. Thus, the antibody becomes the antigen.

need to be purified. Relative titers of different antibodies or antisera can be quantitated using this assay format.

Sometimes it is desirable to separate the analyte from the medium in which it is found prior to carrying out the assay. This assay format therefore can be used in class-capture immunoassays where the immunoglobulin class of interest is selectively removed from the other immunoglobulin classes. The latter may contain antibodies that could interfere with the reaction between the sample antibody and antigen. The immunoglobulin class for which this is the greatest problem is IgM (Kemeny 1992). IgM antibodies are of diagnostic value in identifying recent infection with a number of pathogens. The difficulty is that there may be many times more IgG antipathogen antibodies, which are always present regardless of infection and thus are of no diagnostic value. The problem is compounded by the fact that the IgG antibodies present will tend to be of much higher affinity and may disproportionately inhibit binding of IgM antibodies to the antigen. A similar problem also exists when one is trying to measure IgG subclass antibodies where the antibody of one subclass may be of much higher affinity than another.

In IgM class-capture assays, the solid phase is coated with polyclonal or monoclonal antibody specific for IgM. This anti-IgM-coated solid phase is then used to adsorb most of the IgM in the test sample. Any IgM antibody activity bound to the solid phase then can be determined by the addition of labeled antigen.

Antibody capture assays can also be used to compare the binding sites of two monoclonal antibodies using the antibody competition variation of the assay design (Harlow and Lane 1988). In this approach, the two monoclonal antibodies, one of which is labeled, are added to a solid phase containing the immobilized antigen. If these two antibodies bind to separate and discrete sites on the antigen, the labeled antibody will bind to the same level, whether or not the competing antibody is present. However, if their sites of interaction are identical or overlapping, the unlabeled antibody will compete with the labeled antibody for the available binding sites, thereby resulting in a decreased signal. A basic requirement for using this design is that the labeled monoclonal antibody must be present in a pure form, whereas none of the antibodies used as competitors needs to be purified.

In a third variation of this design, antibody capture assays with antigen competition can be used to detect and quantitate the presence of an antigen in a test sample. A sample of the test solution together with a labeled antibody specific for the antigen are allowed to compete for the antigen binding sites on a solid phase. The strength of the assay signal decreases with increasing amounts of free antigen in the test sample, because less and less labeled antibody will bind to the solid phase.

A major drawback of this assay format is that the labeled antibody may have differing avidities for the bound and free antigen. Problems related to differences in avidity can be identified by comparing standard curves using known concentrations of antigen. Anomalous assay results can be corrected by using lower concentrations of the antigen immobilized on the solid phase, thereby decreasing the local concentration, by premixing the test solution with the labeled antibody and incubating for 30–60 min prior to adding to the solid phase, or by using labeled Fab fragments.

Antibody capture assays for specific antibodies are normally qualitative, and therefore are optimized to minimize false-positive and false-negative results and to provide a low detection limit. Depending on the assay configuration, antibody capture immunoassays can also be used to detect and quantitate the presence of antigens. The sensitivity of these assays depends on the specific activity of antigen and the avidity of the antibody used. The detection limit of antibody capture assays can be as low as 0.1–1.0 fmol, i.e., about 0.01 to 0.1 ng.

The overall development and optimization process for antibody capture assays involves titration of the solid phase and the marker antibody and antigen concentrations with dilutions of known strongly positive and weakly positive sera. Sufficient quantity of these two reagents must be used to ensure that reagent excess conditions prevail for strongly positive samples, while keeping nonspecific binding and the imprecision of low assay response levels as low as possible (Micallef and Ahsan 1994). This maximizes the signal-to-noise ratio and the sensitivity of the assay for weakly positive samples.

Incubation times and temperatures also need to be optimized to ensure that the binding reactions approach equilibrium. Reaction rates can be determined by incubating a few test samples for varying lengths of time and plotting the resultant signals against time. Reaction is typically slower for low concentrations of a specific antibody. Thus, reaction kinetics must be determined using weakly positive samples as well as samples producing a strong assay response.

Antigen Capture Assays

These assays were originally developed by Yalow and her coworkers (Berson et al. 1956; Yalow and Berson 1959, 1960). Antigen capture assays are primarily used for the detection and quantitation of small molecular weight analytes that cannot be detected by two-site immunometric sandwich assays (see below). The sandwich formation is only possible for antigens large enough to bind two molecules of antibodies independently without steric hindrance. Consequently, antigen capture assays are limited-reagent assays for which there are no limitations on the nature of the analyte as long as a specific antibody is available.

In the basic assay format, the amount of antigen in a test sample can be quantitated using a competition between labeled and unlabeled antigen. The unlabeled antibody is immobilized on a solid phase either directly or indirectly through an intermediate protein such as anti-immunoglobulin antibody, protein A, or protein G or via an avidin-biotin bridge. A sample of the test sample and labeled antigen are added, and, after a separation step, the amount of label bound to the solid phase is quantitated. The assay signal is inversely proportional to the level of analyte in the test sample. The relative levels of antigen in the test sample can be quantitated using appropriate calibration curves. Sometimes, conjugation of the analyte to the enzyme label may alter its immunogenic properties. One solution to this is to label it with a small molecule like biotin.

Alternatively, a limited-reagent assay involving analyte immobilized on a solid phase and based on the detection of the bound labeled antibody can also be formatted. A major drawback of this format is that a precise, limiting quantity of antigen must be bound to the solid phase. Furthermore, any dissociated antigen will also act as an inhibitor, thereby limiting the sensitivity of the assay.

The basic requirements for such a labeled analyte, reagent-limited assay format are the availability of a high affinity specific antibody, a marker or label that remains highly detectable when covalently linked to the hapten (or a chemical derivative of it), a calibrant preparation, and usually a method for separation of the free and antibody-bound analyte.

The sensitivity of the antigen capture assays using labeled antigen is determined by: (1) the number of antibody molecules immobilized on the solid phase, (2) the avidity of these immobilized antibodies for the antigen, and (3) the specific activity of the labeled antigen (Harlow and Lane 1988). The first parameter can be easily optimized by testing a number of different concentrations of the antibody solution for immobilization on the solid phase chosen. Sometimes, higher levels of coating can be achieved by changing the type of solid phase used. In contrast, the avidity of the antibody for the antigen can only be altered by changing the antibody source. The specific activity of the label can be monitored by selecting and optimizing the enzyme-substrate system.

Optimization of reagent-limited antigen capture immunoassays involves selection of the appropriate antibody and labeled analyte concentrations by making serial dilutions of both, and testing the combinations of each antibody dilution with every labeled hapten dilution in the absence of any free analyte, as well as at analyte concentrations corresponding to the anticipated highest and lowest standards. The concentrations finally selected are those giving the greatest and most reproducible changes in signal with increasing analyte concentration, as well as yielding the best precision in the clinical diagnostic ranges required. Optimal times and temperatures of incubation also need to be determined.

The optimization of reagent-limited antigen capture immunoassays for maximal sensitivity has been recently reviewed (Ekins 1991; Ezan et al. 1991; Micallef and Ahsan 1994). The concentration range over which acceptable precision may be obtained is narrower for these types of assays than for reagent-excess immunometric sandwich assays. The range of acceptable precision, and hence the limiting sensitivity of these assays, can be shifted to higher or lower concentrations of analyte by changes in the concentrations of antibody and labeled hapten used.

Reagent-limited "competitive" antigen capture immunoassays are rapid and easy to perform. They approach a detection limit of approximately 0.1–1.0 fmol (about 0.01 to 0.1 ng).

Two-Site Immunometric Sandwich Assays

These reagent-excess assay formats are quite popular in the immunodiagnostic industry for the detection and quantitation of large analytes, such as proteins, which have multiple epitopes that are spatially sufficiently well separated. This assay format, therefore, requires two antibodies that are capable of binding to discrete, nonoverlapping epitopes on the antigen. This matched pair of antibodies ideally should have a high combined affinity and specificity.

The basic assay protocol involves incubation of the test sample with a solid phase immobilized "capture" antibody. After a separation step, a labeled "detecting" antibody is added that binds to the immobilized antigen-antibody complex. After a second separation step, the amount of label bound is quantitated. The assay can be optimized in a similar way as described for the other two basic formats. Individual steps in the assay development and optimization process are described in greater detail later in this chapter.

Sandwich assays can also be formatted as single-step assays. This involves simultaneous reaction of the capture antibody and the detecting antibody with the analyte. Although convenient to perform, a major disadvantage of this assay format is that a "hook effect" is observed at very high antigen concentrations, i.e., increasing concentrations of the antigen cause a paradoxal fall in signal (Nomura et al. 1982; Micallef and Ahsan 1994). The signal decreases because at high analyte concentrations there tends to be sufficient analyte to saturate both the labeled and capture antibodies. This in turn prevents sandwich formation. Although the magnitude of the hook effect varies with different analytes, it is always undesirable and may lead to misdiagnosis.

The two-step format described earlier, although less convenient to use, does not produce any hook effect, because the analyte not captured in the first incubation step is physically removed by washing during the separation step.

The common variations of these three basic immunoassay formats, all widely used in the diagnostics industry are illustrated in Figure 9.1.

PREPARATION OF THE SOLID PHASE

The preparation of the solid phase is perhaps the most important step in developing solid-phase enzyme immunoassays, because the solid phase serves as an anchor to which all the subsequent reactants must bind. The performance of the solid phase immunosorbent is usually indirectly inferred from the performance of the complete immunoassays. However, the efficiency of a solid-phase system containing the adsorbed capture antibody can often be monitored by using an anti-immunoglobulin-enzyme conjugate or radiolabeled analyte. Alternatively, the binding of different radiolabeled antigens and their subsequent dissociation can

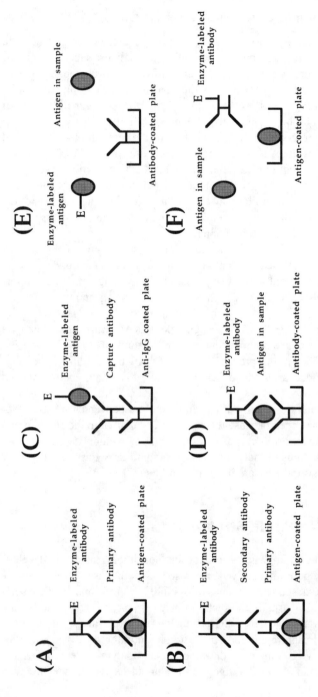

Figure 9.1. Assay formats commonly used commercially for enzme immunoassay products. A, B, and C represent antibody capture formats. In A and B, antigen is bound to the solid phase to capture the desired antibody. The bound antibody is then quantitated using an enzyme-labeled second antibody specific for the primary antibody (A), or by using a "ladder" of secondary antibodies (B) to increase assay specificity. Assay format C is used to capture class-specific antibodies by using a solid phase coated with either a polyclonal or monoclonal antibody specific for the class of immunoglobulins to be captured and detected using an enzyme-labeled antigen. D is an example of an immunometric, reagent-excess sandwich immunoassay commonly used for polyvalent antigens. E and F represent examples of reagent-limited, competitive immunoassays using enzyme-labeled antigen and enzyme-labeled antibody, respectively, on an antibody-coated or antigen-coated solid phase.

284

also be tested. Optimizing the preparation of the solid-phase system is absolutely critical—if it fails, then no matter how well the subsequent stages of the assay work, the assay will still fail.

For evaluating the performance of an immunoassay, perhaps more problems can be attributed to the coating of the carrier used than to any other part of the assay (Kemeny 1992). For example, if an assay fails to yield satisfactory results or gives poor precision, it may be because there is little or no capture antibody or antigen actually bound to the solid phase in the first place, or the immunosorbent may have lost its functional ability because of denaturation or some other effects. In this section, certain basic parameters and the pitfalls involved in the preparation of an ideal solid phase are described.

Basic Considerations

A large body of literature is available describing the adsorptive properties of proteins on solid-phase supports, particularly polystyrene. These solid-phase assays have some properties that are quite unique: (1) they all display a type of prozone phenomenon or "hook effect," (2) the rate of binding has an optimum at which time maximal binding occurs, and (3) there can be a change in the reaction of antigen with antibody, depending upon the form in which the antigen is adsorbed to the polystyrene support (Pesce et al. 1981; Kemeny 1992; Butler et al. 1992).

An example of a type of prozone phenomenon occurring in the sandwich type of enzyme immunoassays is shown in Figure 9.2. When a polystyrene support is coated with increasing amounts of the immunosorbent, followed by its quantitation by any appropriate method, a linear increase in signal is generally observed with increasing amounts of the immunosorbent adsorbed to the solid phase in the lower concentration range of the immunosorbent used. However, beyond a critical concentration, instead of reaching a maximal value, a decrease in signal is generally observed. This phenomenon was actually shown in one of the first publications on ELISA by Engvall and Perlman (1972). Thus, a simple saturation type of analysis cannot be done using such a system (Pesce et al. 1981). One must determine the optimal binding of the antigen or antibody to the solid-phase support.

The reaction between the solid phase and the immunosorbent also influences the sensitivity of the solid-phase immunoassays. The amount of antigen binding to the polystyrene support and its subsequent retention of antigenicity can be measured by its ability to interact with an enzyme-labeled antibody. Generally, the binding reaches a maximum and then decreases with time (Pesce et al. 1977). In the initial phase of the reaction, the sharp increase of IgG bound may be ascribed to the initial interaction of the IgG with the polystyrene support. The second phase of a slow decrease of binding of the enzyme-labeled antibody to the adsorbed IgG reflects a loss of antigenicity of the IgG on the surface. There is apparently a def-

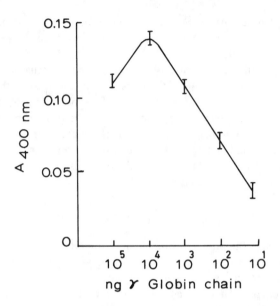

Figure 9.2. An example of a "hook effect" (prozone phenomenon) commonly associated with solid-phase enzyme immunoassays. Polystyrene tubes were coated with different amounts of cord blood containing 70% hemoglobin that was detected with a rabbit antibody specific for the γ-globin in a sandwich assay format. At low concentrations of antigen, i.e., 10^1 to 10^4 ng, a linear increase in absorbance is observed with increasing amounts of antigen coated on the solid phase. However, the absorbance decreases above 10^4 ng antigen concentration, indicating the prozone phenomenon. From: Pesce et al. (1981)

inite structural change in the immunoglobulin molecule with time. It is well known that the binding of proteins to plastic surfaces is saturable and reaches a plateau at high antigen concentrations; thus, this prozone phenomenon must be related in some way to the antibody-antigen interaction.

Pesce et al. (1981) have suggested at least two possible explanations for this prozone phenomenon. The first is that as the plastic surface approaches saturation, steric interference ("crowding") reduces the area of interaction between the plastic surface and the protein. The addition of antibody to a weakly bound antigen molecule might result in its removal from the solid surface during the washing step. The second possibility is that there is a time-dependent denaturation of the antigen on the plastic surfaces which results in the loss of antigenic determinants. These researchers have shown that approximately 90% of the antigen sites are denatured after 10 days on the plastic surface.

Generally, the amount of antibody adsorbed on a polystyrene support is linearly proportional to the amount of antibody added in a log-log plot until a concentration of 5 μg/ml of added antibody has been exceeded, i.e., 600 ng has become adsorbed (Figure 9.3). Increasing the concentration of added antibody

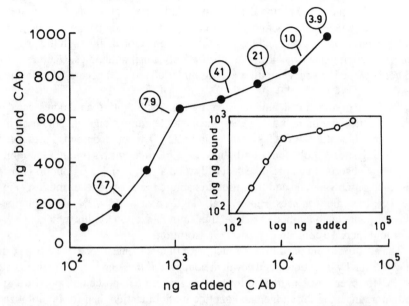

Figure 9.3. The relationship between the amount of added capture antibody (CAb) and the amount adsorbed on Immulon $2^®$ polystyrene microtiter plates. The inset depicts this relationship in a log-log plot. Values encircled are the percentage of added CAb that becomes stably adsorbed at different points on the adsorption plot.
From: Butler et al. (1991)

beyond this point results in a loss of log-log linearity in the binding plot and in the efficiency of adsorption (Butler et al. 1991). The log-log linear binding of antibody over the low concentration range supports homogeneous receptor-ligand interaction up to a point where the primary binding sites are no longer in excess. Adsorption beyond this point could result from adsorption to heterogeneous, lower affinity binding sites, or perhaps the formation of protein aggregates. These data are typical of most immunoglobulins as well as polystyrene microtiter plates from a variety of commercial sources.

Recently, Butler et al. (1991, 1992) investigated the physical and functional behavior of capture antibodies immobilized passively on polystyrene supports. They observed that each protein has a different affinity for polystyrene so that it is impossible to predict the adsorption behavior of a particular protein without having performed such adsorption studies. Moreover, there is a tendency for the affinity of proteins to adsorb on plastics to be correlated with their molecular weights. Furthermore, saturation of plastic binding sites appears to occur approximately at a point at which the surface is covered with a single monolayer of protein.

The functional activity of antibodies changes dramatically once they are immobilized on the plastic surface. The percentage of binding sites of both polyclonal and monoclonal antibodies remaining after adsorption on polystyrene is shown in Figure 9.4. In this study, Buter et al. (1992) found that monoclonal antibodies tend to have fewer binding sites than the polyclonal, although the proportion of surviving sites was low even for the polyclonal antibodies. Capture antibodies adsorbed at a concentration of 5 μg/ml (1000 ng/well) generally performed best.

The loss of functional binding sites by adsorbed capture antibodies can be greatly reduced by immobilizing them via a streptavidin linkage, i.e., the protein-avidin-biotin capture (PABC) system (Figure 9.5). This procedure also preserved the binding sites of polyclonal antibodies. Loss of binding sites through immobilization by adsorption can also be reduced when a polyclonal antiglobulin is first adsorbed on polystyrene and then used to capture the antibodies of interest (Figure 9.5). In some instances, Butler et al. (1992) observed > 70% of the binding sites of monoclonal antibodies active after immobilization by this method, i.e., a > 60-fold increase compared to passive adsorption.

The adsorption-induced loss of function of monoclonal antibodies perhaps can be overcome by directionalized covalent coupling of these antibodies (Mize et al. 1991). However, covalent immobilization may be more important in specific antibody immunoassay for antigen and peptide immobilization than for immobilization of antibodies (Butler 1991). It must be noted that in the Mize et al. (1991) study, none of the reactive polymer surfaces used for the directional immobilization of capture antibodies bound more antibody than polystyrene. Therefore, it ap-

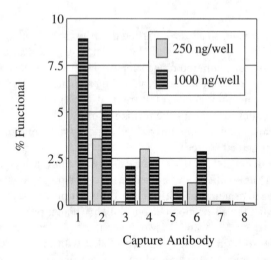

Figure 9.4. The proportion of functional CAb binding sites after the passive adsorption of two affinity-purified polyclonal (nos. 1 and 2) and six monoclonal (nos. 3 to 8) antibodies to fluorescein on polystyrene microtiter plates. The CAbs were tested at two different adsorption concentrations. Redrawn from Butler et al. (1992)

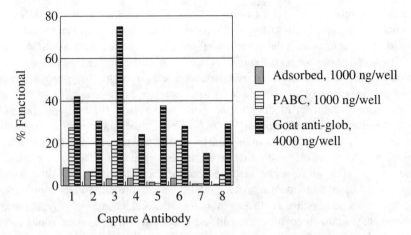

Figure 9.5. The partial prevention of functional binding site loss of CAb due to passive adsorption on a polystyrene microtiter plate. Antibodies 1 and 2 were polyclonal and 3–8 were the different monoclonals tested following either immobilization using the protein-avidin-biotin capture (PABC) system or using polyclonal goat antiglobulin that had been previously adsorbed to the microtiter plate. Data for antibodies 5 and 7 immobilized by the PABC system were unavailable to the authors and not shown here.
Redrawn from Butler et al. (1992)

pears that a hydrophobic surface, such as polystyrene, which binds proteins through many small nonspecific interactions, will tend to bind more protein than hydrophilic surfaces that have been modified to covalently immobilize antibody through a few specific interactions.

Butler et al. (1992) further suggest that immobilization through the use of an adsorbed antiglobulin might appear to be the best procedure if only the "percentage of active sites" is used as a criterion. However, if one assumes that only 5–10% of the sites of the adsorbed antiglobulin are active, then the number of secondary capture antibodies that can be immobilized will also be low, thereby resulting in nearly the same number of or fewer active sites per well as when the capture antibodies are immobilized using the PABC system or merely by passive adsorption. In terms of the "economy of immobilizing," which considers the amount of each capture antibody required to produce an effective solid phase, these researchers found that immobilization using an antiglobulin tends to be less effective than the PABC system and even passive adsorption. The last criterion for evaluating the performance of capture antibody may be most relevant to kit manufacturers because it directly relates to the cost of preparing solid phase capture antibodies.

The studies of Butler et al. (1991, 1992) clearly suggest that polyclonal capture antibodies are superior to monoclonal antibodies when evaluated in terms of their antigen capture capacity and dynamic range. This preferential susceptibility of monoclonal antibodies to surface-induced loss of activity for multivalent antigens could result from at least three, somewhat interrelated phenomena. Although loss of function of polyclonal antibody does occur during adsorption, a small to moderate proportion of denaturation-resistant antibodies of high affinity may survive, whereas resistance or susceptibility, owing to protein homogeneity, is an "all-or-none" phenomenon for monoclonal antibodies. The surviving subpopulations of polyclonal antibodies therefore appear to bind high molecular weight antigens homogeneously at low antigen input (Butler et al. 1991). Secondly, minor differences in glycosylation may destabilize the monoclonal antibody and render it more susceptible to adsorption-induced conformational changes and the corresponding loss of functional activity. Finally, the capture antibodies may be adsorbed on the solid phase as clusters of molecules. Such clusters of polyclonal antibody may allow recognition of different epitopes of the same multivalent antigen; however, those of monoclonal antibody are incapable of forming multiple interactions because of the infrequency of identical determinants. Therefore, one must take special care while working with monoclonal antibodies in solid-phase immunoassays.

Because the polyclonal antibodies consistently perform better and retain more functional sites upon immobilization, Butler et al. (1991) also studied the effect of method of purification of polyclonal antibodies on their retention of functional activity and performance in sandwich immunoassays. Purification of a goat polyclonal antiserum using saturated ammonium sulfate concentrations from 25–50% indicated that the antibody-enriched fraction precipitated at 30% saturation either performed best or as well as any other ammonium sulfate fraction (Figure 9.6). Similar results were also obtained by these researchers when rabbit or swine antisera were tested. Among the methods tested, antigen-affinity purification of polyclonals gave the best capture antibody performance. This method, however, may be impractical if large amounts of capture antibody are required. High-pressure affinity chromatography on protein G column and preparation using caprylic acid gave the next best level of performance (Figure 9.6). Of these two, the caprylic acid method performed the best in all three species tested. Butler et al. (1991) speculated that the short chain fatty acid may in fact stabilize the immunoglobulins against adsorption-induced loss of functional activity.

Most of the foregoing discussion was based on the use of polystyrene as a solid support. Similar observations regarding immobilization of capture antibodies and the subsequent retention of their functional activity cannot be extended to other solid-phase supports, especially membranes and microparticles, which have a high surface area-to-volume ratio. The popularity of these solid phases is primarily due to their higher capacity for immobilization of the capture immunosorbent reagent as compared to polystyrene. The high functional capture antibody concentration on these supports probably reflects their greater surface area, and

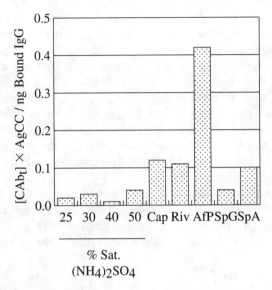

Figure 9.6. The functional activity of polyclonal swine anti-TEPC-15 mouse IgA CAb prepared by various methods. The four bars on the left are the activity of CAbs precipitated from serum using four different concentrations of saturated ammonium sulfate expressed as percent of saturation. The abbreviations under the bars on the right are as follows: Cap = caprylic acid, Riv = Rivanol, AfP = affinity-purified standard, SpG = high pressure protein G affinity chromatography, SpA = high pressure protein A affinity chromatography. CABt refers to functional capture antibody concentration and AgCC is the antigen capture capacity.
From: Butler et al. (1991)

hence, their popularity in immunoassays where broad dynamic ranges are required. However, such an increase in dynamic range also comes at the expense of a greater amount of capture antibody (Brown et al. 1991).

Coating Process

1. Immunoreagents. Most commonly, the reagent immobilized on the solid phase is a protein. Antibodies are immobilized in two-site immunometric sandwich assays or in labeled-analyte competitive assays. For the two-site assays, the antibody with the higher affinity constant should be used for solid-phase adsorption. For specific antibody immunoassays, an antigen is immobilized on the solid phase, whereas for labeled antibody competitive assays, a hapten is immobilized. The hapten may be coupled directly and covalently to the solid-phase material, or indirectly by conjugating the hapten to a protein and then immobilizing the protein.

In the case of proteins that have lipid moieties, delipidation may be required before it can be appropriately bound to the surface (Stein et al. 1986). The delipidation process, however, may denature the protein, thereby creating a series of new epitopes that may not be the same as those that are to be measured in analytes from the test sample. Consequently, the test analyte may also require a series of delipidation reactions before it can be measured as the same molecule present on the solid-phase surface.

Sometimes insoluble proteins, e.g., recombinant proteins, need to be solubilized using appropriate reagents. However, reagents such as SDS or DMSO inhibit solid-phase adsorption. SDS-solubilized proteins can be coated in the presence of 0.1–0.5 mol/l K_2HPO_4, which precipitates SDS as potassium dodecyl sulfate (Porstmann and Kiessig 1992). The precipitate then can be removed by washing the solid phase after protein adsorption.

One also needs to consider the form of the reagent that needs to be immobilized. Pesce et al. (1981) observed considerable differences in the sensitivity of the assay depending upon the form of the antigen immobilized (Figure 9.7). They observed a much better response with the same antibody when a γ-globin purified

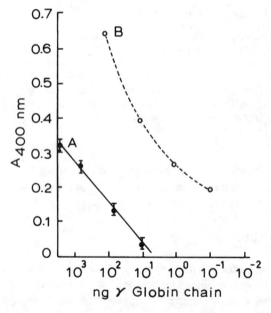

Figure 9.7. The influence of the form of antigen on the sensitivity of enzyme immunoassay. The quantitation of the γ-globin chain using hemoglobin F and the isolated γ-globin polypeptide molecule as the adsorbed antigen is shown here. Varying amounts of globin chain were incubated with the polystyrene tubes. Rabbit antibody specific for the γ-globin chain was added and the bound antibody detected in a sandwich assay format. Curve A shows the color yield of the reaction with hemoglobin F and curve B with the isolated γ-globin. The greater yield of color in the latter case increased the assay sensitivity by 10–50-fold.
From: Pesce et al. (1981)

free of the α-chain is used as the capture antigen as compared to hemoglobin F as the capture antigen. Pesce et al. (1981) observed a 10–50-fold increase in assay sensitivity when the purified antigen was used as the immunosorbent.

Sometimes, several antibodies are available for a given analyte. In such cases, the best antibody can be chosen by using the relative ability of various concentrations of the analyte to displace the antibody at an antibody dilution that corresponds to one-half the maximal binding optical density ($1/2\ B_{max}$) of the assay (Rosen and Fischberg-Bender 1995). The $1/2\ B_{max}$ can be determined by titrating the antibody on an ELISA plate coated with an analyte-protein conjugate (Figure 9.8A). The concentration of the drug that is required to displace enough antibody to lower $1/2\ B_{max}$ by 50% can then be used as an arbitrary value to compare different antibodies. The less analyte that is required to lower $1/2\ B_{max}$ by 50%, the more sensitive the antibody will be in that system. In the example shown in Figure 9.8B, the displacement of two antibodies (MAb1 and MAb2, each with a $1/2\ B_{max}$ of 1.0) by several dilutions of free analyte is shown. Less drug is required to displace MAb2 to 50% of $1/2\ B_{max}$ than MAb1. Thus, the former has a higher affinity for free analyte than the latter, and hence, should be chosen to optimize the assay.

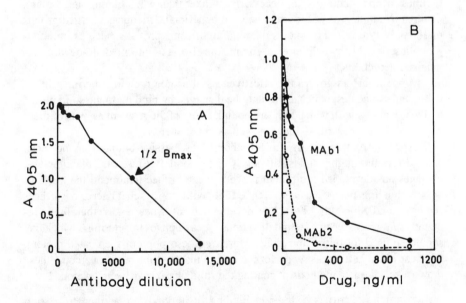

Figure 9.8. Evaluating the sensitivity of different antibodies for a given analyte.
A: Titration of a monoclonal antibody to yield 1/2 maximal binding optical density ($1/2\ B_{max}$)
B: Relative ease of displacement of two monoclonal antibodies (MAb1 and MAb2) by free drug in
  an enzyme immunoassay
From: Rosen and Fischberg-Bender (1995)

For solid-phase adsorption, the diagnostically relevant antigen or specific antibody should be as pure as possible to obtain high specific antibody loading. It also allows the use of a minimum surface area of solid phase in the assay, producing low nonspecific binding of label as well as economic benefits. Contaminating proteins in the solution will start to compete for the limited number of binding sites available at a coating density of 150 ng protein/cm^2 (Pesce et al. 1977). They also lower the precision of the assay. The reagent should also be filtered through a 0.2-micron filter to remove any protein aggregates present.

2. Method of Immobilization.   Three different approaches can be used for the preparation of solid phase. These are briefly described below.

DIRECT OR NONCOVALENT BINDING METHODS.   The easiest method to coat the solid phase is directly by passive adsorption of the capture reagent to the surface of the plastic. Noncovalent binding of proteins to solid supports such as plastics, glass, etc., is a natural and spontaneous process. It is also often a source of difficulty when very small amounts of protein are being handled.

Passive adsorption is mainly due to hydrophobic interactions between the surface and subsurface hydrophobic amino acid residues on the solid surface. The tightness of the binding is not necessarily weaker than when binding is covalent. Gamma irradiation of polystyrene is often beneficial. It induces charged groups on the polystyrene and promotes the binding of some proteins, such as immunoglobulins, possibly by helping to overcome the effects of intermediate-distance repulsive interactions.

Two main factors govern the effectiveness of coated proteins: the integrity of the bound molecules, which should not be damaged by binding to the solid phase, and the firmness with which they are bound. Some of the relevant aspects of these two factors were described above as well as in Chapter 7.

Despite its routine use in clinical laboratories, passive binding has not been studied very thoroughly and is still poorly understood. Because each manufacturer of plates has developed a different procedure for surface treatment, the kinds of passive binding plates have proliferated, all with poorly characterized surfaces (Douglas and Monteith 1994). The development of a new assay, therefore, frequently requires extensive testing of various types of plates to determine which one should be used for optimum results. Moreover, desorption of the coated protein, which adversely affects assay performance, is a major problem. To overcome these drawbacks, alternate methods of coating the solid phase have been developed.

INDIRECT METHODS.   Indirect binding of antibody or antigen to a solid phase usually means that a second antibody, protein A or protein G, or streptavidin is first immobilized noncovalently and then the first antibody is allowed to specifically adsorb to the bound antibody. These methods have advantages in terms of reproducibility, sensitivity, and specificity. They also require less use of the precious first antibody. Moreover, some antibodies tend to perform much better when indirectly bound (Davies 1994).

Indirect binding methods are particularly useful while working with monoclonal antibodies. Some monoclonal antibody preparations perform poorly when bound directly to the plate. Linking them to the plate with antimouse immunoglobulins considerably increases their capacity to bind the primary or the capture antibody (Figure 9.9).

Indirect methods should also be used to screen hybridomas where the antibody is present at very low concentrations. In such instances, it is also impractical to purify the antibody. Furthermore, coating with tissue culture supernatant containing 5–10% fetal calf serum (FCS) would result in FCS rather than the monoclonal antibody being bound to the plate. These problems can be overcome by using antimouse immunoglobulins to first coat the plates (Kemeny 1992).

Among the indirect methods, the PABC approach is especially useful, because it provides a common solid-phase reagent for the immobilization of a variety of capture antibodies or antigens. Because most proteins can be easily biotinylated, the use of avidin- (or streptavidin-) coated plates allows one to immobilize a number of biotinylated reagents. Some indirect approaches, such as the use of antiglobulins and protein A or G, also facilitate directional immobilization of the cap-

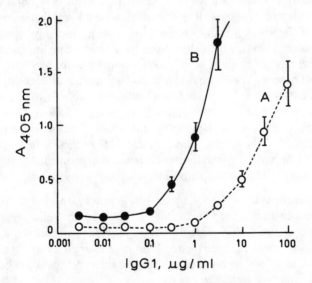

Figure 9.9. The influence of direct and indirect coating of antibody on assay performance. In this example, the capacity of a mouse monoclonal antihuman IgG1 antibody for human IgG was greater if bound to an antimouse IgG coated plate (curve B) than a plate coated directly (curve A) with the monoclonal antibody.
From: Kemeny (1992)

ture antibody via its Fc portion. This enhances the antigen binding capacity and thus the dynamic range of the assay.

The major drawback of such indirect methods is that they require more effort to set up and so should only be adopted if absolutely necessary. In addition, background signal is usually greater with these methods of immobilization.

COVALENT COUPLING METHODS.   Attempts to overcome the limitations of the above two methods and to produce more predictable results have led to the development of polystyrene plates with chemically reactive surfaces (Douglas and Monteith 1994). The concept here is to avoid the problem of equilibrium attachment of the passive binding surface in favor of a chemical link between the capture antibody or the antigen to be immobilized and the surface. By thus permitting the development of complete monolayer coverage at relatively low solution concentration, the need to try a variety of surfaces to find the most appropriate one for a given application is reduced.

Among the various surfaces developed are amino surfaces, which require a pretreatment with glutaraldehyde to react to an immunoglobulin (Inman 1974), a carboxylic acid surface, which requires a carbodiimide (Rassmusen et al. 1991), a hydrazide surface capable of binding to the Fc region of oxidized immunoglobulins (Ngo 1991), and several other surfaces, all of which require a linker agent applied by the user to produce a covalent bond (Cragie et al. 1991; Mize et al. 1991).

In some instances, such as immobilization onto polysaccharide or polyacrylamide particles, which are not naturally adsorptive, the binding is generally covalent. For materials that are naturally adsorptive, covalent coupling can enhance the apparent tightness of binding, or give more control over the binding process.

Covalent coupling methods provide uniformly coated surfaces that provide increased assay signal and improved assay precision. Nonspecific binding is also much lower compared to passive plates (Douglas and Monteith 1994). However, they suffer from the disadvantages of limited stability, the use of hazardous materials as linkers, and the need to modify protocols from those already in use. Manufacturers of such plates generally provide detailed experimental protocols for covalent coupling of immunoreagents to these solid-phase supports.

3. Concentration.   The most obvious and variable aspect of the coated material is the concentration used to coat the plate because the solid phase has only a limited capacity for binding proteins. It is essential to use a concentration that can saturate the surface of the solid matrix. Most microtiter plates are capable of binding approximately 400 ng protein/cm^2. Thus, the concentration can be calculated for any given surface area to be coated.

Typically, protein concentrations of 3 $\mu g/ml$ or more need to be used to saturate the surface. For smaller haptens such as peptides, much higher concentrations will be required. The range of protein concentrations at which there is no interference with binding to the plastic is called the zone of independent binding (Kemeny

1992; Cantarero et al. 1980). In practice, this is typically 1 μg/ml, although as much as 10 μg/ml can be used without difficulty.

For solid-phase systems with high surface area, higher concentrations of the capture reagents are required. The variables that control this process include the surface area to be covered, the affinity of the antibody for analyte, the length of time that the reaction is carried out, the purity of the antibody preparation, and the sensitivity of the detection method. For example, for membranes, application of submicroliter quantities of purified antibody solutions with a concentration range of 100 μg/ml–1 mg/ml in a 1-mm diameter surface area is typical.

Sometimes, the capture antibody or antigen is particularly precious because it is either difficult to obtain or expensive. Such a molecule can be coupled to protein carriers such as BSA. This will not only allow its quantitative binding to the solid phase, but also greater recognition by the detector antibody due to less steric hindrance. Alternatively, such reagents can be coated at suboptimal concentration, or the coating solutions reused in subsequent assays. However, such approaches often lead to poor assay precision.

Generally, if the capture antibody concentration exceeds 1 μg/cm^2, unstable bi- and polylayers will be formed, because there is no space left on the surface of the plastic. Such protein-protein interactions are generally weaker than those between protein and plastic. Thus, analyte molecules bound to these layers will be removed during the separation of the bound and free forms, thereby compromising assay results and performance. Moreover, the binding constant of the antibody continues to decrease with increasing loading density of antibodies because of steric hindrance.

Generally, the optimal coating concentration, conjugate dilution, and minimal antigen concentration can be easily determined by the checkerboard titration method (Figure 9.10). Here, the antigen solution is serially diluted across the plate, while the capture antibody solution is diluted down the plate. An enzyme-labeled secondary antibody is then added at a fixed concentration to all wells. Upon completion of the assay and based on an optimal absorbance value, the optimal concentrations of both antigen and antibody can be derived from the corresponding well. Alternatively, the assay can also be performed with various concentrations of the antigen using such an assay format. As a general rule, the nonspecific response increases with increasing coating and conjugate concentrations. However, the specific response, expressed as the assay response at given analyte concentrations, plateaus at increasing coating concentrations.

Similar methods can also be used to optimize the coating concentrations of capture reagents on other immobile solid-phase surfaces such as coated tubes. Particulate solid phases, such as latex beads or magnetizable particles coated to a fixed density with the capture reagents, can be serially diluted for such testing.

Depending on the assay format used, the following points need to be considered to optimize coating of the solid-phase capture reagent as well as the concentration of the labeled antibody or labeled hapten.

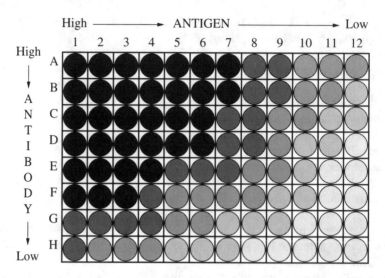

High ——————→ ANTIGEN ——————→ Low

Figure 9.10. An example of a checkerboard titration using polystyrene microtiter plate. See text for details.

For the development of a two-site enzyme immunoassay, the dose range is limited at the low end by the signal-to-noise ratio and reproducibility of the assay response. This is in turn determined by the combined affinity of the two antibodies, the detectability of labeled antibody, and the nonspecific signal produced in the absence of analyte. The assay response increases with increasing amounts of capture antibody and labeled antibodies. However, a limit may be reached eventually beyond which increasing background offsets the benefits of increasing signal magnitude (Micallef and Ahsan 1994). The upper limit of the dose range may be determined by the amounts of antibodies used or by the method used to determine the end point of the assay. Therefore, both the solid phase and labeled antibodies need to be titrated to ensure sufficient amounts of both to operate in excess in the presence of the highest analyte concentrations to be measured, while maximizing precision at critical diagnostic levels.

As compared to the two-site reagent excess immunometric assays described above, the concentration range over which acceptable precision is obtained is much narrower for the reagent-limited, labeled-analyte competitive assays. By changing the concentration of antibody and labeled hapten used, one can change the range of acceptable precision, and hence, the limiting sensitivity of these types of assays. The approximate antibody and labeled-analyte concentrations can be determined by using a checkerboard titration in which each combination of antibody and labeled hapten dilution is tested in the presence of zero analyte and analyte concentrations corresponding to the anticipated highest and lowest standards. The initial optimum reagent concentrations are those giving the greatest and most reproducible changes in signal with increasing analyte concentration. The concentrations finally selected are those giving the best precision in the diagnostic ranges required.

Immunoassays for detecting specific antibodies are generally qualitative in nature, and hence, need to be optimized to give a low detection limit as well as to minimize false positive and false negative results. In this case, the capture antibody and antigen concentrations need to be titrated with dilutions of known, strongly positive and weakly positive sera. Sufficient quantities of both must be used to ensure that reagent excess conditions prevail for strongly positive samples, while keeping nonspecific binding and the imprecision of low assay response levels as low as possible (Micallef and Ahsan 1994). This will allow one to maximize the signal-to-noise ratio and the sensitivity of the assay for weakly positive samples.

4. Coating Buffers and pH. The mechanisms by which proteins bind to plastics, although still poorly understood, certainly involve charge and hydrophobicity. The charge expressed by a protein is a function of the pH of the buffer in which it is dissolved. Adsorption appears to be relatively independent of pH as long as the pH of the coating buffer is at least two units above the pI of the protein. However, the rate and pattern of binding may be dependent on the pH.

Recently, Butler et al. (1992) studied the adsorption pattern of both polyclonal and monoclonal antibodies on microtiter plates as a function of pH (Figure 9.11A). They observed a similar protein loading density at the three different pH values used in this study. However, a typical saturation plot displaying a characteristic linear binding region was observed only at pH 9.6. At both pH 4.5 and 7.0, the percentage of IgG adsorbed, i.e., its affinity for polystyrene, was dependent on the concentration of the capture antibody used. Because no linear binding region was observed, the adsorptive process appears to be cooperative at these two pH values. These data suggest that the pattern of adsorption observed at alkaline pH does permit the construction of a simple standard adsorption plot to be used to determine the amount of capture antibody that is adsorbed on polystyrene relative to the amount added (Figure 9.11B). Ideally, to eliminate pipetting errors in aliquoting the antibody and thus ensure good assay precision, one needs to select the concentration at the top of the linear portion of such a calibration curve.

Typically, a carbonate/bicarbonate buffer system of pH 9.6 is routinely used for coating of the capture reagent onto the solid phase. However, phosphate and Tris buffers of pH 7.0–7.5 can also be used. A slight loss of sensitivity may occur at non-physiological pH (Kemeny 1992). It is also essential that the same conditions be applied to the calibrant as to the sample. Sometimes the pH of the coating buffer can markedly influence the amount of a particular antibody that can bind to the solid phase while having no effect on other antibodies. Hence, it is often desirable to establish optimal pH and the type of buffer for a given application.

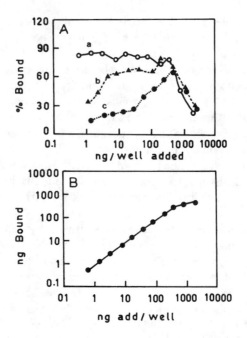

Figure 9.11. Passive adsorption of rabbit IgG as a function of pH on Immulon 2® polystyrene microtiter plates

A: Percent bound plot showing the adsorption characteristics at pH 9.6 (curve a), 7.6 (curve b), and 4.5 (curve c). The coating was done overnight at ambient temperature

B: Standard curve for determining the amount of CAb passively adsorbed at pH 9.6 relative to the concentration of the CAb solution incubated with the polystyrene well

From: Butler et al. (1992)

The concentration of buffers used for coating typically range between 0.05–0.1 M. As long as the capture reagent is stable under these conditions, the molarity of the coating buffer should not have any adverse effects. Generally, buffers of low ionic strength are preferred. The buffers should also be free of any contaminating protein to avoid interference with the coating process.

For some antigens, such as wheat gliadin and some viral and bacterial extracts which are difficult to solubilize, it is necessary to coat in 70% alcohol. Alcohol is reported to provide better coating of such antigens (Kemeny 1992).

5. Coating Conditions.   Stationary or rigid solid phases, such as microtiter plates and tubes, can be coated by aliquoting an appropriate volume of the solution of capture reagent. In contrast, mobile solid phases that are free to move in solution, such as agarose and dextran beads, cellulose discs, magnetizable particles, latex beads, nylon and polystyrene balls, and porous glass beads, can be coated easily by suspending and agitating them in the solution of the coating reagent.

When preparing a large number of solid-phase supports, such as coated tubes, it is often desirable to add a dye to the coating buffer to ensure that all the tubes are filled with the reagent. However, such additives should be carefully tested for any possible adverse effects on assay performance.

Membranes can be coated with capture reagent in one of two ways. In both methods, some mechanism is present to restrict the surface area involved in the application. One method to apply the reagent to a defined surface area of a wetted membrane is to use a manifold filtration device. It permits rapid filtration application of the reagent, while preventing its diffusion beyond the desired surface area (Harvey 1991). This method has the advantage of permitting the rapid application of dilute solutions of proteins, while still controlling the surface area covered. Because the interaction of most proteins with membranes, especially nitrocellulose, is fast, the vacuum application process can be carried out quickly.

A second method of application of capture reagents to a membrane is to apply small volumes (< 1.0 µl) to a dry membrane. This can be carried out with a variety of common laboratory micropipetting devices such as Hamilton syringes.

Extremely accurate application in a very defined surface area of the membrane can be achieved with microprocessor-controlled metering pumps. The advantage of using this type of pump is that it can be used for precise volume placement in a continuous path or repetitious pattern. This type of application is advantageous during device manufacturing as it can be automated. It allows control of the surface area derivatized, the use of a small application volume, and rapid drying (Harvey 1991). One drawback of this method is that the concentration of the capture reagent to be immobilized must be fairly high.

Because passive adsorption of the capture reagent, especially on the plastic surface, is a diffusion-controlled process, sufficient time must be allowed to saturate the surface of the solid-phase supports. The time required to coat a solid-phase support depends on the concentration of the coating solution and the temperature at which this is carried out. Coating at ambient temperatures or at 37°C for 1–2 hr is generally sufficient. Shorter incubation times, however, often result in unequal coating or immobilization of the capture reagent. Ther solution and the temperature at which this is carried out. Coating at ambient temperatures or at 37°C for 1–2 hr is generally sufficient. Shorter incubation times, however, often result in unequal coating or immobilization of the capture reagent. Therefore, it is desirable to coat for at least 16–24 hr at 4°C to obtain the best precision. For large scale preparation of solid phase, these conditions are particularly useful. However, because proteins tend to denature slowly with time, particularly in dilute solutions, optimal coating times must be determined for each new type and lot of capture reagent. The efficiency of the coating process can be monitored as described below.

Evaluation of Coating Efficiency

Uniform coating of reagents to a solid phase is key to the consistent and accurate performance of enzyme immunoassays. For consistent assay performance, the variation in coating levels, as indicated by the coefficient of variation (CV), should not exceed 5%. Additionally, all binding values should be within 10% of the mean (McLean 1992). Assays that accurately measure the binding uniformity of coating reagents to solid phases should therefore be employed to critically evaluate microplates prepared for use in an ELISA.

Nowadays, most automated plate readers come with their own software programs that display an (8×12)-matrix of all the absorbance readings. They also calculate the mean absorbance value, standard deviation, and CV for the entire plate. Any wells with absorbance values $\geq 10\%$ from the mean are also identified. The software then allows the user to select any combination of wells for further analyses. For example, the wells in a strip or at the edge or center of a plate may be compared to other areas of the plate. The mean absorbance value, standard deviation, CV, and percent deviation from the mean are then calculated for such a subset of wells. The continued application of these analytical procedures in quality assurance and stability testing programs will enable one to evaluate the effects of different processes, materials, and storage conditions on coating uniformity of the capture reagent, thereby ensuring the production of a high quality product.

Similar approaches can also be used to evaluate the coating efficiency of other solid-phase systems. For example, coated tubes can be randomly selected from a given manufacturing lot and tested for precision. A high precision indicates uniform coating of the capture reagent.

BLOCKING

Once the primary molecule of interest, such as the capture antibody or the antigen, is immobilized on the solid-phase, many developers of solid-phase immunoassay methods carry out an extra immobilization step to block the unoccupied sites to prevent nonspecific adsorption of assay components (e.g., the enzyme tracer) during the subsequent assay steps. It is generally not necessary to completely dry the solid phase prior to the application of a blocking agent.

An ideal blocking agent should have the following characteristics.

1. It should bind to all the remaining protein binding sites,
2. It should not displace the immobilized component during the blocking step,
3. It should not alter the immunoreactivity of the immobilized components,
4. It should be inert and thus not reactive with any of the subsequent reactants in the assay, and
5. It should not interfere or contribute to the detection signal.

Sometimes it is desirable to coadsorb the blocking agent with hard-to-adsorb proteins. This is generally the case with some monoclonal antibodies. In such instances, coimmobilization with a blocking agent often promotes their attachment to the solid phase. With certain solid-phase systems, such as microparticles or latex beads, coadsorption also helps to space out the antigen or antibody molecules to ensure optimum coverage and to preserve their upright orientation. For such solid phases with high surface area, the blocking agent may be used before, during, or after coating with the primary protein, depending on the ratio of blocker molecules to desired protein and the strategy for their use.

Blocking agents that have found widespread use and popularity in diagnostic kit development can be broadly classified into three categories: proteins, nonproteins, and detergents or surfactants. These are listed in Table 9.2. Proteins are the most effective blocking agents, because they need to be used only once and their effects seem to be permanent. Of these, bovine serum albumin (BSA) is by far the most popular and commonly employed blocking reagent. It yields satisfactory results for most systems, and is commonly available and relatively inexpensive. Gelatin and casein (or nonfat dry milk) are also commonly used. Generally, 0.1–1.0% concentration of the protein agent is used for blocking, although concentrations as high as 5% can be required with some solid-phase supports.

Polyvinyl alcohol and polyvinyl pyrrolidone are nonprotein blockers. They have the advantage of not being contaminated with interfering proteins. For example, the presence of immunoglobulins in BSA preparations, or endogenous peroxidase activity in a variety of different protein blockers can result in serious background contribution.

Detergents, especially of nonionic type, can also be used as blocking agents to eliminate nonspecific interactions. However, unlike the other two classes of

TABLE 9.2. Blocking Agents Commonly Used in Solid-Phase Immunoassays

Reagent	Concentration (%)
Protein Agents	
BSA	1–5
Casein	1–2
Gelatin	1–3
Nonfat dry milk	1–5
Ovalbumin	1
Nonprotein Agents	
Polyvinyl pyrrolidone	2
Polyvinyl alcohol	1
Detergents	
Tween®-20	0.05–0.5
Triton®-X-100	0.05–0.5

blocking agents, detergents do not block protein binding sites on the solid phase, but rather reduce the ability of subsequent reactants to interact with the solid phase. Tween®-20 and Triton®, generally at a concentration of less than 0.2%, are the most effectively employed detergents in diagnostic kit development.

The use of detergents should be employed with some caution as they may inhibit not only the interaction of reactants with the solid phase immunosorbent, but may also reduce the interaction of reactants with specific analytes. In addition, some detergents may remove the immobilized protein from the solid phase. Detergents are also more labile and will easily desorb if the equilibrium solution concentration is changed. Therefore, they must be added to all wash solutions and protein diluents if they are to be kept on the surface.

It is essential to select a blocking agent that will not interfere with the assay performance. Depending on the assay system, sometimes both increased and decreased background activity may be seen as a result of blocking. For immunoblot assays, an unsuitable blocking reagent may in fact aggravate the sample-dependent high background; thus, if the use of a blocking protein proves to be necessary, the protein should be carefully chosen and tested (Craig et al. 1993). Thus, it may be necessary to perform detailed studies using a variety of detergents and other blocking agents over a range of concentrations for each particular immunoblot assay in order to arrive at the correct set of conditions. Blocking appears to be an essential step for solid-phase supports with high surface area. However, not all blocking reagents work equally well and the size of the protein used to block, for example, may influence this step. Generally, small protein molecules provide a more efficient barrier than large ones (Kemeny 1992).

Although, traditionally, a blocking step is routinely carried out to minimize nonspecific binding of assay components in solid-phase immunoassays, in most normal circumstances, this is not necessary and other approaches to reduce non-specific binding may prove to be more important (Mohammed and Esen 1989). The constitution of the assay and wash buffers should discourage nonspecific binding. For example, assay buffers may contain detergent with or without an inert protein. In addition, only labels that have a low tendency to bind nonspecifically should be used. Thorough washing at every step is also essential. Blocking also may not be needed when working with solid-phase supports to which the protein can be covalently linked, provided the remaining reactive groups on the solid phase after the coating process are thoroughly quenched.

ENZYME LABELS

The enzyme labels used must have high immunoreactivity, high specific activity, and low nonspecific binding. Each tracer needs to be carefully tested for the presence of any unique instability, inadequate optimization, or any special interferences that might be related to the biochemical characteristics of the enzyme itself.

The optimum concentration of the label required must be determined by experimentation. The checkerboard titration method described earlier often gives a fairly accurate estimate of the label concentration. For competitive immunoassays, the enzyme label should be used at a limited concentration. This can be determined experimentally in the absence of analyte by adding sufficient quantity to give an optical density as close as possible to the maximum that can be obtained with the given solid phase. This will normally be in the steep part of the binding curve and will give the widest detectable range practical. For noncompetitive assays, the enzyme label should be used in excess. If there are a number of reagents used in the detection stage of the assay, it is essential to optimize the first layer and then the subsequent steps.

SAMPLE AND STANDARD PREPARATION

Although great emphasis traditionally has been placed on the accuracy and precision of immunoassays, some of the largest potential sources of error concern sample collection and preparation, and handling methods. Fortunately, one of the major advantages of the immunoassay as compared to traditional analytic techniques is that extremely specific assays can be performed on relatively crude samples. However, it is still essential that a well-thought-out sample preparation method be developed for effectively getting the analyte out of the sample matrix and into the assay at an appropriate concentration for testing and free of sample matrix interferences.

The sample preparation step can also be used to concentrate and dilute a sample depending on the sensitivity requirements of the application. Ideally, the structure of the analyte after extraction from the sample matrix should reflect the immunogen structure to which the antibodies were developed. Generally, there is a limit to the amount of sample that can be measured. In practice, this is rather similar for most immunoassays over a concentration range of 1–1000 ng/ml (Kemeny 1992). Very low concentrations require special modifications and higher concentrations need dilution. High concentrations of sample, particularly in clinical settings, also cause other problems such as a greater tendency of low affinity antibodies to bind to the solid phase as well as enhanced nonspecific binding. It is therefore rarely practical to assay serum samples at a dilution of less than 1/50. The high samples need to be diluted appropriately so that they can be measured in the most precise and accurate part of the standard curve.

The sample extraction or dilution buffer thus is an important element of the immunoassay, both in providing efficient extraction of the analyte from the sample into a liquid phase, and in minimizing background noise due to nonspecific reactions. For clinical samples, the main components affecting the primary antigen-antibody reaction appear to be proteins—their concentration and, in some cases, the specific distribution of individual proteins. The pH, ionic strength, and other components such as lipids and specific ions also influence the diluent matrix. In patients, abnormal levels of specific proteins, immunoglobulins, or enzymes also change the matrix.

The selection of a suitable diluent for high samples therefore requires careful consideration of several factors. A variety of buffered protein solutions are used for this purpose. The commonly used diluents include assay buffer containing protein, zero standard hormone-free serum, stripped serum, hypo-serum pool, heterologous serum from horse or pig, and simulated serum containing albumin and thyroid binding globulin (Feldkamp and Smith 1987).

The best diluent will vary from kit to kit, and therefore, needs to be explicitly tested. Because standards are often prepared in analyte-free human serum, many commercial protocols specify that the zero standard should be used to dilute high samples. In such cases, enough zero should be provided for this purpose. When the method is to be used over a very wide range of concentrations and samples will frequently need dilution, the manufacturer should specify and supply the proper diluent. Even a specified diluent must be tested in actual case.

Sometimes, a clinical sample must be diluted with a pool of normal serum, the patient's own baseline, or prestimulated serum rather than with the zero standard which has been chemically treated or lyophilized (Feldkamp and Smith 1987). Examples of cases in which the selection of the best diluent significantly affected the accuracy of the final assay results are shown in Figure 9.12. Information on subject preparation and sample collection and handling in clinical settings for immunodiagnostic purposes was recently reviewed (Wilde 1994).

In environmental diagnostics, water-miscible solvents such as methanol, isopropanol, or acetonitrile are sometimes used to solubilize analytes from a sample. These solvents can be tolerated in immunoassays up to levels of 10–20% without affecting the primary binding reaction or enzyme activity.

The sample preparation employed should complement the immunoassay to provide an integrated system. Simple and rapid methods of sample preparation are needed to enable an on-site assay format to be effectively utilized. Likewise, a throughput assay format such as the 96-well microtiter plate requires a sample preparation protocol that is capable of similar throughput and speed.

Similar considerations must also be given to preparing reference standard or calibrator solutions. Information on this topic is comprehensively reviewed in Chapter 10.

The sample and standard preparatory procedures should be performed in a reproducible way day after day. Appropriate time should also be allowed for this purpose. The reagents and samples must be stored in well-marked containers, in

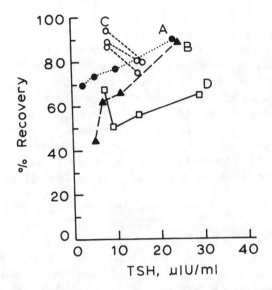

Figure 9.12. Effect of matrix on the recovery of thyrotropin (TSH). A high-TSH kit standard and a commercially available TSH (Calstan, Calbiochem) were diluted in the kit zero (curve A), a low-TSH serum pool (curves B and D), and an individual patient sample (curve C). The recovery and parallelism varied with the diluent selected. Highest recovery (curve C) was observed when a low-TSH serum from a single patient was used.
From: Feldkamp and Smith (1987)

standardized concentrations and volumes to avoid simple mistakes such as erroneous dilution or a mix-up of vials. Reagents should also preferably be stored in the concentration in which they are to be used in the assay, or in a way such that they require only minimal treatment prior to use.

It is also important that both reagents and samples be equilibrated to the appropriate assay temperature before being combined. The same consideration should also be given to the solid phase.

ASSAY CONDITIONS

Successful binding of the test analyte, either antibody or antigen, to the solid-phase-immobilized capture reagent is critical for the design, optimization, and success of the immunoassay. Immunoassay kinetics are generally limited by diffusion. Despite macroscopic mixing of a reaction, molecules take time to travel by Brownian motion across microscopic distances in order to interact. This "diffusion to capture" process, i.e., the overall process by which ligands move towards and successfully bind to receptors or binding proteins, has been reviewed (Berg

1984; Geurts 1989; Stenberg et al. 1988). The rate of diffusion to capture is indirectly proportional to the size (radius) of the diffusing species and to the square of the distance between them. If a small species, such as a hapten, interacts with a larger one, the overall rate is dominated by the smaller partner. Particular attention must also be given to reaction kinetics in assays that involve binding to the walls of a well, cuvette, or test tube, because large distances may be involved (Nygren and Stenberg 1989; DeLisi 1980). Such assays are quite sensitive to mixing, which effectively reduces this distance. If such systems are not deliberately mixed, thermal convection can adversely affect assay reproducibility.

Several factors thus need to be considered to optimize assay conditions in order to obtain desirable results within given assay specifications. These are briefly described below.

Incubation Temperature

Increasing assay temperature generally increases the rate of the reaction between the antigen and the antibody. The affinity constant K_a of this reversible reaction, however, decreases with increased temperature. Therefore, the overall avidity of the antibody, and, hence, the potential sensitivity of the assay often increase at lower temperatures. Similarly, reproducibility and precision of the assay are also improved by extended incubation in the cold.

Most naturally occurring binding proteins are much more sensitive to temperature changes than are the antibodies. For example, the K_a for cortisol-binding globulin (transcortin) increased 20-fold as the temperature was lowered from 37°C to 4°C (Chan and Slaunwhite 1977), and that for thyroxine-binding globulin increased 3-fold from 37°C to room temperature. In contrast, the affinity of T_4 for its antibody remained essentially unchanged. Proteolysis and other adverse reactions can sometimes occur at higher temperatures, thereby affecting assay performance.

Generally, in most normal standard procedures with nearly all assays, the higher temperatures will not have any adverse effects unless the immunoreactants are temperature labile and long incubation times are used. Lower temperatures are therefore essential while performing assays for temperature-sensitive analytes.

Incubation Time

To achieve maximal signal sizes and to minimize drifts in results, it is usual to incubate for a sufficient time to allow all reactions to achieve or approach equilibrium. Thus, any small differences in the length of time that a particular sample has been incubated will have little or no effect on the amount of signal generated. The time required may vary from a few minutes for a modern particulate or membrane immunoextraction step at 37°C to several hours for some other assay systems at 4°C. The rate of reaction may also be different for the samples and standard.

Shorter incubation times can be used for assays of analytes that occur in relatively high concentrations, e.g., human placental lactogen (hPL), thyroxine, cortisol, and digoxin. In contrast, low analyte concentrations require longer times to reach equilibrium.

The rate of equilibrium of the reaction can be determined by incubating a few test samples and standards of different concentrations for varying lengths of time and plotting the resultant signal against time. The incubation time and temperature chosen should be sufficient so that the signals for samples and standards for all concentrations approach stable maxima.

Sometimes, sequential assay designs, in which the unlabeled antigen is added before the labeled antigen, can be used to increase assay sensitivity and to reduce the incubation time. This is particularly useful for assays where the labeled antigen is sensitive to degradation.

Currently, an increasing number of assays are performed as nonequilibrium assays. In such cases, exact incubation time becomes more critical. The exactness of timing of the various steps of the assay also becomes important. Therefore, care should be taken to maintain the same time lag between the different steps of the assay. Otherwise, the characteristics of most nonequilibrium assays do not differ from those of the equilibrium assays, provided that the timing of the reactions is appropriately controlled.

Noncompetitive immunoassays are affected by temperature and incubation time similarly to competitive assays. Prolonged incubation time thus also increases both sensitivity and specificity in noncompetitive assays. When sensitivity is not the most essential assay feature, assays may be performed with a shorter incubation.

The incubation times of assays used in clinical practice are often influenced more by the need for urgent results than for optimal sensitivity or precision. The assays used for decision making in acute medicine, such as digoxin, estriol, or drug assays, must be performed within a short period of time if the results are to be clinically useful. In nonacute clinical settings, it is often practical to perform the assays with longer incubation times.

Buffers and Incubation pH

The composition of the buffer and its pH influence the antigen-antibody reaction, particularly if the binding is pH dependent. For example, pH 8.0 is optimal for the binding of cortisol to transcortin, the binding protein for cortisol in serum. In designing an immunoassay, one can take advantage of the temperature and pH effect. The binding of cortisol to transcortin is highly temperature dependent, whereas the binding to antibody is not (Chan 1987). By incubating at tempera-

TABLE 9.3.   Buffer Systems Commonly Used in Enzyme Immunoassays

- Phosphate-buffered saline (PBS), pH 7.2, 0.15 *M*
- Carbonate buffer, pH 9.6, 0.05 *M*
- Borate buffer, pH 7.4, 0.2 *M*
- Borate-buffered saline, pH 8.3–8.5, 0.1 *M*
- Tris-buffered saline (TBS), pH 7.5, 0.1 *M*
- Citrate buffer, pH 3.0–7.0, 0.1 *M*
- Veronal®-buffered saline

tures greater than 37°C and at pH 5.0, cortisol will be released from transcortin because of the low K_a at high temperature and low pH; however, its binding to antibody will be about the same. Therefore, the extraction of cortisol from its binder and the binding to antibody can be carried out in a single step, thereby eliminating the pretreatment of serum by heat denaturation.

Buffers of high or low pH values can sometimes prevent antigen-antibody reactions. Ideally, immunoassay buffers should include factors that can stabilize the immunoreactants and promote their interactions, increase signal-to-noise ratio, and possibly also eliminate the chance for false positives.

The efficiency of the binding reaction therefore depends on its environment (pH, ionic strength), and the presence of additives (carrier proteins, detergents, protease inhibitors, and antibacterial chemicals). Several major buffer systems that have been commonly employed in immunoassays are listed in Table 9.3. The buffer salts maintain a restricted pH range and establish the ionic strength. Other additives function in important roles. For example, Tween®-20 is added as a detergent to reduce nonspecific binding. Sodium azide prevents bacterial growth during buffer storage and prolonged assay conditions. Bovine and human serum albumin and other proteins are added to block nonspecific binding sites. In assays where small particles can cause interference, the buffer needs to be filtered through a 0.22-micron filter. Among the most widely used enzyme labels, alkaline phosphatase is incompatible with phosphate buffers, while sodium azide inhibits horseradish peroxidase. Hence, appropriate buffer systems need to be selected while working with these two enzymes.

Finally, one must remember that the optimal pH or buffer to be employed in any immunoassay system cannot be predicted empirically. Therefore, multiple buffer conditions must be examined for each antigen and antibody combination prior to settling on the "optimal" conditions.

Reagent Mixing

The immunochemistry and kinetics of solid-phase antigen-antibody reactions differ significantly from classical liquid-phase reactions involving haptens and antibodies in solutions. Two factors characterize these reactions on solid-phase supports. First, the immobilized reactant can no longer diffuse and its contribution to the collision rate is thus eliminated. Second, the average nearest distance between antigen and antibody is greatly increased. The kinetics of these reactions therefore become important in the design, optimization, and performance of a solid-phase immunoassay based on equilibrium assay principles.

In a solid-phase system, immunologically active reactants, i.e., antigen and antibody, are not uniformly distributed. The distance between the analyte and the capture reagent is not the same for all molecules; analyte molecules close to the solid phase collide and react faster with the capture reagent than those in solution in the middle of the microtiter well (Franz and Stegemann 1991). Therefore, several compartments exist in the reaction vessel, and the overall binding of analyte with capture molecules consists of: (1) their transport to the solid-phase surface, (2) the binding between antibodies and antigens, (3) the dissociation of reactants, and (4) the transport of dissociated reactants back into the solution.

The rate of bimolecular reaction is a time-limiting process at low concentrations of capture reagent immobilized on the solid phase. In contrast, at high concentrations, the reaction may occur so fast that the transport of analyte molecules to the solid phase becomes rate limiting. When the microtiter plate is vigorously agitated, enough energy for the transport of analyte molecules to the solid-phase capture reagent is provided so that diffusion is of less importance.

Generally, equilibrium can be attained by agitation at high speed (1200 rpm) within 1 hr, as compared to 16–32 hr nonagitated systems (Franz and Stegemann 1991). These times are likely to vary depending on the affinity of the antibody used in a given assay system. At high shaker rotation speeds, a second mechanism appears to facilitate the rapid attainment of equilibrium. The geometry of the fluid within the reaction vessel is altered. Using high speed photographic techniques, Franz and Stegemann (1991) have shown that agitation at 1200 rpm results in a larger area of the surface of the microtiter well being exposed to the soluble reactants. Thus, the area covered by 50 μl of PBS was increased by 50%, from 75 mm^2 to 115 mm^2, assuming a cylindrical geometry of the microtiter well. As the volume of the fluid in the well remains unchanged, the diffusion distances are greatly lowered.

Mixing of the reactants generally reduces incubation times considerably, because the reaction rate is proportional to the diffusion distance raised to the inverse third power. Uniform speed of mixing is particularly important for nonequilibrium immunoassays to obtain good assay precision.

Separation Steps

The choice of separation method also influences the binding reaction, or at least how this reaction is quantitated after the performance of the separation. Certain separation techniques directly affect the primary reaction between the antigen and its antibody. With solid-phase immunoassays, three factors are important: wash solution composition, gentle versus vigorous washing, and aspirating versus hand decanting.

The wash solution should ideally contain a physiological or enzyme friendly buffer. Thus, phosphate buffers should be avoided for alkaline phosphatase labels, while the use of sodium azide should be avoided in wash solutions when working with horseradish peroxidase. Detergents such as Tween®-20 should also be included in wash buffers to block sites that may become available due to protein desorption during washing.

The nature of the washing step is also critical. Vigorous washing can strip off the bound complex as well as inactivate the bound enzyme label. In contrast, gentle washing can separate the free without stripping off the bound complex from the solid phase. If background levels appear to be high, one can simply increase the number of washes.

Hand decanting of the solid phase, although tedious and slow, often gives excellent results. For automated microtiter plate washers, the results can be more variable. Generally, a vacuum of around 400 mm Hg is optimal for the 96-well microtiter plates. A high vacuum can cause excessive drying, resulting in enzyme inactivation, while a lower vacuum may leave residual wash solution in the wells, thereby affecting assay precision.

Color Development

An understanding of the kinetics of the bound enzyme label is essential during the course of assay optimization. Optimization of this step of an enzyme immunoassay involves selection of suitable substrates, timing of the reaction, and selection of appropriate development conditions.

Appropriate selection of the substrate can often influence the assay results. For example, TMB (3,3',5,5'-tetramethylbenzidine) is ideally suited for kinetic analyses because of its rapid reaction rates with peroxidase. It also shortens the incubation time and results in greater assay sensitivity. In contrast, ABTS [2,2'-azino-di-(3-ethylbenzothiazoline-6-sulfonate)], because of its slower reaction rate, is ideally suited for endpoint analyses. It also offers a greater dynamic range for the assay.

Kinetic analyses are often preferred in monitoring homogeneous enzyme immunoassays. Here high specific enzyme activities and pronounced alterations of activity in the course of the immune reaction are necessary to guarantee sensitive assays with short substrate reaction times. In contrast, end point analyses are widely used in heterogeneous enzyme immunoassays testing large panels of samples and whenever high sensitivity and a greater dynamic range are required. Here the color development reaction is terminated by the addition of stopping reagents (in most cases acid or alkaline solutions), which often leads to a bathochromic or hypsochromic shift of the absorption maxima and an increased absorption coefficient of the formed product.

The optimal time of reaction can be easily determined by incubating the bound enzyme with excess of substrate. Incubations can be carried out at 2-min intervals for up to 30 min. The absorbance readings then can be plotted as a function of incubation time. Color development increases linearly as a function of time during the initial stages and reaches a plateau. Ideally, to eliminate errors due to timing of steps, the incubation time should be at least 3 min long.

In addition to time, the effects of temperature and light, if the chromophore is light sensitive, also need to be investigated. Similarly, mechanical mixing results in even distribution of the substrate and color formation as well as efficient inactivation of the bound enzyme label at the termination of the reaction.

In general, the following points need to be kept in mind when optimizing this step of the immunoassay.

1. For successful measurement of the enzyme activity in these assays, there must be tight control of temperature, pH, ionic strengths of the buffer, and various cofactor concentrations necessary for the enzyme to convert substrate into product.

2. Because enzymes are proteins and can be subject to rapid denaturation under improper incubation conditions, close attention must be paid to preserve the enzyme activity.

3. Improper storage of immunochemical reagents must be avoided to preserve the enzyme activity of the labeled reagents.

4. Substrate depletion may become a problem. Often, if a high quantity of the enzyme label is present on the solid phase, substrates can be depleted very rapidly. One must take care to have sufficient substrate so that full color development can occur. The upper limits of linearity should be carefully defined.

5. Improper pH and ionic strength of the reaction buffer can adversely influence color development.

ASSAY EVALUATION

Technical evaluation is the first step in the assessment of the performance of a newly developed immunoassay system. Evaluation of immunoassays requires an efficient protocol and objective criteria. Various specific experimental approaches can be used to test the performance of a method and to determine whether it meets the defined goals.

The methods of evaluation may vary somewhat with the assay format. Nonetheless, the assay should be tested with regard to precision, sensitivity, accuracy, and specificity (in that order); should be compared with another method, preferably a reference method using actual clinical/field samples; and should be tested for lot-to-lot variation. Furthermore, potential sources of errors of measurement must be understood and evaluated for a given immunoassay system.

These errors can be systematic, consistently deflecting all repeated measurements of the same quantity of analyte from the true value, thereby constituting the source of assay bias; and random errors, affecting individual measurements inconsistently and causing a scatter of results about some average value. The latter effect is often referred to as the assay precision profile.

Several of the above parameters are also used routinely for quality assurance and quality control (QA/QC) for daily use of the procedures, FDA approval of a test kit, intralaboratory comparison of test methodologies, and interlaboratory evaluation. These parameters for the evaluation of immunoassays are briefly described below.

Precision

Precision is probably the most important technical aspect of an immunoassay performance. Precision, also sometimes referred to as reproducibility, is a statistical measure of the variation between repeated determinations on the same sample, either within the same run or from day to day, i.e., between runs. Imprecision is the opposite of precision.

Precision is typically tested early in the evaluation protocol because, if a method cannot give reproducible results in a given laboratory, it is unlikely that other aspects of the evaluation will be valid or that the method will be acceptable for routine use. A well-planned evaluation with pooled samples prepared to run many times can efficiently accumulate adequate data within and between runs to assess the precision of the kit. Several protocols offer convenient approaches to establish precision in clinical chemistry methods cost effectively (NCCLS 1984, 1989).

The statistics conventionally used to express the precision profile of an assay are the mean $(\overline{X})$, standard deviation (SD) and the coefficient of variation, CV [% $CV = (SD/\overline{X}) * 100$] at a particular analyte level.

1. Precision Profile.   The precision profile is a way of visualizing the performance of an immunoassay over the full range of analyte concentrations. This is particularly useful for immunoassays that are usually less precise at both low and high concentrations (competitive assays) and principally at low concentrations (immunometric assays). The precision profile can be constructed by plotting CV versus concentration (Figure 9.13). For competitive assays, the precision profile resembles a U-shaped curve, whereas immunometric assays tend to have a flat precision curve. At the extreme low concentration, approaching the detection limit, the sensitivity of the assay affects the precision. At the other end of the concentration curve, where the absorbance is high for the enzyme immunoassays, the precision also suffers.

To construct a precision profile of the assay, ideally at least 10–20 replicates should be run at each standard concentration, as the estimate of true precision im-

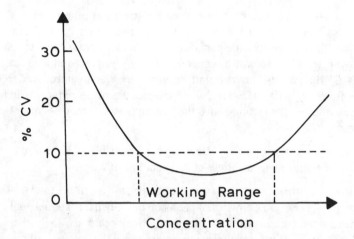

Figure 9.13. The precision profile of an immunoassay illustrates nonuniform error, expressed as % coefficient of variation (CV), in the dose estimate as a function of analyte concentration. The working range of an assay can be established as the range where imprecision is below a preset level, such as 10% CV in this example.

proves with the number of replicates. Although it is desirable to use as many standard concentrations as possible, at least three concentrations should be selected representing low (80% B_0), midrange (50% B_0) and high (20%B_0), where B_0, the maximum binding, represents the binding of the zero standard. In addition, clinically important ranges must be included in such analyses.

The precision profile prepared at the time of assay development contains useful information on the "working range" of the assay. The working range is that range of analyte concentrations over which measurements are of a precision sufficient for a given application. This minimal acceptable precision must be defined for each immunoassay kit. High precision is particularly important in the concentration range useful for critical clinical decisions. Hence, the assay design, i.e., the concentration of antibody and enzyme label, should be optimized to result in excellent precision in these concentration ranges. Usually, a CV of 10%, often much less, is obtainable with current immunoassay procedures (Figure 9.13).

2. Measurement of Precision.   Measurement of precision may be partitioned to allow evaluation of its causes. These are briefly described below.

WITHIN-RUN PRECISION.   Within-run precision is defined as the precision of the same sample run several times within the same assay. At least three levels of samples (80, 50, and 20% B_0) should be used to cover the useful range of the assay. Ideally, at least 20 replicates should be run. Samples should be spaced equidistantly throughout the assay rather than sequentially. The data also need to be analyzed for any outliers.

BETWEEN-RUN PRECISION.   Between-run precision is an index of the ability of the assay to reproduce a result on the same sample from run to run, and from day to day. It is a more realistic performance indicator because real test samples are analyzed from day to day. In general, between-run precision tends to have a higher % *CV* than within-run precision, because variables are introduced from run to run or from day to day. However, when samples are run as replicates and precision assessed using the mean results, the between-run *CV* may be lower than the within-run *CV*.

BETWEEN-LOT PRECISION.   Between-lot precision is an estimate of the variability of results using a variety of different lots of reagents.

WITHIN-METHOD, WITHIN-METHOD GROUP, AND ALL-LABORATORY PRECISION.   These are estimates of precision described by external quality control schemes for control samples run in many laboratories. Only the within-method category is of practical relevance in assessing the precision of an individual method.

3. *Factors Affecting Precision.* Because precision is the most important aspect of the technical performance, identifying and controlling the source of imprecision is quite important. Assay imprecision is caused by the combined effects of several sources of variation (Table 9.4). Some of these factors are briefly described below.

TABLE 9.4.   Sources of Imprecision in Immunoassays

Reagents
Capture antibody or antigen, antibody or hapten enzyme conjugate, calibrator, diluent, wash solution, substrate set, quality control

Pipetting Errors
Accuracy, precision, carryover, calibration, setting, reproducibility

Binding Reaction
Timing of reagent additions, incubation time and temperature, equilibrium vs nonequilibrium assay, high dose hook effect

Color Reaction and Detection
Timing, temperature, quenching, stability of substrate and colored product, spectrophotometric error

Interfering Substances
Add color or reduce color (e.g., bilirubin, drugs), act as enzyme (e.g., hemoglobin, endogenous alkaline phosphatase), inhibit enzymes (metals), scatter light (lipids, macromolecules)

Data Processing
Curve fitting techniques

ERRORS IN THE PRIMARY REAGENTS. With the exception of reference standards, imprecision due to some basic fault in the reagent solutions is very rare, unless the assay was not optimized properly to begin with. Generally, factors that affect the rate of antigen-antibody binding reaction also influence the assay precision. These include antibody concentration, temperature and time:temperature profile, pH and ionic strength of the reaction mixture, and variation in effective antibody coating density if a solid phase adsorbed antibody is used (Davies 1994). Thus, precision of an assay is influenced by the equilibrium constant of the reaction, the rate of reaction, and whether equilibrium has been reached.

The range of acceptable precision, and hence the limiting sensitivity of limited-reagent competitive assays, can be shifted within limits to higher or lower concentrations of analyte by changes in the concentrations of antibody and the labeled hapten used. The effect of changes of antibody concentration on the precision profile when other factors are held constant is illustrated in Figure 9.14. Similar effects can be obtained for changes in labeled hapten concentration.

ERRORS DUE TO SEPARATION PROCEDURE. The basic design of a separation procedure can also have an important influence on precision. The bound and free forms should be separated completely with a minimum of misclassification errors and without affecting the primary binding reaction. The wash step should be designed so that the exact amount of separating wash solution is not critical. If this volume is noncritical over a reasonably wide range, one potential source of

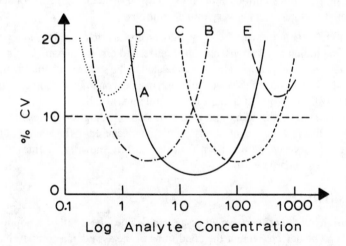

Figure 9.14. Precision profile of immunoassay changes with antibody concentration. Curve A represents an optimal precision profile for a given competitive assay. Curves B and C represent a shift in the precision profile to a lower or higher concentration of analyte, respectively, with a decrease or increase in antibody concentration. Curves D and E illustrate examples of how extremes of antibody concentration can cause a marked deterioration of assay precision.

imprecision is eliminated. However, incomplete decantation of wash solution, drops of supernatant on the sides of decanted tubes, and reaction volumes that are too small to permit effective liquid separation will all lead to poor precision (Davies 1994). Automated microtiter plate washers are also a common source of poor precision, particularly if not adequately maintained or if too high a vacuum is used for aspiration of wells.

ERRORS DUE TO DISEQUILIBRIUM.   This may become a potential source of imprecision. If the time taken for separation represents a substantial part of the total assay time, then significant differences may occur between the first and last tubes of the assay. Problems can also arise if the incubation temperatures are poorly controlled or uneven, leading to the so-called edge effects in microwell assays.

ERRORS DUE TO STANDARDS.   Preparation of standard material often is a major source of imprecision in immunoassays. Individual sets of standards should always be prepared as aliquots from a large pool made at each concentration. Each standard concentration should also be prepared independently using a master standard solution. Doubling dilutions and pipetting on the edges of the tubes should also be avoided to minimize pipetting errors.

Because every step of sampling and dispensing has an inbuilt error, standards should always be prepared in the largest possible batches. Lyophilized standards additionally have two sources of error—at the initial dispensing, and when they are reconstituted before use (Chard 1978). Similarly, protein standards should always be prepared in solutions containing a carrier protein, e.g., BSA or serum, to avoid denaturation effects in very dilute solutions.

MISCELLANEOUS ERRORS.   Assay imprecision is also caused by poor pipetting, errors in the timing of additions, and operator-related errors. Enzyme labels can also contribute to assay imprecision by factors related to the spectrophotometric measurement itself. Thus, timing of the incubation and reagent conditions, stability of the final color, and instrument quality and quality control must also be taken into account.

The precision of a binding assay also varies according to the dose level measured. Hence, analyte concentrations should be measured in the central portion of the standard curve where precision is at its best. It gradually declines at the extremes of the standard curve.

Sensitivity

Sensitivity is the property of an assay that defines the smallest amount of analyte that can be detected under the conditions of the assay. In a competitive assay, the dose-response curve is represented by the following equation:

$$(B/F)^2 + (B/F) (K_a \cdot Ag_T - K_a \cdot Ab_T + 1) - K_a \cdot Ab_T = 0 \qquad (1)$$

where B/F is the ratio of bound-to-free antigen, K_a is the affinity constant, Ag_T the total antigen concentration, and Ab_T the total antibody concentration.

There are two different approaches for determining the theoretical sensitivity and optimal concentrations of reactants in competitive assay designs. Yalow and Berson (1970) defined sensitivity as the maximal slope of the standard curve independent of the experimental error. Ekins and his coworkers (1968, 1970, 1974) defined sensitivity as the lower limit of detection, i.e., the least or minimum detectable dose (LDD or MDD). These two approaches in designing and optimizing the sensitivity of the immunoassays are briefly described below.

1. Yalow and Berson Model.   In this model, the optimal conditions are $Ag_T = 0$ and $Ab_T = 0.5/K_a$. Therefore, equation (1) becomes

$$(B/F)^2 + (B/F)(0 - 0.5 + 1) - 0.5 = 0 \tag{2}$$

$$(B/F)^2 + 0.5(B/F) - 0.5 = 0 \tag{3}$$

$$(B/F + 1)(B/F - 0.5) = 0 \tag{4}$$

$$B/F = -1; 0.5 \tag{5}$$

$$B/Total = 1/3, \text{ percent bound} = 33\% \tag{6}$$

The maximal slope is obtained with a initial percent bound of 33% when Ag_T approaches zero. Thus, assay sensitivity in this model is defined as the smallest measurable change in the slope of the dose-response curve. For sensitive assays, the concentration of antibody binding sites must be low. In the Yalow and Berson model, maximal sensitivity, therefore, can be obtained by using (1) a tracer with very high specific activity so that the initial antigen concentration approaches zero, (2) an antibody with a sufficiently large affinity constant at a concentration equal to $0.5/K_a$, and (3) an initial percent binding of 33%.

2. Ekins' Model.   The treatment of Ekins differs fundamentally from the Yalow and Berson model. Instead of considering sensitivity as the smallest measurable change in the response variable with a small change in analyte concentration (i.e., the smallest measurable change in the slope of the dose-response curve), Ekins has defined sensitivity as the quantity of unlabeled analyte that will change the distribution of radioactivity by an amount that is just equal to the standard deviation of the experimental determination of the ratio of free to bound activity at zero concentration of the analyte.

In the Ekins' model, the optimal conditions are $Ag_T = 4/K_a$ and $Ab_T = 3/K_a$. Therefore, equation (1) becomes

$$(B/F)^2 + (B/F)(4 - 3 + 1) - 3 = 0 \tag{7}$$

$$(B/F)^2 + 2(B/F) - 3 = 0 \tag{8}$$

$$(B/F + 3)(B/F - 1) = 0 \qquad (9)$$

$$B/F = -3; 1 \qquad (10)$$

$$B/\text{Total} = 1/2, \text{ percent bound} = 50\% \qquad (11)$$

Ekins' model thus implies that maximal sensitivity can be achieved with an initial percent binding of 50%. The lower limit of detection expressed as the LDD is inversely proportional to the square root of (1) the specific activity of the tracer, (2) the affinity constant, (3) the reaction volume, and (4) the counting time. Furthermore, this relationship is affected by the experimental error incurred in the measurement of bound and free antigens. An increase in the specific activity of the tracer will reduce the LDD, i.e., the assay sensitivity will be increased, probably through a reduction in counting error. The affinity constant K_a and the experimental error are the limiting factors in achieving maximal sensitivity. Other factors such as reaction volume and counting time can provide limited compensation for the assay conditions (Ekins 1981).

It should be remembered that the original calculations in the Ekins' model refer to one standard deviation of the error at zero dose and not the 95% confidence limit using two standard deviations. The model also assumes that there is no nonspecific binding. It also assumes that the sensitivity is that of the assay reaction and not of the sample, and that experimental errors are limited to a *CV* of 1%. This model also assumes that there is homogeneity of labeled and unlabeled antigen, that there is a single equilibrium constant, and that the reaction proceeds to equilibrium. However, few, if any, assays will comply with all of these assumptions in practice.

Disagreement regarding the concept of assay sensitivity has led to controversy regarding optimal immunoassay design. The Yalow and Berson model disregards errors in the measurement of the assay response, and is therefore unrelated to the statistical concept of precision described earlier. Nonetheless, this model continues to profoundly influence the immunoassay protocols both in research laboratories as well as in commercial settings.

Theoretical sensitivity of the immunometric assay designs can also be calculated in two possible ways. Ekins and coworkers have derived unified models of dose-response relationships and errors for these assays based on the same concepts of response/error relationships that were derived for competitive assays. Detailed descriptions of these models can be found in several articles (Jackson et al. 1983; Davies 1994; Ekins 1991).

In immunometric assays, the low-end sensitivity is very much affected by the accuracy of the lowest standard and often the matrix of the standards. Moreover, for these assays, the precision is also poorest at the low end because of a low signal-to-noise ratio. The sensitivity of these assays is directly proportional to antibody concentration and inversely proportional to nonspecific binding. The inherent biochemical characteristics of immunometric reactions and the availability of suitable monoclonal antibodies, however, have allowed newer assays to have excellent sensitivity.

A graphic representation of these two models for determining the sensitivity of competitive assays is shown in Figure 9.15.

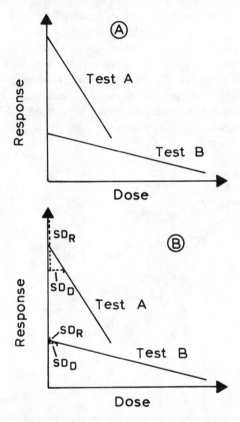

Figure 9.15. Two differing concepts of sensitivity underlying immunoassay design theories.

(A) Yalow and Berson model defines sensitivity as the maximum slope of the standard curve. Thus Test A is more sensitive than Test B. Although widely used in designing and optimizing competitive immunoassays, this model must be viewed in the context of the precision with which the assay response can be defined. A large imprecision will adversely affect the sensitivity of the immunoassay.

(B) Ekins model defines sensitivity as the least amount of analyte that can be clearly distinguished from zero concentration of the standard. Thus Test B, because of better precision, is more sensitive than Test A. This model thus accounts for sensitivity as the precision of the analyte concentration as that concentration approaches zero. SD_R defines the standard deviation of measurement of assay response at a given dose approaching zero; a smaller value as indicated by SD_D defines the least concentration of the analyte that can be measured with statistical certainty.

The LDD, commonly used to define sensitivity, is measured by assaying at least 10–20 replicates of the zero standard and calculating the mean and standard deviation. The mean is used for the standard curve, and the response, mean –2 standard deviations (plus for immunometric assays), read in dose or mass from the standard curve is the LDD, i.e., the smallest dose that is not zero with 95% confidence (Figure 9.16). This equates to the probability that a single replicate of sample does not form part of the distribution of zero analyte values. This is also commonly referred to as the "analytical sensitivity" of the assay.

Knowing the true precision of an immunoassay also enables one to assess the probability that a given concentration differs from a specified value and thus to define the minimal distinguishable difference in concentration (MDDC) (Figure 9.16). Thus, each point on the standard curve is associated with a measurement error (mean response ± 2 standard deviations). Sometimes, the MDDC is also referred to as the "resolution." This "accuracy" of the measurement will be described later in this section.

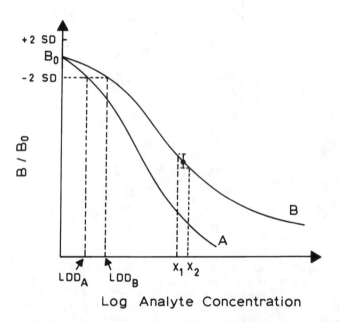

Figure 9.16. Immunoassay sensitivity expressed as least detectable dose (LDD). The precision of B_0 is measured by assaying replicates ($n \geq 10$) of the zero standard. LDD is calculated as the analyte concentration corresponding to $B_0 - 2$ standard deviations (SD) (LDD_A and LDD_B for the two curves shown). The "resolution" or "minimal distinguishable difference in concentration" (MDDC) of the assay is defined by $X_2 - X_1$. For the two assays with the same precision at zero standard, the steepest curve (curve A) is the most sensitive.

The above approach generally gives the best (lowest) LDD possible because of the better precision within the same run. It is also the accepted industry standard. However, the LDD calculated in this manner is unrealistic and usually not reproducible from day to day. Two other approaches can be used to arrive at a better estimation of the sensitivity of a given immunoassay. One can determine the sensitivity between days. This approach is similar to LDD except that the zero standard is run as an unknown between days for a period of time. The mean plus 2 standard deviations of the actual value is calculated from the zero standard as an unknown. This approach is more realistic because the claimed sensitivity should be achievable from day to day. Typically, this value is higher than the LDD by 50–100% (Chan 1992).

Another approach to the evaluation of sensitivity is to dilute a patient's serum with the assay diluent to below the detection limit of the assay. The sensitivity is then determined at the dilution where the percent recovery found, calculated from the found/expected concentration plot, is no longer close to 100% of the expected value.

A final point in the definition of sensitivity is the units in which it is specified. Sensitivity should always be expressed in terms of concentration, i.e., weight/volume, and should refer to the biological fluid for which the assay is intended.

Sensitivity depends on both the slope of the standard curve and the precision of individual measurements at or near zero. For two assays with the same precision at zero, the steepest curve is the most sensitive, i.e., it has the greatest change in measurable response per small change in dose (Figure 9.16). For two assays with the same slope, the more precise is the more sensitive.

Understanding how the various assay components contribute to sensitivity is helpful to more effective assay optimization and troubleshooting. Testing some of the parameters listed in Table 9.5 at the time of assay evaluation will demonstrate how robust the assay is to small changes in reaction conditions. The sensitivity of immunoassays can be improved in certain assay formats. For example, sample size can be increased, but the effect of the changed proportion of sample to reagent must be tested for matrix effects. Additional low concentration standards may be necessary to better define the standard curve (Feldkamp 1992). Increased incubation time, increased temperature, and improved mixing can be predicted to increase analyte binding and precision in some assays. Sequential assay designs also give a better sensitivity than assays with simultaneous addition of the sample and the enzyme tracer.

Sometimes it is also possible to arrange an assay to suit a specific clinical purpose. The basic principle here is that the most precise part of the standard curve (usually the central portion) should be set so that it coincides with the range of practical interest (Chard 1978). This may be either the whole of a normal range for analytes for which both high and low values may be of diagnostic significance (Figure 9.17A), or one extreme of a normal range for analytes for which either low

TABLE 9.5.   Influence of Assay Variables on the Sensitivity of Immunoassays

To Improve Assay Sensitivity
 1. Reduce the amount of tracer. For any given set of assay conditions, however, there is a limiting concentration of tracer below which further reduction leads to no significant changes in assay sensitivity.
 2. Reduce the amount of antibody. Dilutions will shift the standard curve down and to the left. However, for increasing sensitivity using this approach, assay precision must be taken into consideration.
 3. Increase the incubation time for equilibrium assays. This may affect the low end of the standard curve more than the high end. In the initial development of an immunoassay, however, every effort should be made to reduce the incubation time.
 4. Reduce the incubation time for nonequilibrium assays. However, accuracy may suffer.
 5. Sequential saturation assay designs, in which the sample is allowed to react with the antibody prior to the addition of tracer, often improve the slope of the standard curve and, therefore, the assay sensitivity. However, such assay designs may increase the number of steps and reduce assay precision, and are extremely sensitive to time.
 6. Increase sample size. The effects of matrix and volume changes, however, must be taken into account. Generally, if the total concentration of the sample in the incubation mixture is < 10%, nonspecific effects are minimal.
 7. Increase assay temperature. Such an approach enhances the speed of an assay. Edge effects, however, must be carefully investigated.
 8. Increase the number of replicates. Better assay precision is critical for improved sensitivity.
 9. Extract and concentrate the analyte from the sample. Assay sensitivity depends on the concentration of the analyte in the sample. This approach is often useful for analyzing basal circulating levels of several important biological materials, such as hormones.
 10. For immunometric sandwich assays, increase antibody concentration, time, and/or sample size.

To Desensitize an Assay for Targeting the Standard Curve for a Range of Clinical and Biological Interest
 1. Increase the amount of antibody and the tracer. This implies high level of binding of the zero standard and improved precision, thereby increasing the slope of the standard curve. It also shortens incubation time, because equilibrium is reached rapidly.
 2. Reduce sample volume to a minimum compatible with accurate and reproducible pipetting. Assay precision, however, may be lowered.
 3. Use low affinity antibodies, particularly because increasing antibody concentrations lowers assay sensitivity.
 4. The use of above conditions often increases the convenience of performing the assay in terms of ease, speed, and reliability.

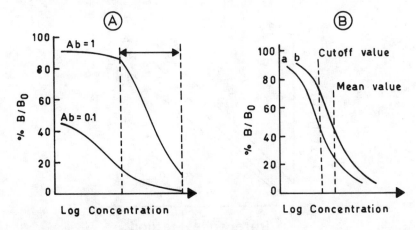

Figure 9.17. Using immunoassay sensitivity to meet specific needs.

A. Although a lower concentration of antibody (Ab = 0.1) yields a very sensitive curve, it covers only a part of the range of physiological or clinical interest (denoted by an arrow between the two dotted lines). Desensitizing the assay by using a higher antibody concentration (Ab = 1) produces a test that covers the range of interest by providing a much steeper slope in this concentration range.

B. A similar concept can also be used to separate normal and abnormal populations based on a test measurement. The midpoint (i.e., 50% B/B_0 ratio) of a standard curve should be chosen as the mean value of the normal population (curve b). To separate the normal from an abnormal population, this ratio should be the value that best separates these two populations (curve a).

(Figure 9.17A), or one extreme of a normal range for analytes for which either low values or high values, but not both, are of diagnostic significance (Figure 9.17B).

Accuracy

Accuracy is the fundamental ability of any assay to measure the true value of an analyte. In analytical measurements, it is defined as how close the "average" measured value is to the true value. Similar to precision, in practice accuracy is evaluated in terms of "inaccuracy" or "bias." Bias is a measure of the difference between the measured and the true value. It often represents a systematic error inherent in a method, or caused by some artifact of the measurement system. Thus, an accurate method is one capable of providing precise and unbiased results.

Assay bias may be both positive and negative. It is said to be "proportional" when the assay reads a constant percentage higher or lower than the true value, or "constant" when the assay reads a constant concentration higher or lower (Figure 9.18). Both types of bias may occur together, and therefore, the overall bias may vary over the assay range (Davies 1994). The relationship between precision, accuracy, and bias is shown in Figure 9.19. This aspect of analytical measurements is described further later in this chapter under Measurement Errors.

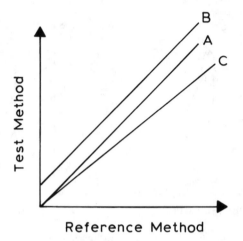

Figure 9.18. Examples of bias in immunoassay measurements. Curve A represents perfect correlation between the test and reference method. Curves B and C represent constant and proportional errors, respectively, in measurements.

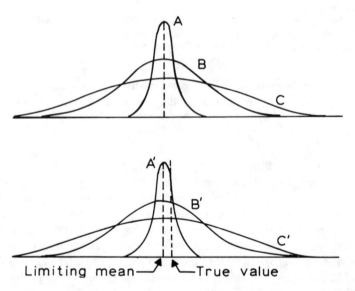

Figure 9.19. The relationship between accuracy, precision, and bias in immunoassay measurements. The top panel illustrates immunoassay tests with no bias measurements. The precision decreases in the order A > B > C. The dotted line indicates the same limiting mean and the true value of measurement for all the three tests. However, because of its large imprecision, test C is relatively inaccurate compared to test A. The bottom panel illustrates biased measurements. All the three tests in this panel are biased, and hence inaccurate, because the limiting means do not coincide with the true value of measurement in each case. However, most of the results for the A' will be more accurate than for test C and even B (top panel) because of its high precision.

Accuracy can be affected by every component of the immunoassay. This includes systematic errors (cross-reactivity or poor recovery), poor tracer, lack of specificity of the antibody, and poor precision (technical errors, low affinity antibodies, or poor quality tracer).

If a definitive method exists to obtain the true value of the analyte of interest, the accuracy of the test method can be determined by comparing the results to such a reference method. However, in most instances, only an indirect assessment of the accuracy is possible. Several indirect measurements of the accuracy are commonly employed. These include recovery studies, parallelism, calibration standards, specificity (or cross-reactivity), and interferences. With the exception of specificity of the assay, which is described in the following section, these techniques are briefly described below.

1. Calibration Standards.   Accuracy of the measurements is greatly influenced by the characteristics of the standard itself used in the assay. An ideal standard should be chemically well defined, homogeneous, and available in pure form. Furthermore, it must be immunologically identical to the analyte of interest in a universally valid matrix (Feldkamp and Smith 1987). However, it is often difficult to obtain standards of high purity and quality for several compounds to assess the true accuracy of an immunoassay system.

2. Recovery Studies.   Recovery is an indirect assessment of accuracy. It is the ability of a test to recover, or measure, a known incremental amount of an analyte from a sample matrix. The experimental protocol for a such study includes adding a known amount of analyte (A) to a base (B) and measuring the concentration (C). The percentage of recovery can be calculated by using the formula

$$(C - B)/A * 100 \qquad\qquad (12)$$

An implicit assumption in such studies is that the standards themselves, against which recovery samples are to be compared, are accurate, i.e., they are pure and properly prepared. This is possible only for certain analytes such as drugs, vitamins, steroid and small peptide hormones, and thyroid hormones. International standards are available for some analytes, but they are not necessarily pure and are assigned values based on activity.

The volume of the spike used in these studies should also be as small as possible in order to avoid major changes in the base matrix. Hence, high concentrations of the spikes should be used. The selection of the base material or matrix is also important. Ideally, it should contain very little or none of the analyte of interest. Recovery studies should normally be performed using at least three or more base samples spiked with at least three concentrations of the analyte representing the working range of the assay.

The recovery is affected by the precision of the assay. Therefore, the spikes should be of sufficient concentration relative to the standard curve and relative to the base such that the precision is adequate to resolve the difference from the base.

Recently, Davies (1994) has reviewed the relative drawbacks of using recovery as an objective criterion for assessing accuracy. Because the calculations are subject to considerable errors in measurement, no recovery calculations can be assessed objectively without their accompanying 95% confidence intervals. Moreover, recovery studies can only detect the proportional bias of the system, because the constant bias is subtracted from both the base and recovery measurements equally. Davies (1994) also points out that the assessment of recovery in assays for serum antibodies, given their differing affinity and avidity, is of questionable value. Accuracy in such assays, therefore, is relative rather than absolute, and is best ascertained on the basis of clinical sensitivity and specificity rather than on technical criteria. Relative recovery should also be used for assaying substances that are difficult to obtain in pure form.

3. Parallelism.   This is one of the most useful experimental approaches to demonstrate the relative recovery and thus the accuracy of an assay. An accurate assay will be independent of sample size. This characteristic can be tested by dilution experiments. It is demonstrated by diluting a sample or a standard with an appropriate diluent prior to running the assay and correcting the results for the dilution factor used. Several different samples should be tested using at least three different dilutions per sample to demonstrate linearity. The initial samples and dilutions chosen ideally should cover the entire working range of the assay. Fractional, instead of serial or double dilutions, are preferred in dilution experiments.

Results of the dilution experiments can be displayed by several methods to evaluate parallelism. The simplest approach is to calculate the final concentration by multiplying by the appropriate dilution factor. The results can be plotted against dilution or sample size (Figure 9.20). A parallel response is inferred from a horizontal line. Statistical parameters can be used to test the significance of least-squares correlation of the measured value against dilution.

Because assays of antibody reflect both the concentration and affinity of the antibody, they will show parallelism only if the affinity of the antibodies in the sample corresponds exactly with those in the assay calibrators (Davies 1994). Thus, samples containing high affinity antibody will show over-recovery on dilution, i.e., higher values in the assay. In contrast, samples with antibodies of low affinity will show disproportionately lower levels on dilution. Such nonlinearity can lead to difficulties in clinical interpretation where serial monitoring of antibody titers is performed and where the activity of the antibody is such as to require different dilutions on each occasion to bring the results into the working range of the assay. One approach to minimize this problem is to use a reference preparation that contains antibodies of representative average affinity. This will ensure that patient samples dilute correctly on "average," i.e., approximately half of them will show apparent under-recovery on dilution, and half over-recovery (Davies 1994).

For dilution experiments, selection of the appropriate diluent is critical. It should be selected carefully to ensure a consistent matrix among the dilutions. The

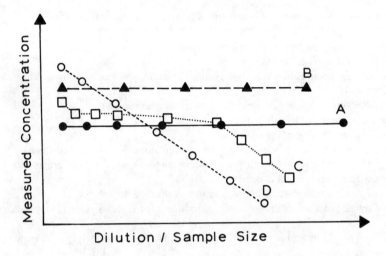

Figure 9.20. Types of curves that can result from parallelism studies. Curve A represents ideal data where the measured concentration is independent of sample size or dilution used. Curve B results when a linear, parallel relationship is displaced by a constant amount; curve C shows nonparallelism, usually due to interferences, such as pH, salts, and endogenous binding proteins in the sample, that cause a disruption in antigen:antibody binding kinetics; and curve D illustrates nonparallelism caused by cross-reacting substances in the assay.

differences in the matrix of the diluted standard and the standard curve matrix are a major cause of a lack of parallelism. It also suggests that the assay may be sensitive to matrix effects. Furthermore, these experiments are also subject both to measurement and pipetting errors; the latter may cause systematic and cumulative errors in dilution. Other causes of nonparallel response include a lack of dose response at the extremes of the curve, imprecision, and inaccurate standards.

4. Interferences. The final consideration in accuracy evaluation is ascertaining that the assay is free from nonspecific interferences that may increase or decrease the apparent result of a sample. Interferences thus are factors other than the true cross-reacting substances that can cause bias in an assay result. Most of the manifestations of interferences are generally classified as "matrix" effects. Davies (1994) recently reviewed literature on various aspects of this topic. Mechanisms by which interferences or matrix effects influence immunoassay results are summarized in Table 9.6.

Specificity (Cross-Reactivity)

The accuracy of an assay depends on the specificity of the assay. The specificity of an assay can be defined as the ability to assay or detect only the analyte of interest in a heterogeneous mixture. It depends on the properties of the antibody used. An antibody has a unique specificity for a particular epitope. This speci-

TABLE 9.6.   Mechanisms of Interferences or Matrix Effects in Immunoassays

1. Alteration of effective analyte concentration
   - Removal or blocking of the analyte
   - Displacement of analyte from physiological binding protein (e.g., thyroid hormone assays)
   - Alteration of antigen conformation due to complexation
2. Interference with antibody binding
   - Physical masking of the antibody
   - Alteration of the antibody-binding site conformation due to sample-induced changes in the ionic strength and/or pH, or the presence of residual organic solvent
   - Low dose hook effect in competitive assays
   - High dose hook effect in one-step immunometric sandwich assays
   - Presence of heterophilic antibodies in immunometric assays yielding false positive results
   - Presence of autoantibodies yielding a low concentration of analyte in immunometric assays
   - The presence of complement and rheumatoid factor leading to an effective reduction in measured analyte concentration
3. Interference in the binding of the capture antibody to the solid phase in protein-avidin-biotin capture (PABC) assay formats
4. Interference in the separation step
5. Interference with detection systems
   - Endogenous signal generating systems
   - Enzyme inhibitors
   - Enzyme catalysts or cofactors

ficity towards the principle analyte is inherently limited by the heterogeneous nature of the antigenic response in a polyclonal antiserum. However, even when monoclonal antibodies are used, the structural similarity of epitopes on different antigens will create a true site-specific competition between these antigens and the tracer-labeled analyte used in the assay. This competition is commonly referred to as cross-reaction. Cross-reactivity is thus a measurement of antibody response to substances other than the analyte of primary interest.

The specificity and cross-reactivity are important parameters for the evaluation of immunoassays. Some frequently encountered situations include: (1) existence of endogenous molecules that are structurally similar to the principal analyte, (2) in vivo production of metabolites of the principal analyte that have common cross-reactive epitopes, and (3) simultaneous administration of other structurally similar analytes, as in medications. In these cases, the accuracy of the immunoassay results will depend on eliminating or at least minimizing the cross-reactive components in the analyses (Miller and Valdes 1991). Even if all available potential cross-reactants do not show a response in the assay, the assay specificity is still not unequivocally established. This can be partly attributed to the unavailability of potential cross-reactants in a pure form, or they may be unknown. These studies should therefore be as comprehensive as possible.

The nature of potential cross-reactants varies with the nature of the analyte for which the assay is designed. One such example is the assay for thyroid-stimulating hormone (TSH). This glycoprotein produced in the pituitary consists of α and β subunits. The α subunit, however, is also common to luteinizing hormone (LH), follicle-stimulating hormone (FSH), and human chorionic gonadotropin (hCG). The β subunit, which differs among these hormones, is responsible for the specificity of the biological action of these hormones. Therefore, to evaluate the specificity of a TSH immunoassay, potential cross-reactivity studies should include these other hormones, isolated α and β subunits, as well as other pituitary hormones, such as prolactin and human growth hormone. These all are possible contaminants of the pituitary extract material commonly used in the immunogen (Micallef and Ahsan, 1994). Similarly, cross-reactivity studies for immunoassay of a new drug ideally should include all structurally related molecules (including any binding protein blocker analogs used for assays without an extraction step) and their metabolites. The decision of what possible cross-reactants should be tested thus varies from assay to assay. The determination of potential cross-reactive substances is particularly important for steroid assays, some drug assays, and gonadotropin assays.

The specificity and cross-reactivity of an assay system can be examined and expressed in several ways. In a limited-reagent competitive assay, it is usually defined as the percentage ratio of the concentrations of analyte (or standard) and cross-reactant that produces a 50% (some researchers use 10%) displacement of the tracer. Sometimes, it is also expressed as the apparent concentration of the analyte resulting from the assay response due to an arbitrary concentration of the cross-reactant. The use of these different methods may give quantitatively different estimates of cross-reactivity. Recently, Miller and Valdes (1992) have proposed a new protocol for evaluating and expressing cross-reactivity data involving the construction of interferographs for analytically important cross-reactants.

In competitive assays, the most common experimental approach is to spike a pure sample of the cross-reacting substance into an analyte-free matrix to give a suitable wide range of concentrations. Dose-response curves are then constructed using the analyte and the potential cross-reactant under identical assay conditions. Cross-reactivity can be defined as the point where the reduction in assay signal corresponds to 50% of that achieved in the absence of analyte (i.e., B/B_0 of 50%), as a percentage of the analyte concentration giving the same fall in signal (Figure 9.21). Thus,

$$\% \text{ cross-reactivity} = \frac{\text{concentration of standard at 50\% } B/B_0 \text{ (S)}}{\text{concentration of cross-reactant giving 50\% } B/B_0 \text{ (C)}} \times 100 \quad (13)$$

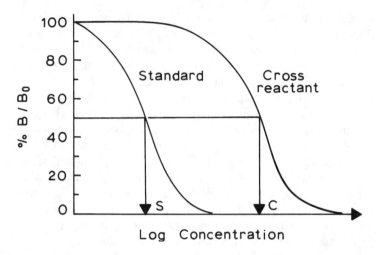

Figure 9.21. Cross-reactivity in immunoassays. A displacement curve for a cross-reactant is compared to the standard curve. The concentration at 50% displacement of the standard (S), divided by that of the cross-reactant (C) is the fractional cross-reactivity in mass units.

It should be remembered that the displacement curves generated using such a protocol may not be parallel. Thus, the effect of cross-reacting substances may differ according to the relative concentrations of the analyte of interest and the cross-reactant. This is often the case when polyclonal antibodies are used in designing the assay. Furthermore, the specificity of a polyclonal antiserum often changes with subsequent bleeds from the same animal. In such instances, the observed cross-reactivity reflects the cross-reactivity of the particular antibody clone having most influence in the binding reaction at that analyte concentration. At low analyte concentrations, this will predominantly be high affinity antibodies; at high concentrations, the weaker clones will have the most influence. One useful technique in the development of steroid assays using polyclonal antisera is to add deliberately a small quantity of the cross-reactive steroid in order to "swamp" a minor, but particularly cross-reactive, antibody clone (Davies 1994).

The determination of assay specificity and cross-reactivity is particularly important in the development of immunometric assays. For example, a cross-reactant that binds to only one of the two antibodies will prompt little or no assay response when used alone as a pure standard, but may significantly reduce analyte binding if simultaneously present in sufficiently high concentrations, causing false low results. Even when binding of cross-reactant by both antibodies occurs, some cross-reactions may be canceled by the high dose hook effect, if present (Micallef and Ahsan 1994). True cross-reactivity can thus only be assessed in this type of assay by measuring the cross-reactivity of each antibody individually in a competitive assay format as described above. These individual measurements, however, by themselves are of little practical use except as an assay design tool. In spite of these limitations in determining cross-reactivity, for many clinical assays, specificity has been greatly improved by the use of immunometric assay designs and formats where only molecules that react with two different antibodies are detected.

Recently, Miller and Valdes (1992) and Davies (1994) have suggested a novel approach to determine the cross-reactivity and specificity of immunoassays, irrespective and independent of assay format and design. Routine samples provide a more realistic matrix than a chemically or physically stripped analyte-free matrix, and there may be differences in their behavior when the cross-reactant or analyte is bound to protein. Thus, the cross-reacting substance can be spiked directly into such samples containing endogenous analyte. The cross-reactivity is then calculated at some multiple of the upper reference limit for the cross-reacting substance, rather than at some arbitrary point on the dose-response curve, and expressed as the apparent percent change in the endogenous analyte concentration. This approach provides clinically a more useful way of assessing the likely degree of interference than would be encountered in routine practice.

For several small analytes, certain empirical approaches have proved useful in maximizing the specificity of the assay for the principal analyte. These approaches include isolating the analyte of interest from the cross-reactants before immunoassay, altering the thermodynamic conditions used in the assay, or developing more specific antibodies (Miller and Valdes 1991). The efficacy of these techniques depends on the differences in physicochemical properties between the cross-reacting substance and the analyte of principal interest. For large molecules or small molecules tightly bound to large proteins, separation by ultrafiltration or chromatography is effective. However, for molecules similar in size to the principal analyte, separation is much more difficult. In such instances, altering the kinetic or thermodynamic conditions of the immunoassay have proved effective. Several such approaches to improve the specificity of the immunoassays are summarized in Table 9.7.

Ruggedness Testing

The critical operational parameters and the tolerances for their control must be known in order to judge the trueness of the concentration measured by every immunoassay procedure. Such parameters, for example, include assay temperature, time of incubation, pipetting volumes for various reagents, separation procedures, the order of reagent addition, and signal detection. Functional relationships for these should be known, or their reliable effects on measured values should be estimated. Such information is useful to establish tolerances within which the parameters must be maintained in order to obtain results within acceptable limits. Although it is possible to optimize such variables, it may not be

TABLE 9.7.   Various Approaches to Improve Assay Specificity

1. Separation of analyte before immunoassay
   • Ultrafiltration, protein precipitation, chromatography
2. Use of highly specific antibodies
   • Use a highly pure immunogen to remove irrelevant materials in the ligand
   • Use a fragment of the analyte that does not contain the common antigenic site to eliminate nonspecificity due to the presence of common antigenic sites on the analyte and cross-reacting substances, e.g., glycoprotein hormone (TSH, LH, FSH, hCG) assays
   • Use novel hapten-protein conjugates for immunization
3. Manipulation of assay parameters
   • Nonspecificity due to a population of low affinity antibodies can be minimized if the antiserum is used at very high dilutions
   • Increasing incubation time decreases the degree of cross-reaction with structurally similar compounds
   • Because increase in the incubation time means attainment of reaction equilibrium, increasing incubation temperature helps in more rapid attainment of equilibrium
   • Use a highly purified analyte for tracer labeling
   • Use antigen-affinity purified antibodies in the assay. Alternatively, remove the unwanted populations of antibody by adsorption with cross-reacting substances
   • Select the most specific monoclonal antibody clone by screening for cross-reactivity
   • Use two-site immunometric sandwich assay design wherever possible with antibodies that can recognize two independent epitopes on the analyte

---

feasible in many cases so it may be that only empirical operational values and their tolerances are known. When developing a new method, it is important to know whether the tolerances established are effective and/or whether unsuspected interactions exist.

Youden (1984) recommended the use of a novel experimental protocol, based on $n + 1$ measurements to test the effect of n variables, to determine whether established tolerances are effective, for example, if $\pm t$ represents the tolerance for temperature control. In essence, he suggested that measurements be conducted in which the various critical parameters are maintained at either their upper limit or their lower tolerance limit in combinations as illustrated by the matrix presented in Table 9.8. The example presented is designed for the study of seven variables or factors using eight measurements. Thus, measurement 1 is made with all variables at their maximum values (denoted by capital letter). Measurement 2 has variables A, B, and D at their maximum values while the remainder are at their minimum values.

Such ruggedness testing is recommended when developing a new method, particularly if it is intended for on-site testing of samples. Ideally, ruggedness testing should precede every collaborative test of a proposed standard method.

TABLE 9.8. An Example of a Matrix Consisting of Eight Combinations of Seven Factors Used to Test the Ruggedness of an Immunoassay Method

Factor Value	Combination or Determination Number							
	1	2	3	4	5	6	7	8
A or a	A	A	A	A	a	a	a	a
B or b	B	B	b	b	B	B	b	b
C or c	C	c	C	c	C	c	C	c
D or d	D	D	d	d	d	d	D	D
E or e	E	e	E	e	e	E	e	E
F or f	F	f	f	F	F	f	f	F
G or g	G	g	g	G	g	G	G	g
Observed result	s	t	u	v	w	x	y	z

Procedure:
1. Choose 4 minus 4 combinations of s to z to get 4 caps minus 4 l.c. of desired letter, e.g., $[(s + t + u + v)/4 - (w + x + y + z)/4] = A - a$.
2. Rank the seven differences to identify problems.
3. Calculate standard deviation from the eight results.

Nonspecific Binding (NSB)

As described earlier in Chapter 7, nonspecific binding (NSB) is conventionally used to measure the efficiency of separation of the free from bound form. NSB tubes are therefore often included in immunoassays to correct for any apparent binding of the tracer that is not due to the specific reaction with the antibody. In a competitive assay, NSB is measured by running a control tube once per assay which contains all the reagents (including a test sample) except the antibody. If the antibody is normally bound to a solid phase, a "blank" antiserum reagent, consisting of the normal solid phase without the antibody, should ideally be used. To be absolutely consistent, nonimmune animal serum is often substituted for the specific antibody to maintain the same matrix. In an immunometric assay, NSB is measured by using zero concentration samples. In this case, the NSB is the binding of the labeled antibody to the solid phase in the absence of analyte. The NSB level is assumed to have a constant effect at all concentrations, and is typically subtracted from all tubes to correct the final assay results.

In solid-phase immunoassays, NSB can be caused by two mechanisms: direct binding of the enzyme label, and serum-mediated NSB. These are discussed in the following sections.

1. Direct Binding of the Enzyme Label. Excess enzyme conjugate may bind to the surface of the solid phase, thereby producing elevated signals in a uniform manner regardless of the sample being tested. An excessive direct NSB diminishes the ability of an assay to distinguish a 'true' signal derived from the analyte from the background signal. The assay sensitivity is therefore reduced.

Direct NSB is often a major problem with solid phases with high surface area. This type of NSB can generally be minimized by blocking the remaining binding sites on the solid phase after the coating of the capture reagent with an appropriate blocking agent. Addition of detergents, such as Tween®-20, to all subsequent assay reagents also helps in preventing such NSB of the enzyme label. Increasing the number of washes to separate the free from bound complex also alleviates this problem to a large degree. Similarly, only labels that have a low tendency to bind nonspecifically should be used.

2. Serum-Mediated NSB.   In this case, the enzyme label is bound to the solid phase by a serum component other than the analyte of interest (Gorman et al. 1991). Such a serum component may be common to all samples or found only in a small subset. In the former case the assay sensitivity is lowered in a manner analogous to that of direct NSB. However, in the latter case, erroneous but believable results that are often difficult to detect are produced.

Serum-mediated NSB may be caused by enzyme polymorphism and is sometimes seen in assays that use alkaline phosphatase as the enzyme label. The selection of appropriate enzyme label is thus an important criterion to eliminate this type of NSB.

As described earlier, heterophilic antibodies are yet another source of serum-mediated NSB in immunoassays (Kricka et al. 1990; Boerman et al. 1990; Levinson and Goldman 1988; Thompson et al. 1986). Generally, serum or immunoglobulins from the same species can be included as antibody reagents in the assay to neutralize such heterophilic antibody activities and effectively remove NSB (Boscato and Stuart 1988; Howanitz et al. 1982).

Rheumatoid factor, which reacts with the Fc portion of the antigen-antibody complexes, has long been known as a common source of serum-mediated NSB. Such NSB can be avoided by the use of fragmented immunoglobulins [Fab or (Fab')2] for one or both of the specific antibodies in the immunometric assays. Similarly, in some immunoassays designed for plasma samples, coagulation related processes (e.g., fibrin formation) often produce false elevated results, especially if a high surface area solid-phase support is used. This problem is presumably caused by nonspecific binding of the enzyme conjugate to fibrin or other coagulation intermediates deposited on the solid phase surface. The use of anticoagulants has been found to be effective in preventing NSB by this mechanism; however, their effects on assay performance must be thoroughly investigated (Gorman et al. 1991).

Ideally, immunoassays should be designed to have a very low NSB. Thus, some or all of the above strategies should be considered to remove this potential interference. Nonetheless, final design decisions must be based on a proper balance of performance needs and manufacturing costs.

Method Comparison

A newly developed immunoassay method should produce results similar to those of other reliable, clinically validated immunoassays, as well as independent methods of assay. Where possible, comparison should be made with an established reference method, such as isotope dilution-gas chromatography-mass spectrometry (ID-GCMS) for hapten analysis (Tunn et al. 1990; Gosling et al. 1993). For protein analysis, a well-validated and well-established clinical test should be used, if possible one that has given accurate and precise results in external quality assessment schemes.

To perform the method comparison, one should randomly select at least 30 (ideally 100–200) patient/field samples from a variety of clinical/field conditions and with analyte values ranging throughout the entire calibration curve. These samples are then analyzed by both methods and the results compared. A graph of the two methods plotted against one another with identical axes, along with the bisecting line of perfect agreement, allows a good visual assessment of the degree of agreement between the two methods. Simple linear regression analysis is widely used to estimate the Pearson correlation coefficient.

A prerequisite for such a study is that the samples should be spread across the entire working range of the assay. If most samples are clustered narrowly in one end of the concentration range with very few points at the other extreme of the calibration curve, the statistical parameter used to evaluate method comparison will be biased with those few samples. Similarly, if samples are distributed in two groups at both ends of the calibration curve, the slope of the linear regression will be determined by the mean of those two groups. In both cases, a falsely elevated coefficient may be obtained together with an incorrect slope of the linear regression.

Ideally, such a method comparison study should yield excellent linear correlation between the two methods with a slope equal to 1. However, the two methods may display a proportional bias (slope $\neq$ 1) as well as a constant bias (intercept $\neq$ 0). In such cases, a simple adjustment of normal ranges is not enough, and further studies to verify important clinical decision points are often necessary.

The use of simple regression techniques in method comparison has several drawbacks (Bland and Altman 1986; Pollock et al. 1992; Davies 1994). Some of these include the following:

1. The relationship is assumed to be linear, even though this may not be the case.
2. The correlation coefficient nearly always ten the face of poor agreement.
3. It assumes that the x-variable is without error, which is not true when two immunoassay methods are compared.
4. It assumes a normal distribution of data points, which is frequently not the case. Most studies include disproportionately higher numbers of low and high samples to cover the entire working range of the assay. The statistical parameters are thus greatly biased in favor of such clustered data points.

5. Although an excellent correlation coefficient may indicate that the two methods are comparable, the new method may not work within the clinical range of the assay with the desired sensitivity, precision, and accuracy.

6. The slope and intercept from such a comparison can be used to predict the new method results from the old, but not vice versa.

Some of these drawbacks can be avoided by using Deming's regression method (Deming 1943) which gives the same values for slope and intercept regardless of which assay method is used for the *x*-axis data.

At present, the method of Bland and Altman (1986) is preferred to assess how clearly results from the two methods agree. This technique uses percentage rather than absolute differences between measurements, because in most immunoassay systems the standard deviation is concentration dependent. The differences in the two assay results are then plotted against the mean result to determine the presence of any concentration-dependent assay bias.

Lot-to-Lot Variation

The technical performance of multiple lots of reagents used in an immunoassay system should be evaluated for consistency. The degree of inaccuracy will depend on the QC tolerance limit set by the manufacturer. The tolerance limit generally varies from 2–10%, or 1–2 standard deviations of the difference between lots. Ideally, at least three lots of reagents shoul be evaluated. The most frequent lot change is often the enzyme conjugate. In most instances, such changes are trivial and should not affect the performance of the assay (Chan 1992). However, this may produce rather dramatic changes in the absolute absorbance.

The calibrator lots also need to be evaluated for proper performance of the assay. In fact, the absorbance of the calibrator is often a good indicator of lot-to-lot variation of the conjugate. In addition to the absolute absorbance, the differences in the absorbance of the calibrator affect both the dynamic ranges and the sensitivity of the assay. For qualitative assays, this is especially important, as no quantitative results are produced. Thus, a positive or negative result will greatly depend on the differences in the absorbance of the zero and the positive cutoff calibrator. To avoid false-positive results, a minimum acceptable absorbance value should be set for such assays.

Measurement Errors

Analytical errors and undesirable performance of an immunoassay method are discovered as the result of a laboratory's own quality assessment efforts, from its participation in collaborative test programs, or from external quality assessment schemes. Every analytical measurement is associated with an estimate of error without which the significance of the measurement itself cannot be assessed.

Measurement errors can be classified into three groups. Systematic errors are always of the same sign and magnitude and produce biases. They are constant no matter how many measurements are made. Random errors vary in sign and magnitude and are unpredictable. In other words, they occur by chance. Random errors average out and approach zero if enough measurements are made. In contrast, blunders are simply mistakes that occur on occasion due to carelessness, and produce erroneous results. Measuring the wrong sample, adding the wrong reagents, and incorrectly reading values are examples of blunders. They produce outlying results that may be recognized as such by statistical procedures, but they cannot be treated by statistics.

The importance of systematic and random errors in immunoassay measurements is briefly described below.

1. Systematic Errors. As described earlier, systematic errors or "biases" can be constant or proportional (Figure 9.18). Systematic errors are generally in principle (but not always in practice) attributable to factors within the control of, or measurable by, the operator. It is important to understand that usually the bias error and the imprecision of a measurement behave independently of one another (Figure 9.19). Thus, a highly precise measurement may still be very inaccurate because of the presence of a bias error. Similarly, even in the absence of bias, the measurement may still be inaccurate due to poor precision of the assay.

Factors that contribute to constant systematic error are independent of the analyte concentration, because it is of a constant magnitude throughout the concentration range of the analyte. It is generally caused by an interfering substance that gives rise to either a low (negative error) or high (positive error) false signal. Lack of specificity and interfering substances are a major source of this type of measurement error. Constant systematic errors also occur in enzymatic assays that depend on oxidase-peroxidase-coupled detection systems. Here, the substrate intermediate, hydrogen peroxide, is destroyed by endogenous reducing agents such as ascorbic acid. Improper blanking of the sample or the reagents can also introduce such error.

Constant error or bias in an assay can be obtained from studies in the presence of cross-reacting or interfering substances, and are later confirmed by comparison with a reference method (Westgard et al. 1978a). It is compared directly to the allowable error for the appropriate decision level. If the constant error is less than the allowable error, it is judged acceptable. This decision is based on clinical limits, rather than a statistical test of significance.

In contrast, proportional error is caused by incorrect assignment of the amount of standard in the calibrator. If the calibrator has more analyte than is labeled, all the unknown determinations will be low in proportion to the amount of error in the calibration and vice versa. Erroneous calibration thus is the most frequent cause of proportional error. It may at times also be caused by a side reaction of the analyte. The percentage of analyte that undergoes a side reaction will be the percentage of error in the method.

Proportional errors can be determined from recovery experiments, and later by comparison with a reference method (Westgard et al. 1978a). The difference between the calculated percentage recovery and its ideal value (100%) is deemed as the percent of proportional error. However, one needs to convert this to concentration units at the clinical decision level to decide its acceptability. Similar to constant error, this decision is based on medical requirements rather than statistical tests of significance.

2. Random Errors.   Even when all sources of systematic error are removed from an immunoassay system, there still remains the possibility of random error attributable to factors that affect the reproducibility of the measurement. These include instability of the instrument; variation in the temperature, reagents, and calibrators; variabilities in handling techniques such as pipetting, mixing, and timing; and variabilities in operators. These factors superimpose their effects on each other at different times; some cause rapid fluctuations, while others occurr over a longer time.

Random errors can be described in terms of precision, imprecision, reproducibility, and repeatability of an assay. In each case, they refer to random dispersion of results or measurements around some point of central tendency. Random errors can be estimated from within-run and between-run precision studies for both short and long term evaluation of the assay (Westgard et al. 1978a). These replication studies should be first performed with an aqueous solution of calibrator or standard, and then repeated with samples whose matrix is as similar as possible to that of the intended test samples. The concentrations to be studied should be at or near the clinical range concentrations for the analyte. Generally, 95% confidence limits are used for the allowable random error in the assay.

3. Estimation of Total Error.   The total error of a test method, obtained by combining the estimates of systematic and random errors, is a way of assessing a kit or instrument and making a judgement of acceptable performance. This is the most severe criterion for the test method to meet. Several approaches to estimate total analytical error are described in the literature (Westgard et al. 1974; Krouwer 1991; NCCLS 1989). Some estimates include total imprecision and fixed bias, and others attempt to include estimates of protocol-specific bias, such as sample carryover and drift, whose estimated contribution to error depends on how they were measured.

The National Committee for Clinical Laboratory Standards (NCCLS) protocol EP10-T (NCCLS 1989) describes a convenient method for initially estimating within-run and between-run precision and linearity (parallelism) with a minimum number of standards. The rationale is that if the total error is acceptable, individual components need not be tested. This protocol, however, does not address specific matrix or other interferences, or selection of standard concentrations, data reduction, or sensitivity. Although samples in a run are arranged to detect drift and carryover, the number of samples included in a run may not be adequate to detect some drift that will appear later. However, failure of these initial tests certainly will alert the laboratory or the manufacturer of potential problems and lay the groundwork for collection of more extensive data on the kit of choice. Long term performance should be assessed continually through QC programs as well as devices such as the precision profile (Krouwer 1991).

Several factors may influence the final estimate of the total analytical error of the system. For example, there will be instances in which the combined error will be exactly equal to the systematic error. At other times the combined error for a given result will be less than the average systematic error by some amount because of the random error of the method. Sometimes, the combined error will be greater than the systematic error, again by some amount caused by the random error of the method. The operator has no way of knowing what the various components of error are or when they will cause a large error. Hence, the total error of the system must be defined for the worst-case combination. If it is less than the allowable error, the method's overall performance will be acceptable.

Several excellent reviews are available describing the occurrence of these errors and the relevant statistics to evaluate their significance in immunoassay measurements (NCCLS 1984, 1986, 1987, 1989; Krouwer 1991; Westgard et al. 1978a–e, 1974; Tonks 1968; Westgard and Hunt 1973; ASTM 1968; Waakers et al. 1975; Cornbleet and Gochman 1979; Channing Rodgers 1981).

QUALITY ASSURANCE/QUALITY CONTROL (QA/QC)

Quality assurance/quality control (QA/QC) is an essential function of every laboratory and manufacturing process. It is an objective means of evaluating one's assay systems against the clinical requirements of the test. Because of the inherent nature of immunoassays, the comparability of the results of measurements carried out in different places and over long periods of time may be much less than is required for consistently accurate diagnosis. A comprehensive discussion of the QA/QC process is beyond the scope of this chapter. Hence, only the salient features are described.

Internal Quality Control

The internal QC system must be designed to ensure that results are within acceptable limits of accuracy and precision, with particular attention to the concentration ranges that have the most critical clinical significance. Such a program should routinely assess equipment performance, reagent stability, technique, assay conditions, sample handling, and any other factors that can influence assay performance. QC samples representing the concentration ranges where decisions are made that affect the clinical significance of the test results thus need to be included in every assay or at regular intervals. Such samples produce a retrospective check of each assay using such parameters as within-assay and between-assay precision; bias or assay drift; and any trends, cycles, or patterns. Consistency of data or a lack of it, as judged by comparison of results with those of previous runs/lots, is then used to evaluate the long term performance of an immunoassay kit.

The most useful and common form of presentation of such data and related statistical parameters is the Shewhart or Levey-Jennings control charts (Westgard et al. 1981; Blockx and Martin 1994). These charts require an estimate of the mean and standard deviation for each control in use. The results from different lots or batches are plotted against a vertical axis showing the deviation in units of between-batch standard deviation.

Control charts are an excellent method for detecting problems such as bias or poor precision. When a batch fails to meet a control rule, the cause of the failure should be investigated. Failure to pass any of the checks suggests a failure of control and such a lot therefore should be rejected.

A series of well-established and statistically sound decision rules, known as the "multirule Shewhart chart" or "Westgard analysis," is available to interpret data from control charts (Westgard et al. 1981). Westgard analysis focuses on the decision of accepting or rejecting data from an assay run, in some cases taking data from previous runs into account. The term "multirule" is often used to describe such analyses, because several different control rules are used in this QC procedure.

Because results from the Shewhart/Levey-Jennings type of control charts are not very sensitive to small changes in mean values, Cusum charts are often used to identify a constant bias from the target mean (Jeffcoate 1981). This chart uses a similar format except, instead of plotting the individual values for a control, the difference of each new value from the target mean is added to the cumulative sum of all the previous differences. However, because of the difficulty in interpreting such data, Cusum charts have not found wide application in the immunodiagnostic field.

In addition to controls, which are the most useful monitor of assay quality, several other parameters can be used to monitor assay performance and as indicators of assay problems. These include: (1) percent binding at zero standard concentration, (2) percent binding at the concentration of the highest calibrator, (3) nonspecific binding (NSB), (4) curve-fit parameters, such as goodness of fit, number of iterations, variance ratio, slope and intercept, and (5) estimated dose at 20%, 50%, and 80% binding. Each of these variables can give an indication of the quality and the behavior of the reagents, and the influence of the reaction conditions used (Blockx and Martin 1994). For example, high NSB could indicate that the tracer has degraded, or a low percent binding could be caused by an incubation at too low a temperature.

In addition to the use of such internal QC data for day-to-day assay control, the cumulative data should be periodically reviewed to monitor longer term trends in assay performance. Such a practice helps in identifying gradual shifts in bias and/or precision.

External Quality Assessment Schemes (EQAS)

External quality assessment schemes (EQAS) provide an independent assessment of a laboratory's performance, as well as complementing internal QC in quality assurance. In such a program, the organizer sends the same control samples to all the participating laboratories for testing at regular intervals. No target values are provided and the samples are tested in the same way as patient samples. The participants are required to return the results quickly. This permits remedial actions to be taken in a realistic time frame when unwanted trends in performance occur.

Many EQAS provide data at three levels, describing laboratory, method-group, and all-laboratory performance. The scheme protocols and statistical procedures used to derive these data differ between schemes and have been reviewed (Seth 1987). A well-organized EQAS, therefore, can provide a unified, reliable source of information, advice, and education on all aspects of the assays, and an independent check of the effectiveness of the laboratory's internal QC programs.

Both internal and external quality assessment results are thus essential elements in evaluating quality, efficiency, and productivity. The internal QC data provides a good assessment of precision; accuracy is most reliably confirmed by EQAS.

Several reviews are available in the literature detailing the role of both internal QC and EQAS in evaluating the quality of immunoassay methods (Gosling and Basso 1994; Blockx and Martin 1994; Perlstein 1987; Seth 1987, 1991; WHO 1981; Uldall 1987; De Verdier et al. 1981; Whitehead 1977; Seth et al. 1991; Bergmeyer 1991; Taylor 1987).

ASSAY VALIDATION

Clinical Diagnostics

The effectiveness of a newly developed immunoassay intended for a specific clinical purpose must be assessed in terms of meeting the clinical objectives. The main aim of a clinical evaluation is to assess the ability of a system to provide accurate test results in a timely fashion for the clinical need. The need could be for disease screening or monitoring, diagnosis, or management. A thorough discussion of the procedures involved in clinical validation of immunoassays is beyond the scope of this book. However, several excellent reviews describe the various protocols involved in such a validation process (Feldkamp 1992; Zweig and Robertson 1987; NCCLS 1987; Chan 1992; Davies 1994; Emanuel and Perelmutter 1994). Several important parameters described under Assay Evaluation, however, are also important in clinical validation of immunoassays.

Several aspects of validation apply to the evaluation and selection of an immunoassay or immunoassay kit for clinical use. The appropriate studies and extent of the evaluation depend on whether the assay defines a new test, or an application not previously established in the medical literature, or whether a laboratory is evaluating an FDA approved kit for an analyte with a well-established use in clinical medicine (Feldkamp 1992). In the former situation, the developing laboratory or manufacturer must establish not only the range of values of normal subjects but also the range of values in important clinical conditions to which the test is applied. In addition, both the clinical sensitivity and specificity must be proven. In contrast, for clinical validation of a kit to correctly assess the validity of test results, the laboratory must have an independent knowledge of the patient's condition.

The evaluation and in-house validation of a kit for a well-known analyte is quite a different process. It primarily focuses on the determination of the normal range and on correlation or comparison of values with those obtained by another valid method or reference method. This process is mainly concerned with identifying any analytical biases, resolving any observed discrepancies in individual samples, and understanding differences in clinical sensitivity and specificity of the two methods.

Some of the important concepts involved in a clinical validation process of immunodiagnostic tests are briefly described below.

1. Normal Range. More accurately defined as "reference value," "reference range," or "reference interval," the normal range must be established for every new system. Three specific issues must be addressed to define the normal range: (1) how to define the normal population, (2) what the sample size would be, and (3) what statistics to apply.

Traditionally, normal range or reference values are defined as the range within which 95% of the normal healthy population will fall. Because most biological substances are subject to some kind of homeostatic mechanism that ensures that their concentrations only vary within a particular range of values, the central point of this variation or the mean concentration is referred to as the "homeostatic set point." This value is specific for a given individual.

The establishment of a normal range for a given analyte is affected by several sources of variation. These include within-subject biological variation, between-subject biological variation, the collection of the sample for analysis or the preanalytical variation, and the analytical variation defined by the between-batch variation of the test system. The normal range thus reflects the sum of variations from all these sources (Davies 1994; Cotlove et al. 1970; Harris 1979).

Several excellent reviews deal with the question of sample size and statistical treatment of the data in validation studies (Dixon 1953; Reed et al. 1971; Solberg 1987; Taylor 1987). For a 95% confidence interval, the theoretical lower limit of the sample size to estimate the 2.5% to 97.5% of normal population is $1.0/0.025 = 40$. The precision of the estimate increases with larger sample population. Ideally, the practical limit should be at least three times the theoretical limit, i.e., $n = 120$ or greater.

Statistical analysis using the mean ± 2 standard deviations for a population with Gaussian (normal) or log Gaussian distribution is the most frequently used method for data analysis. Nonparametric statistical methods, which do not assume any specific distribution of data, are usually recommended when the sample size is ≤ 120. Statistical analyses are used to set the tolerance interval, which includes a predetermined proportion of the normal population with a predetermined probability.

2. Predictive Value of a Diagnostic Test.   The predictive value of a diagnostic test is defined in terms of clinical sensitivity, specificity, and efficiency. These terms are briefly described below.

CLINICAL SENSITIVITY.   Clinical sensitivity measures how well the test detects those patients that have the disease. When the patient is known to suffer from a given disease and a diagnostic test for that disease is positive, the result is referred to as true-positive (TP). False-negatives (FN) are the number of patients suffering from the disease that are misclassified by the test. Clinical sensitivity or the positivity in disease then can be calculated as:

$$\text{clinical sensitivity} = [TP/(TP + FN)] * 100 \qquad (14)$$

CLINICAL SPECIFICITY.   Clinical specificity measures how well the test correctly identifies those patients who do not have the disease. Thus, false-positives (FP) are the number of nondiseased or healthy patients who are misclassified by the test, and true-negatives (TN) are those patients correctly classified by the test. Clinical specificity is then defined as

$$\text{clinical specificity} = [TN/(TN + FP)] * 100 \qquad (15)$$

The concept of clinical sensitivity and specificity using these four parameters is illustrated in Figure 9.22.

Based on the above relationships, three other terms can be defined as follows.

The predictive value of a positive test gives the percentage of patients suffering from the disease with positive test results and is given by the relationship:

$$[TP/(TP + FP)] * 100 \qquad (16)$$

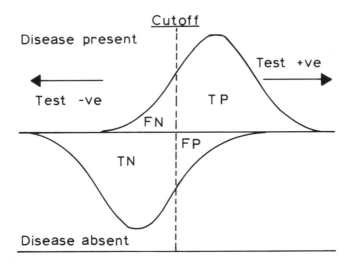

Figure 9.22. Clinical concepts of sensitivity and specificity illustrating the influence of cutoff values on the predictive value of a diagnostic immunoassay test. TP, FP, TN, and FN refer to true-positives, false-positives, true-negatives, and false-negatives, respectively, measured by the test. The clinical specificity of a diagnostic test increases at the expense of sensitivity.

The predictive value of a negative test is defined as the percentage of healthy patients with negative test results using the following relationship:

$$[TN/(TN + FN)] * 100 \tag{17}$$

The overall efficiency of a diagnostic test, defined as the percentage of patients correctly classified as diseased or healthy, is given by the following relationship:

$$\text{efficiency} = [(TP + TN)/(TP + FP + TN + FN)] * 100 \tag{18}$$

A perfect test should exhibit 100% sensitivity, specificity, and efficiency. However, in real life situations this is not the case, and values for these parameters depend on the decision level (also referred to as "decision point," "upper limit of normal," "cutoff value," "reference value," etc.) chosen. The decision level or cutoff value for an assay should not be based solely on statistical considerations. Thus, factors such as relative medical, ethical, psychological, and financial costs associated with an FP and FN result must also be considered (Davies 1994; Zweig and Robertson 1987). For example, quite different cutoffs are used when screening for spina bifida using maternal serum α-fetoprotein (AFP), where a positive result leads to the offer of amniocentesis and further tests, compared to the subsequent choice of cutoff used in assessing amniotic fluid AFP, where a positive result may lead to an offer of termination of pregnancy.

A high cutoff value reduces the sensitivity of a diagnostic test while increasing its specificity, and vice versa (Figure 9.22). High clinical sensitivity is essential in, for example, screening tests for HIV or hepatitis infection in blood or organ donation. Such tests may inevitably produce some false positive results, but this may be acceptable if all the samples tested as positive can be further investigated by means of more specific tests (Micallef and Ahsan 1994). In contrast, high clinical specificity is especially important when treatment of the disease diagnosed is potentially dangerous, for example, chemotherapy or radiotherapy. False negative results may be identified by other means, but the possibility of a healthy person being subjected to treatment is minimized.

The two parameters, clinical sensitivity and specificity, thus occur in pairs. A given test will have one set of such sensitivity-specificity pair in one clinical situation, but may have a different set of pairs when applied to another clinical situation where the group being tested is different. These two parameters can be evaluated in terms of the receiver operating characteristic (ROC) curve. The ROC curves can be constructed by plotting the relationship between the false-positive rate (i.e., 1 – specificity) and the sensitivity or detection level at various cutoff levels.

Theoretical ROC curves are shown in Figure 9.23. These curves display assay performance over the entire decision levels. Curve A in this example shows a diagnostic test with no discrimination, i.e., it has the same true and false positive rates for any given decision level. Tests with increasing clinical utility are characterized by curves B, C, and D, while the diagnostic test represented by curve E

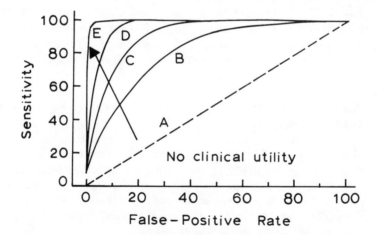

Figure 9.23. Theoretical receiver operating characteristic (ROC) curves for diagnostic tests. Curve A illustrates a diagnostic test with no discrimination; the true-positives and false-positives identified by this test are same at all the decision levels. The clinical utility of the tests increases as shown by the arrow from tests B to E. Curve E illustrates a diagnostic test with 100% specificity (no false-positives) and sensitivity

has virtually 100% specificity and sensitivity. If the "cost" of an FN result is the same as that of an FP result, then the optimal cutoff point will maximize the sum of the specificity and the sensitivity, i.e., the point nearest the top left corner (Davies 1994).

The ROC curves can also be constructed as a plot of TP rate (sensitivity) versus TN rate (specificity). This produces a mirror image of the curves shown in Figure 9.23.

The ROC method is of limited applicability in assessing tests where the specificity is very high. However, it is extremely valuable in assessing the clinical utility of screening tests where the specificity is relatively low in comparison to sensitivity, and for the comparison of two competing tests. Thus, by superimposing the ROC curves of more than one test method, one can select the best method for a given clinical need. A better test is one that displays a higher TP rate and a lower FP rate.

The ROC curve thus provides a comprehensive picture of the test's performance capabilities at all possible decision levels. It does this without the need to choose a decision level or establish a normal range in advance. Furthermore, it allows complete comparisons of any number of tests to one another over all possible decision levels. The determination and importance of ROC curves in the clinical evaluation and comparison of analytical methods has been comprehensively reviewed (Zweig and Robertson 1987; Zweig and Campbell 1993; Henderson 1993).

Thus, several criteria are used for the clinical validation of immunoassay diagnostic tests. Such a study should evaluate the clinical usefulness of a test to establish clearly and explicitly a given clinical goal. It is essential to identify what issue of consequence to patient management is to be addressed by the test. Zweig and Robertson (1987) have suggested the following guidelines for a clinical test evaluation or diagnostic trial.

1. Choose study subjects who are representative of the clinical population to which the test is ultimately be applied.
2. Perform all tests being evaluated on all the subjects; perform all tests on an individual subject at the same point in the subject's clinical course.
3. Classify the subjects as affected versus unaffected or diseased versus healthy by rigorous and complete means so that the true diagnoses or outcomes are approached closely. Diagnostic maneuvers that extend beyond routine clinical practice may be required for the purpose of the evaluation. All diagnostic criteria should be independent of the test or tests being studied.
4. Evaluate and compare test performance at all decision levels by using ROC curves.
5. Select decision levels for the test(s) being evaluated based on the ROC curve, the intended use of the test, the prevalence of the condition, and the relative costs of false-positive and false-negative results.

In the United States, clinical validation of immunoassay tests comes under the guidelines set forth by the FDA. The FDA has also developed a protocol to verify conformance of in vitro diagnostic devices' performance to that claimed by the manufacturers. Performance is tested separately at 10%, 50%, and 90% of a method's claimed linear range (Lee et al. 1982). A chi-square test is used to compare the total standard deviation observed in an analysis of variance experiment to the manufacturer's claimed standard deviation. In the comparison-of-methods experiment, 20 patient specimens with analyte concentrations within 10% of each of the above concentrations (total 60 specimens) are selected and run in duplicate by the test and comparative methods. The bias is calculated for the 20 specimens at each level, and is compared to the bias claim by use of a one-sided t test at each level. Because the data used to make a decision at each concentration are within 10% of the decision level concentration, the problems of the bias and the t test are eliminated. It is assumed that if the bias is within a manufacturer's claim at these three levels, it is also within claimed limits at all intermediate levels. This method does not use data as efficiently as one that uses a linear regression approach, but it avoids the problems caused by violating the underlying assumptions of linear regression.

Environmental Diagnostics

Assay validation and certification for immunoassay kits targeted for environmental monitoring are under the authority of the U.S. Environmental Protection Agency (EPA). Such attributes as selectivity, sensitivity, portability, and rapid turnaround time make immunoassay methodology particularly useful for on-site testing. It is especially useful in lowering the costs of cleanup projects because equipment does not have to lay idle while awaiting the results of laboratory analyses.

The earlier immunoassay methods development efforts in this field were largely unsuccessful. The chemistry utilized in these kits was not sufficiently rugged for use on real world environmental matrices, although these methods did perform well on clean water matrices and spiked samples (Lesnik 1994). This poor initial performance of immunoassay kits resulted in a decline of interest at the EPA Program Office.

The first viable immunoassay test kit for the analysis of pentachlorophenol in both soil and water matrices was demonstrated in 1992 to EPA's Office of Solid Waste (OSW) by Ensys, Inc., Research Triangle Park, North Carolina. Since then the OSW has been working with several manufacturers to develop and validate a whole range of immunoassay test kits for both individual analytes as well as classes of analyte. The following summary of EPA's requirements for the development and validation of immunoassay kits for environmental monitoring is largely based on Lesnik's (1994) review. The process is divided into three parts: the development of screening methods based on immunoassay technology, assay validation, and the regulatory approval process. These guidelines generally are along similar lines to FDA's 510(K) Premarket Notification for clinical diagnostic products.

1. Development of Screening Methods. The major application for immunoassay methods in environmental monitoring is for quantitative screening purposes, where a positive response indicates that the target analyte is present at or above the action level (usually the regulatory action level) and a negative response indicates that it is either absent or present below the level of regulatory concern. Manufacturers of immunoassay based screening methods can obtain a letter from OSW that provides guidance on what general validation criteria should be applied to a screening method for its potential inclusion in SW-846. SW-846 provides official documentation for approved test methods for solid waste. These official methods are used to measure the concentration of specific pollutants or to establish whether a waste stream demonstrates a hazardous characteristic, such as ignitability, corrosivity, reactivity, or toxicity. The EPA does not require immunoassay screening procedures to be fully quantitative as long as they measure the presence or absence of target analytes at or below regulatory action levels.

The initial demonstration of the method involves measuring the percentage of false-negatives and false-positives generated using the procedure for a single sample matrix with reference to an appropriate SW-846 quantitative method. A "false negative" here is defined as a negative response for a sample that contains up to two times the stated detection level of the target analyte(s). A "false positive" is a positive response for a sample that contains analytes at one half the detection level. Ideally, a screening method should produce no false negatives and no more than 10% false-positives.

The maximum permissible false-negative rate for a screening method is 5% at the action level specified. The rate for false-positives at the claimed action level is not specified by the OSW. However, the manufacturer is responsible for providing such data in the product package insert.

The OSW also requires evaluation of interferences and matrix effects on the performance of the screening method. Both positive and negative interferences resulting from target and nontarget analytes need to be investigated. Method performance and matrix effects in a variety of the Resource Conservation and Recovery Act (RCRA) matrices also need to be demonstrated. The types of matrices typically include solids, concentrated waste, ash, groundwater, leachate, or waste waters. Initial method evaluations do not require analyte recovery data from spiked samples, although field trial data with reference to SW-846 methods are recommended.

Thus, initial evaluation of an immunoassay screening procedure for environmental monitoring requires data on the percentage of both false-positives and false-negatives, sensitivity to method interferences, and matrix-specific performance data. At this stage, the manufacturers of the immunodiagnostic kits must also provide a description of the assay procedure and the package insert provided with the test kits.

2. Validation.   The OSW has also formulated validation criteria specifically applicable to immunoassay methods. These must be followed whether the kit is the basis for a new method or is an alternative kit being added to existing methods. The following data are required for validation of immunoassay methods.

1. Cross-reactivity with similar analytes,
2. Cross-reactivity with dissimilar analytes that may be reasonably expected to be found at waste sites,
3. False-negative/false-positive rates,
4. Extraction efficiency for soil test kits,
5. Performance data on spiked samples in environmental matrices validated against standard SW-846 analytical methods, and
6. Performance data on actual environmental field samples validated against standard SW-846 analytical methods.

Interferences can be a major problem in environmental analyses; thus, it is important to demonstrate that the analytes of concern can be identified in the presence of similar analytes or dissimilar analytes that may be present in environmental samples.

The false-negative/false-positive rate for a particular immunoassay kit is also important. OSW screening methods are designed to generate 0% false-negatives and up to 10% false-positives at the regulatory action level (Lesnik 1994). Slightly higher false-positive rates (up to 25%) are tolerable. False-positive rates higher than 25% negate the cost effectiveness of the technique because of the excessive numbers of confirmatory tests they require. In contrast, > 5% false-negative rates at the regulatory action level eliminate the potential use of the method for regulatory purposes.

The extraction efficiency data are also important for setting the appropriate action level for soil analyses. In such instances, recoveries are the primary determining factor for ensuring that the analyte of concern can be detected at the regulatory action level and for minimizing false negative/false positive rates.

The performance data generated from environmental samples spiked with the target analytes gives a good indication as to whether or not an immunoassay method will work. However, the performance generated in the field on real environmental samples is the key determining factor on whether or not the immunoassay method is sufficiently rugged to be included in SW-846 as an analytical method.

All these criteria need to be extensively tested and validated by the manufacturers. EPA validation is confined primarily to confirmation of the manufacturers' results and performing some additional testing on well-characterized environmental samples, which are more easily available to EPA regional laboratories.

3. Regulatory Approval Process.   The regulatory process by which methods are incorporated into SW-846 consists of the following phases.

1. Technical workgroup review,
2. Proposed regulation for public comment through a Federal Register notice,
3. Response to public comments, and
4. Promulgated final regulation through a Federal Register notice.

The first and the most critical phase in the method approval process is the only EPA technical review during the approval process. The methods are carefully reviewed by a panel made up of EPA chemists from across the agency representing program offices, ORD, regional laboratories, and enforcement to determine whether they demonstrate appropriate performance and applicability to address RCRA requirements. After workgroup approval, a "draft" method is prepared. The subsequent stages in the approval process by the EPA are largely administrative in nature. Once the method is approved, it can be used for any appropriate RCRA application including those where the use of SW-846 methods is mandatory.

Under the OSW mandate, the two primary applications of immunoassay methods in the RCRA Program are mapping contamination at well-characterized sites slated for cleanup, and monitoring the effectiveness of cleanup activities. In addition to OSW, other program offices, such as OPPTS and Office of Water, are looking at immunoassay methods to address some of their analytical requirements in the Pesticide Registration Program and the Drinking Water and Wastewater Program, respectively.

The future appears to be bright for the environmental application of immunoassay methodology. In addition to the EPA, many other federal agencies with massive cleanup problems, e.g., the Department of Energy and the Department of Defense, have become interested in this technology for its overall utility in significantly lowering the cost of cleanup operations. Although the current emphasis is on the development and approval of screening methods for environmental monitoring, it will not be long before quantitative methods are routinely used in environmental applications.

Food Diagnostics

The U.S. FDA has delegated the responsibility of validating immunoassay kits intended for food diagnostic purposes to the AOAC Research Institute, Arlington, Virginia. The validation protocols are essentially similar to those used in both clinical and environmental diagnostics. Currently, immunoassay kits have been approved and regularly used for the detection of certain antibiotic and drug residues in milk and milk products and in meats, for which tolerance levels have been set by the regulatory agencies.

The AOAC RI have defined the following parameters for the evaluation of a test kit.

Positive: A result due to the presence of a drug at or above the tolerance level in ppb.

Negative: A result obtained with a sample containing no or less than the tolerance level of the drug.

Specificity: Ability of a test to detect a true-negative.

Selectivity: Ability of a test to detect a true-positive.

Limit of detection: The lowest concentration of a drug that can be distinguished with a statistical confidence from a negative.

Cross-reactivity: The degree to which a procedure will detect a drug or a compound other than the one for which the test kit is designed or intended to detect.

Predictive value: The fraction of positive results that are true-positives.

Tolerance level: The concentration in ppb of a drug in a milk or meat product that results in a violative level, and at which level the product may not be sold for human consumption.

Confirmatory method: A well-accepted analytical method, usually HPLC or GCMS, that can be used to identify with better than 99% certainty the drugs that may be present. For most drugs, no such methods as yet exist that can surpass the sensitivity of the immunoassay methods.

At least 30 samples need to be used for detecting the sensitivity and selectivity of an immunoassay protocol. Dose-response curves, cross-reactivity studies, ruggedness testing, and matrix effects (e.g., frozen versus fresh samples, raw versus pasteurized milk, presence or absence of bacteria and somatic cells in milk, etc.) also need to be investigated to characterize a given assay protocol. In addition, the AOAC RI also requires an incurred residue testing study, and validation of certain portions of the manufacturer's study by an independent testing laboratory. Confidence intervals of 95% are used to evaluate the ability of an immunoassay kit to correctly identify positive samples.

Similar parameters are also used for the evaluation and validation of immunoassay kits designed for the presence and identification of food pathogens, such as *Salmonella* and *E. coli* 0157:H7 in various food samples.

REFERENCES

ASTM. 1968. *ASTM Standard E178–68. Standard recommended practice for dealing with outlying observations.* American Society for Testing and Materials, Philadelphia, PA.

BERG, H. C. 1984. *Random Walks in Biology.* Princeton Univ. Press, Princeton, NJ.

BERGMEYER, H. U. 1991. Immunoassay standardization. *Scand. J. Clin. Lab. Invest.* 51 (Suppl. 205):1–2.

BERSON, S. A.; YALOW, R. S.; BAUMAN, A.; ROTHSCHILD, M. A.; and NEWERLY, K. 1956. Insulin-I[131] metabolism in human subjects. Demonstration of insulin binding globulin in the circulation of insulin-treated subjects. *J. Clin. Invest.* 35:170–190.

BLAND, J. M., and ALTMAN, D. G. 1986. Statistical methods for assessing agreement between two methods of clinical measurement. *Lancet* i:307–310.

BLOCKX, P., and MARTIN, M. 1994. Laboratory quality assurance. In T*he Immunoassay Handbook,* ed. D. Wild, pp. 263–276, Stockton Press, New York.

BOERMAN, O. C.; SEGERS, M. F. G.; POELS, L. G.; KENEMANS, P.; and THOMAS, C. M. G. 1990. Heterophilic antibodies in human sera causing falsely increased results in the CA 125 immunoradiometric assay. *Clin. Chem.* 36:888–891.

BOSCATO, L. M., and STUART, M. C. 1988. Heterophilic antibodies: A problem for all immunoassays. *Clin. Chem.* 34:27–33.

BROWN, W. R.; DIERKS, S. E.; BUTLER, J. E.; and GERSHONI, J. M. 1991. Immunoblotting. Membrane filters as the solid phase for immunoassays. In *Immunochemistry of Solid Phase Immunoassay,* ed. J. E. Butler, pp. 151–172, CRC Press, Boca Raton, FL.

BUTLER, J. E. 1991. *Immunochemistry of Solid Phase Immunoassay.* CRC Press, Boca Raton, FL.

BUTLER, J. E.; JOSHI, K. S.; and BROWN, W. R. 1991. The application of traditional immunochemical methods to evaluate the performance of capture antibodies immobilized on microtiter wells. In *Immunochemistry of Solid Phase Immunoassay,* ed. J. E. Butler, pp. 221–231, CRC Press, Boca Raton, FL.

BUTLER, J. E.; NI, L.; NESSLER, R.; JOSHI, K. S.; SUTER, M.; ROSENBERG, B.; CHANG, J.; BROWN, W. R.; and CANTARERO, L. A. 1992. The physical and functional behavior of capture antibodies adsorbed on polystyrene. *J. Immunol. Meth.* 150:77–90.

CANTARERO, L. A.; BUTLER, J. E.; and OSBORNE, J. W. 1980. The adsorptive characteristics of proteins for polystyrene and their significance in solid phase immunoassays. *Anal. Biochem.* 105:375–380.

CHAN, D. W. 1987. General principle of immunoassay. In *Immunoassay: A Practical Guide,* eds. D. W. Chan and M. T. Perlstein, pp. 1–23, Academic Press, Orlando, FL.

CHAN, D. W. 1992. *Immunoassay Automation. A Practical Guide.* Academic Press, San Diego, CA.

CHAN, D. W., and SLAUNWHITE, W. R. 1977. The chemistry of human transcortin. II. The effects of pH, urea, salt and temperature on the binding of cortisol and progesterone. *Arch. Biochem. Biophys.* 182:437–442.

CHANNING RODGERS, R. P. 1981. *Quality Control and Data Analysis in Binder-Ligand Assay. Volumes 1 and 2,* Scientific Newsletters, Inc., Anaheim, CA.

CHARD, T. 1978. *An Introduction to Radioimmunoassay and Related Techniques.* North-Holland, Amsterdam.

CORNBLEET, P. J., and GOCHMAN, N. 1979. Incorrect least-squares regression coefficients in method-comparison analysis. *Clin. Chem.* 25:432–438.

COTLOVE, E.; HARRIS, E. K.; and WILLIAMS, G. Z. 1970. Biological and analytical components of variation in long-term studies of serum constituents in normal subjects. III. Physiological and medical complications. *Clin. Chem.* 16:1028–1032.

CRAIG, W. Y.; POULIN, S. E.; COLLINS, M. F.; LEDUE, T. B.; and RICHIE, R. F. 1993. Background staining in immunoblot assays. *J. Immunol. Meth.* 158:67–76.

CRAGIE, R.; MIZUUCHI, K.; BUSHMAN, F. D., and ENGELMAN, A. 1991. A rapid in vitro assay for HIV DNA integration. *Nucleic Acid Res.* 19:2729–2733.

DAVIES, C. 1994. Concepts. In *The Immunoassay Handbook,* ed. D. Wild, pp. 83–116, Stockton Press, New York.

DELISI, C. 1980. The biophysics of ligand-receptor interactions. *Q. Rev. Biophys.* 13:201–230.

DEMING, W. E. 1943. *Statistical Adjustment of Data.* John Wiley & Sons, New York.

DESHPANDE, S. S. 1994. Immunodiagnostics in agricultural, food and environmental quality control. *Food Technol.* 48(6):136–141.

DE VERDIER, C. H.; GROTH, T.; and WESTGARD, J. O. 1981. What is the quality of quality control procedures? *Scand. J. Clin. Lab. Invest.* 41:1–14.

DIXON, W. J. 1953. Processing data for outliers. *Biometrics* 9:74.

DOUGLAS, A. S., and MONTEITH, C. A. 1994. Improvements to immunoassays by use of covalent binding assay plates. *Clin. Chem.* 40:1833–1837.

EKINS, R. P. 1968. Limitations of specific activity. In *Protein and Polypeptide Hormones, Part 3 (Discussions),* ed. M. Margoulies, pp. 612–616, Excerpta Medica, Amsterdam.

EKINS, R. P. 1974. Basic principles and theory. *Br. Med. Bull.* 30:3–11.

EKINS, R. P. 1981. Toward immunoassays of greater sensitivity, specificity and speed: An overview. In *Monoclonal Antibodies and Developments in Immunoassay,* eds. A. Albertini and R. P. Ekins, pp. 3–21, Elsevier/North-Holland Biomedical Press, New York.

EKINS, R. P. 1991. Immunoassay design and optimization. In *Principles and Practice of Immunoassay,* eds., C. P. Price and D. J. Newman, pp. 96–153, Stockton Press, New York.

EKINS, R. P.; NEWMAN, G. B.; and O'RIORDAN, J. L. H. 1970. Saturation assays. In *Statistics in Endocrinology,* eds. J. W. McArthur and T. Colton, pp. 345–378, MIT Press, Cambridge, MA.

EMANUEL, I. A., and PERELMUTTER, L. L. 1994. Performance characteristics in the evaluation of in vitro assays. In *In Vitro Testing,* ed. I.A. Emanuel, pp. 31–42, Thieme Medical Publishers, New York.

ENGVALL, E., and PERLMANN, P. 1972. Enzyme-linked immunosorbent assay. III. Quantitation of specific antibodies by enzyme-labeled anti-immunoglobulin in antigen coated tubes. *J. Immunol.* 109:129–135.

EZAN, E.; TIBERGHIEN, C.; and DRAY, F. 1991. Practical method for optimizing radioimmunoassay detection and precision limits. *Clin. Chem.* 37:226–230.

FELDKAMP, C. S. 1992. Evaluation and clinical validation of immunoassays. In *Immunochemical Assays and Biosensor Technology for the 1990s,* eds. R. M. Nakamura, Y. Kasahara, and G. A. Rechnitz, pp. 83–109, Amer. Soc. Microbiol., Washington, D.C.

FELDKAMP, C. S., and SMITH, S. W. 1987. Practical guide to immunoassay method evaluation. In *Immunoassay: A Practical Guide,* eds. D. W. Chan and M. T. Perlstein, pp. 49–95, Academic Press, Orlando, FL.

FRANZ, B., and STEGEMANN, M. 1991. The kinetics of solid phase microtiter immunoassays. In *Immunochemistry of Solid Phase Immunoassay,* ed. J. E. Butler, pp. 277–284, CRC Press, Boca Raton, FL.

GEURTS, B. J. 1989. Diffusion limited immunochemical sensing. *Bull. Math. Biol.* 51:359–379.

GORMAN, E.; HOCHBERG, A.; KNODEL, E.; LEFLAR, C.; and WANG, C. 1991. An overview of automation. In *Principles and Practice of Immunoassay,* eds. C. P. Price and D. J. Newman, pp. 219–245, Stockton Press, New York.

GOSLING, J. P., and BASSO, L. V. 1994. Quality assurance. In *Immunoassay: Laboratory Analysis and Clinical Applications,* eds. J. P. Gosling and L. V. Basso, pp. 69–81, Butterworth-Heinemann, London.

GOSLING, J. P.; MIDDLE, J.; SIEKMANN, L.; and READ, G. 1993. Improvement of the comparability of results from immunoprocedures for the measurement of hapten analyte concentrations: Cortisol. *Scand. J. Clin. Lab. Invest.* 53 (suppl. 216):3–41.

HARLOW, E., and LANE, D. 1988. *Antibodies: A Laboratory Manual.* Cold Spring Harbor Laboratory, Cold Spring Harbor, New York.

HARRIS, E. K. 1979. Statistical principles underlying analytical goal setting in clinical chemistry. *Am. J. Clin. Pathol.* 72:374–382.

HARVEY, M. A. 1991. *Optimization of Nitrocellulose Membrane-Based Immunoassays.* Schleicher and Schuell, Keene, NH.

HENDERSON, A. R. 1993. Assessing test accuracy and its clinical consequences. A primer for receiver operating characteristics curve analysis. *Ann. Clin. Biochem.* 30:521–539.

HOWANITZ, P. J.; HOWANITZ, J. H.; LAMBERSON, H. V.; and ENNIS, K. M. 1982. Incidence and mechanism of spurious increases in serum thyrotropin. *Clin. Chem.* 34:427–431.

INMAN, J. K. 1974. Covalent linkage of functional groups, ligands, and proteins to polysaccharide beads. *Methods Enzymol.* 34:38–42.

JACKSON, T. M.; MARSHALL, N. J.; and EKINS, R. P. 1983. Optimization of immunoradiometric assays. In *Immunoassays for Clinical Chemistry,* eds. W. M. Hunter and J. E. T. Corrie, pp. 557–575, Churchill-Livingstone, Edinburgh.

JEFFCOATE, S. L. 1981. *Efficiency and Effectiveness in the Endocrine Laboratory.* Academic Press, London.

KEMENY, D. M. 1992. Titration of antibodies. *J. Immunol. Meth.* 150:57–76.

KRICKA, L. J.; SCHMERFELD-PRUSS, D.; SENIOR, M.; GOODMAN, B. P.; and KALADAS, P. 1990. Interference by human anti-mouse antibody in two-site immunoassays. *Clin. Chem.* 36:892–894.

KROUWER, J. S. 1991. Multifactor designs. IV. How multifactor designs improve the estimate of total error by accounting for protocol-specific biases. *Clin. Chem.* 37:26–29.

LEE, H. T.; DANIEL, A.; and WALKER, C. D. 1982. Conformance test procedures for verifying labeling claims for precision, bias, and interferences in in vitro diagnostic devices used for the quantitative measurement of analytes in body fluids. *Bureau of Med-*

ical Devices Biometrics Report, U.S. FDA Pub. No. 8202, U.S. Govt. Printing Office, Silver Spring, MD.

LESNIK, B. 1994. Immunoassay methods. The EPA approach. *Environ. Lab* 6(3):37–44.

LEVINSON, S. S., and GOLDMAN, J. 1988. Interference by endogenous immunoglobulins with ligand binding assays. *Clin. Immunol. News* 9:101–104.

MCLEAN, L. 1992. Testing polystyrene microplates for coating uniformity. *Amer. Biotech. Lab.* 10(12):22.

MICALLEF, J., and AHSAN, R. 1994. Immunoassay development. In *Immunoassay: Laboratory Analysis and Clinical Applications,* eds. J. P. Gosling and L. V. Basso, pp. 51–68, Butterworth-Heinemann, London.

MILLER, J. J., and VALDES, R. 1991. Approaches to minimizing interference by cross-reacting molecules in immunoassays. *Clin. Chem.* 37:144–153.

MILLER, J. J., and VALDES, R. 1992. Methods for calculating crossreactivity in immunoassays. *J. Clin. Immunoassay* 15:97–107.

MIZE, P. D.; NAQUI, A.; O'CONNELL, C. M.; WALLER, J. N.; FESLER, M.; MYATICH, R. G.; and KEATING, W. E. 1991. Studies on the covalent attachment of antibodies to controlled surfaces. In *Immunochemistry of Solid Phase Immunoassay,* ed. J. E. Butler, pp. 207–219, CRC Press, Boca Raton, FL.

MOHAMMAD, K., and ESEN, A. 1989. A blocking agent and blocking step are not needed in ELISA, immunostaining dot-blots and Western blots. *J. Immunol. Meth.* 117:141–145.

NCCLS. 1984. *NCCLS tentative guidelines EP5-T. User evaluation of precision performance of clinical chemistry devices.* National Committee for Clinical Laboratory Standards, Villanova, PA.

NCCLS. 1986. *NCCLS propsoed guidelines EP9-P. User comparison of clinical laboratory methods using patient samples.* National Committee for Clinical Laboratory Standards, Villanova, PA.

NCCLS. 1987. *NCCLS proposed guidelines GP10-P. Assessment of clinical sensitivity and specificity of laboratory tests.* National Committee for Clinical Laboratory Standards, Villanova, PA.

NCCLS. 1989. *NCCLS tentative guideline EP10-T. Preliminary evaluation of clinical chemistry methods.* National Committee for Clinical Laboratory Standards, Villanova, PA.

NGO, T. T. 1991. Immunoassay. *Current Opinion in Biotechnol.* 2:102–109.

NOMURA, M.; IMAI, M.; USUDA, S.; NAKAMURA, T.; MIYAKAWA, Y.; and MAYUMI, M. 1982. A pitfall in two-site sandwich 'one-step' immunoassay with monoclonal antibodies for the determination of human alpha-fetoprotein. *J. Immunol. Meth.* 56:13–17.

NYGREN, H., and STENBERG, M. 1989. Immunochemistry at interfaces. *Immunology* 66:321–327.

PERLSTEIN, M. T. 1987. Immunoassays. Quality control and troubleshooting. In *Immunoassay: A Practical Guide,* eds. D. W. Chan and M. T. Perlstein, pp. 149–163, Academic Press, Orlando, FL.

PESCE, A. J.; FORD, D. J.; and MAKLER, M. T. 1981. Properties of enzyme immunoassays. In *Enzyme Immunoassay,* eds. E. Ishikawa, T. Kawai, and K. Miyai, pp. 27–39, Igaku-shoin, Tokyo.

PESCE, A. J.; FORD, D. J.; GAIZUTIS, M.; and POLLAK, V.E. 1977. Binding of protein to polystyrene in solid phase immunoassays. *Biochim. Biophys. Acta* 492:399–407.

POLLOCK, M. A.; JEFFERSON, S. G.; KANE, J. W.; LOMAX, K.; MACKINNON, G.; and WINNARD, C. B. 1992. Method comparison: a different approach. *Ann. Clin. Biochem.* 29:556–560.

PORSTMANN, T., and KIESSIG, S. T. 1992. Enzyme immunoassay techniques. An Overview. *J. Immunol. Meth.* 150:5–21.

RASSMUSEN, S. R.; LARSEN, M. R.; and RASSMUSEN, S. E. 1991. Covalent immobilization of DNA onto polystyrene microwells. The molecules are only bound at the 5' end. *Anal. Biochem.* 198:138–141.

REED, A. H.; HENRY, R. J.; and MASON, W. B. 1971. Influence of statistical method used on the resulting estimate of normal range. *Clin. Chem.* 17:275–284.

ROSEN, S. M., and FISCHBERG-BENDER, E. 1995. Murine monoclonal antibodies to drugs: Strategies for immunization and screening. *IVD Technology,* July, pp. 20–25.

SETH, J. 1987. The external quality assessment of hormone assays. *J. Clin. Biochem. Nutr.* 2:111–139.

SETH, J. 1991. Standardization and quality assurance. In *Principles and Practice of Immunoassay,* eds. C. P. Price and D. J. Newman, pp. 154–189, Stockton Press, New York.

SETH, J.; STURGEON, C. M.; AL-SADIE, R.; HANNING, I.; and ELLIS, A. R. 1991. External quality assessment of immunoassays of peptide hormones and tumor markers. Principles and practice. *Ann. Ist. Super. Sanita* 27:443–452.

SOLBERG, H. E. 1987. Establishment and use of reference values. In *Textbook of Clinical Chemistry,* ed. N. Tietz, pp. 371–378, W.B. Saunders, Philadelphia, PA.

STEIN, E. A.; DIPERSIO, L.; PESCE, A. J.; KASHYAP, M.; KAO, J. T.; SRIVASTAVA, L.; and MCNERNEY, C. Enzyme-linked immunosorbent assay of apolipoprotein AII in plasma with use of a monoclonal antibody. *Clin. Chem.* 32:967–971.

STENBERG, M.; WERTHEN, M.; THEANDER, S.; and NYGREN, H. 1988. A diffusion limited reaction theory for a microtiter plate assay. *J. Immunol. Meth.* 112:23–29.

TAYLOR, J.K. 1987. Quality Assurance of Chemical Measurements. Lewis Publishers, Chelsea, MI.

THOMPSON, R.J., JACKSON, A.P., and LANGLOIS, N. 1986. Circulating antibodies to mouse monoclonal immunoglobulins in normal subjects: incidence, species specificity, and effects on a two-site assay for creatine kinase-MB isoenzyme. Clin. Chem. 32:476–481.

TONKS, D. 1968. A quality control program for quantitative clinical chemistry estimations. *Can. J. Med. Technol.* 30:38–54.

TUNN, S.; PAPPERT, G.; WILLNOW, P.; and KREIG, M. 1990. Multicenter evaluation of an enzyme immunoassay for cortisol determination. *J. Clin. Chem. Clin. Biochem.* 28:929–935.

ULDALL, A. 1987. Quality assurance in clinical chemistry. *Scand. J. Clin. Lab. Invest.* 47 (Suppl. 187):83–91.

WAAKERS, P. J. M.; HELLENDOORN, H. B. A.; OP DE WEEGH, G. J.; and HEERSPINK, W. 1975. Applications of statistics in clinical chemistry. A critical evaluation of regression lines. *Clin. Chim. Acta* 64:173–184.

WESTGARD, J. O., and HUNT, M. R. 1973. Use and interpretation of common statistical tests in method-comparison studies. *Clin. Chem.* 19:49–57.

WESTGARD, J. O.; BARRY, P. L.; and HUNT, M. R. 1981. A multi-rule Shewhart chart for quality control in clinical chemistry. *Clin. Chem.* 27:493–501.

WESTGARD, J. O.; CAREY, R. N.; and WOLD, S. 1974. Criteria for judging precision and accuracy in method development and evaluation. *Clin. Chem.* 20:825–833.

WESTGARD, J. O.; DE VOS, D. J.; and HUNT, M. R. 1978a. Concepts and practices in the selection and evaluation of methods. Part I. Background and approach. *Am. J. Med. Technol.* 44:290–300.

WESTGARD, J. O.; DE VOS, D. J.; and HUNT, M. R. 1978b. Concepts and practices in the selection and evaluation of methods. Part II. Experimental procedures. *Am. J. Med. Technol.* 44:420–430.

WESTGARD, J. O.; DE VOS, D. J.; and HUNT, M. R. 1978c. Concepts and practices in the selection and evaluation of methods. Part III. Statistics. *Am. J. Med. Technol.* 44:552–571.

WESTGARD, J. O.; DE VOS, D. J.; and HUNT, M. R. 1978d. Concepts and practices in the selection and evaluation of methods. Part IV. Decision on acceptability. *Am. J. Med. Technol.* 44:727–742.

WESTGARD, J. O.; DE VOS, D. J.; and HUNT, M. R. 1978e. Concepts and practices in the selection and evaluation of methods. Part V. Applications. *Am. J. Med. Technol.* 44:803–813.

WHITEHEAD, T. P. 1977. Quality Control in Clinical Chemistry. John Wiley & Sons, Chichester, England.

WHO. 1981. *External Quality Assessment of Health Laboratories.* Report on a WHO Working Group. EURO Reports and Studies 36, Regional Office for Europe, World Health Organization, Copenhagen.

WILDE, C. 1994. Subject preparation, sample collection and handling. In *The Immunoassay Handbook,* ed. D. Wild, pp. 243–255, Stockton Press, New York.

YALOW, R. S., and BERSON, S. A. 1959. Assay of plasma insulin in human subjects by immunological methods. *Nature* (London) 184:1648–1649.

YALOW, R. S., and BERSON, S. A. 1960. Immunoassay of endogenous plasma insulin in man. *J. Clin. Invest.* 39:1157–1175.

YALOW, R. S., and BERSON, S. A. 1970. Radioimmunoassays. In *Statistics in Endocrinology,* eds. J. W. McArthur and T. Colton, pp. 327–344, MIT Press, Cambridge, MA.

YOUDEN, W.J. 1984. *Experimentation and Measurement. NBS SP 672.* National Bureau of Standards, Gaithersburg, MD.

ZWEIG, M. H., and CAMPBELL, C. C. 1993. Receiver-operating characteristics (ROC) plots. A fundamental evaluation tool in clinical medicine. *Clin. Chem.* 39:561–577.

ZWEIG, M. H., and ROBERTSON, E. A. 1987. Clinical validation of immunoassays. A well-designed approach to a clinical study. In *Immunoassay. A Practical Guide,* eds. D. W. Chan and M. T. Perlstein, pp. 97–127, Academic Press, Orlando, FL.

10

Reagent Formulations
and Shelf Life Evaluation

INTRODUCTION

The control of assay quality and the degree of quality control required begin with the source and the accurate preparation of reagents and control sera used in the assay procedure. The quality of reagents used in certain steps of the immunoassay procedure is often more critical than the quality of those used for blocking and washing steps. Because several different reagents may be used in performing the immunoassay, some in-laboratory checks are mandatory, and sometimes essential even for commercial immunoassay kits.

With an assay that has been developed totally in-house, new reagents must be thoroughly characterized. Once this is done, the controls can be restricted to those used for monitoring assay performance, in addition to regular checks that the properties of the reagents have not deteriorated. This is particularly important with respect to enzyme label and the standard/calibrator reagents. These two generally have a tendency to deteriorate more rapidly and, hence, require testing quite frequently, often as part of every assay run.

As compared to assays developed in-house, those based on kit material might appear to require less quality control, as one purchases a certain quality from the manufacturer. However, as long as the commercial kit procedures do not provide complete test formats of the individual reagent batches, such material may require more exhaustive testing than that actually necessary for "home-made" reagents.

An important aspect of quality control in the use of commercially available immunodiagnostic kits is the need to ensure that the reagents supplied have not been damaged during transportation. Each newly delivered kit thus should be tested before use. This is particularly important for clinical samples. Such evaluation can be easily done by the "kit overlapping" procedure. Here individual reagents from the new kit are tested against the reagents in the currently run kit. The antibody and the enzyme label from the new kit are substituted in the assay procedure for the reagents used in the currently run kit. Similarly, individual standard concentrations, usually at the midpoint of the standard curve, are assayed with the new kit materials. In this way, a new kit will be appropriately evaluated with respect to standards and antibody and enzyme label preparations before its incorporation into clinical routine.

Reagents supplied with the commercial kits also need to be reconstituted carefully. It is important to follow the instructions in the package insert. This is particularly true for the reconstitution of freeze-dried preparations. Because such vials are at a negative pressure, they should be opened carefully, allowing air to slowly enter the vial to avoid material being forced out by the incoming rush of air. The reconstitution should be carried out by using a calibrated pipette with the correct solvent at the recommended temperature. The mixing of the solution should also be carried out cautiously to avoid foaming and subsequent denaturation of protein reagents. Similarly, extended contact between the solution and the vial stopper is best avoided, as occasionally substances on rubber stoppers can leak into the solution and interfere with the assay. This may occur with commercial controls, where the manufacturer of the control cannot test the effects of the stopper on all the assays with which the control may be used.

When conserving reagents and control preparations, it is important to adhere to the manufacturer's instructions. Blockx and Martin (1994) have recommended the following general guidelines for this purpose.

1. Liquid and freeze-dried reagents should be stored at 2–8°C.
2. Reagents transported in dry ice should be stored at –20°C.
3. After reconstitution, freeze-dried reagents should be kept at 2–8°C for short periods, or as aliquots in the freezer for longer periods.
4. Thawed reagents should not be refrozen.

In this chapter, general guidelines are provided for the formulation, preparation, and shelf life evaluation of both liquid and solid reagents for use in immunodiagnostic kits.

LIQUID FORMULATIONS

Several reagents such as washing buffers, enzyme label, substrate buffers, chromophores, and stop reagents can be formulated and supplied as liquid reagents in commercial immunodiagnostic kits. Because most such reagents are aqueous solutions, the quality of water used plays a critical role in the stability and performance of these reagents in the immunoassay procedure. Untreated water contains a number of impurities, including inorganics, organics, dissolved gases, suspended solids, colloids, microorganisms, and pyrogens. Depending on the application, some or all of these will have to be removed.

Deionization and reverse osmosis are used for preparing a quality source of water for laboratory use. Cation- and anion-exchange resins can be used to remove all dissolved ionizable substances from the primary water source. This process can also be coupled with specific ion-exchange resins or activated carbon filters to remove organic or colloidal matter. In contrast, reverse osmosis uses a semipermeable membrane that separates the water from most of the impurities. The method can remove approximately 95% of the dissolved ions, 97% of dissolved solids, and up to 99% of bacteria, viruses, high molecular mass organics, and pyrogens. If required, both techniques can be combined, reverse osmosis being used as a pretreatment to deionization.

According to Blockx and Martin (1994), the ideal water quality for making up reagents and calibrators is:

Resistivity:	10–18 $M\Omega$/cm at 25°C
Organics (TOC):	20–100 ppb
pH:	7.0

However, most immunoassays do not require such high water quality. Most can be performed with water having a resistivity not lower than 5 $M\Omega$/cm at 25°C, with an organic content of less than 2 ppm. Depending upon the applications, it is therefore sufficient to have a deionization technique, controlled by a conductivity meter, to obtain the water purity necessary for assays and for rinsing labware needed for the preparation of calibrators and reagents.

The preparations of some of the important solutions used in the development of commercial immunodiagnostic kits are briefly described below.

Coating Buffers

Coating buffers used for adsorbing antibodies or antigens to the solid phase should have a pH at least 1–2 units higher than the pI of the protein. Similarly, the ionic strength of the coating buffer should be low. It must be ensured that this buffer contains only the primary coating molecule of interest to bind to the surface. Any contaminating molecules in the buffer will lower the coating efficiency and result in poor assay performance. Phosphate (pH 7.0–7.2) and carbonate (pH 9.2–9.4) buffers of 0.02–0.05 molarity are widely used for coating solid-phase matrices.

Blocking Buffers

Blocking buffers should also be as pure as possible, containing only the buffer salts and the blocking agent. The blocking agent should be chosen so as not to cross-react in the later steps of the assay. Thus, it should not have any affinity for the antigen, antibody, or the enzyme label. If blocking is not done as a separate step, the other buffers will have to be very good to eliminate high background in the assay.

Incubation Buffers

The rates of antigen-antibody reactions have a rather broad pH maximum around 7.0. The primary reaction is little affected by the ions used in most common buffer systems. Generally, assays can be performed with any standard buffer, such as phosphate, Tris, borate, barbital (Veronal®), glycine, or bicarbonate. The addition of 0.15 M NaCl to make them "physiologic" in osmolarity is not essential in most assay systems. These buffers ideally should have a neutral pH value, a low to medium-high ionic strength, and a composition that ensures binding of the analyte to the antibody. Sometimes it is desirable, although not necessary, to include the blocking agent in incubation buffers to improve assay performance.

Wash Buffers

Ideally, wash buffers should have a pH value around 7.0 and a low to medium-high ionic strength. The incorporation of detergents, such as Tween-20, in wash buffers is generally recommended to avoid cross-reactivity problems and high background signal in the assay.

Substrate Buffers

The substrate buffers are extremely important in enzyme immunoassays. Their use and formulation are mandated by the enzyme label and the substrate system used in the assay protocol. Hence, standard protocols should be referred to while preparing substrate buffers.

Aqueous Protein Formulations

Enzymes and antibodies, being proteins, are relatively fragile substances that have a tendency to undergo inactivation or denaturation if not handled properly. Hence, primary consideration should be given to their handling.

Generally, high temperatures and acid or alkaline solutions should be avoided. Most enzymes are inactivated above 35–40°C and in solutions of pH less than 5 or greater than 9. Similarly, in adjusting the pH of protein solutions, one must be careful not to create a zone of destruction around a drop of reagent added, which would tend to inactivate some of the protein in solution. Solutions of enzymes should be well stirred, while avoiding foaming, and acids or bases added dropwise along the side of the tube to avoid any denaturation in adjusting the pH.

The shelf life of most proteins can be greatly prolonged by cold storage. For most enzymes, storage at 2–5°C is sufficient for long term stability in the dry state. Other enzymes are unstable even at these temperatures and must be stored in a freezer below –20°C. Certain enzymes, such as alkaline phosphatase, can be stabilized at high concentrations of salts and can be kept for long periods as a suspension in ammonium sulfate. Such solutions can often be stored in a refrigerator or freezer for months without loss of activity. Repeated freezing and thawing of enzyme and antibody solutions, however, should be avoided.

Organic solvents, such as alcohol, acetone, or ether, denature most enzymes at room temperature except at low concentrations (< 3%). Care therefore must be taken in changing the composition of a solution from aqueous to partly nonaqueous.

Many enzymes are denatured at surfaces. Therefore, vigorous shaking of enzyme solutions should be avoided. Because traces of metal ions can inactivate certain enzymes, it is necessary that the water used in preparation of enzyme reagents should be carefully purified.

Enzyme labels can be supplied as liquid formulations. However, certain guidelines must be adhered to maintain the stability of such preparations. Formulations can be prepared in phosphate buffered saline of pH around 7.0–7.2. Polyols must be used to cryoprotect the enzyme. Glycerol and polyethylene glycols can be used at around 40% concentration for this purpose. The use of sugars at high concentrations must be avoided in such formulations. They tend to crystallize out when stored for long periods. Because the primary reagent, the enzyme label in this case, is often present in very low concentration, the total protein content of such formulations must be raised to at least 4–5 mg/ml. Generally for enzyme-antibody conjugates, the use of immunoglobulin depleted serum proteins from an animal species identical to the antibody donor is recommended. In most instances, BSA is also an ideal protein for this purpose. Depending on the type of enzyme used, other stabilizers may need to be used in such formulations. Similarly, the use of chelating agents with enzymes that contain metal ions must be avoided. The incorporation of preservatives in such formulations is generally recommended. Those widely used in the immunodiagnostic reagent formulations are described later in this section.

It is sometimes necessary to include an agent with some of these solutions and diluents that prevents the adsorption of reagents to walls of tubes, pipettes, etc. Most solid materials have the ability to adsorb constituents of solution to their surfaces. The extent of this adsorption varies with different materials. At the extremely low concentrations of active reagents used in immunoassays, even low adsorbing surfaces tend to remove significant fractions of the reagents from the solution. Moreover, certain analytes can be sticky. For example, drawing a solution of adrenocorticotropic hormone (ACTH) that does not contain any adsorption-preventing material in and out of a micropipette may cause a major loss of ACTH, some of which may remain on the inside walls of the pipette. The addition of proteins to the solution in concentrations that are much higher than those of the active reagents will reduce the adsorption markedly, probably by covering the surface with a nonadhesive film. Human or bovine serum albumin at 0.1–0.5% concentrations is most commonly used for this purpose. Because albumins are known to bind some active molecules and also serve as transport proteins in plasma, their use should be avoided with certain assays. In such instances, gelatin can be a good substitute.

If the manufacturer wishes to supply some of these reagents, such as wash buffers, with their kits, it is generally recommended to supply them in a concentrated form so that the end user can dilute them appropriately prior to running the assay.

Preservatives

Preservatives such as thimerosal (Merthiolate®) at 1:10, 000 or 0.02–0.1% sodium azide have been widely used in diagnostic reagents. However they have proven less than ideal for this purpose. Thimerosal is expensive, blocks some chromogenic substrates, and because it contains mercury, is classified as toxic for disposal. Its use is also banned in some countries.

Sodium azide is primarily biostatic at the levels normally used, and thus might not kill all microorganisms present in a reagent. It also inhibits a number of biological assays and enzymes used in immunoassays. Sodium azide powder is also both hazardous and inconvenient to work with, and, like thimerosal, its use imposes restrictions on product disposal.

Recently, Rohm & Haas Co., Spring House, Pennsylvania, has introduced Pro-Clin™ 300, a new biocide for in vitro diagnostic reagents. It is a highly effective preservative with broad-spectrum biocidal activity, good compatibility and stability, and low toxicity at in-use levels (Chapman 1994). At unusually low concentrations, it eradicates bacteria, fungi, and yeast cells for prolonged periods. It does not interfere with the functioning of most enzyme- or antibody-linked reactions or assay indicators. Furthermore, there are no disposal restrictions on the material when used at recommended levels.

The active components in the preservative are two isothiazolones, 2-methyl-4-isothiazolin-3-one and 5-chloro-2-methyl-4-isothiazolin-3-one, which are formulated with an organic stabilizer (alkyl carboxylate) in a modified glycol. Within minutes of contact with microorganisms, these active compounds in the preservative inhibit growth, macromolecule synthesis, and respiration.

Long term protection in reagents is afforded by 6–20 ppm active components (0.02–0.07% ProClin™ 300 preservative as supplied), with 9–15 ppm typical. The most effective level for a particular reagent, however, must be determined by experimentation. Because its active components are volatile, ProClin™ 300 is not recommended for freeze-dried solid formulations.

The preservative has proven compatible with most enzymes tested, even at concentrations far above normal in-use levels. Similarly, levels of up to 300 ppm of active components appear to have no effect on antigen-antibody binding reactions.

ProClin 300™ is compatible with a wide range of chemicals and pH values, making it suitable for preserving a variety of diagnostic reagents. Nucleophiles, especially secondary amines, and amino buffers (Tris, HES, HEPES) lower its effectiveness. When using the preservatives with these buffers, stability can be often improved by reducing the pH, to protonate the amines and diminish their nucleophilic character.

SOLID FORMULATIONS

Solid reagent formulations are often preferred for diagnostic reagents that are either unstable or insufficiently stable during the product life span for distribution and use in aqueous solutions. In such instances, freeze-drying or lyophilization is the method of choice for preparing solid reagent formulations for use in commercial immunoassay kits. It is a process in which the solvent, often water, is first frozen and then removed by sublimation and desorption in a vacuum environment to produce a stable product. Freeze-drying provides extended stability and shelf life to the product, minimizes chemical decomposition and/or loss of biological activity because of the low temperatures used in the drying process, provides a sterile and low particulate process, and, when carried out in vials, facilitates accurate and sterile dosage.

Compared to other methods of drying, freeze-drying is normally considered less destructive to protein products such as enzymes and antibodies. In addition, control of sterility and foreign particulate matter is relatively easy in the freeze-drying process. The process, however, is inherently expensive from the viewpoints of plant costs and manufacturing. Nevertheless, it is usually dwarfed by the cost of raw materials; both costs being inevitable when a high quality product is the goal.

In this section, relevant aspects of this technology as applied to manufacturing of quality diagnostic reagents are briefly described. Because most reagents are solubilized under aqueous conditions, some of the basic properties of water in relation to this technology are also described. In addition, the effects of low temperature on certain properties of proteins are also discussed.

Basic Properties of Water

Water contributes the bulk of liquid reagents. However, under identical conditions, its properties differ greatly from those of pure water. Some of the phase transition properties of pure water under standard conditions are summarized in Table 10.1. These properties are greatly influenced by the type and the amount of dissolved solids in water.

TABLE 10.1.   Phase Transition Properties of Water

Melting point at 1 atm	0.00°C
Boiling point at 1 atm	100.00°C
Critical temperature	374.15°C
Critical pressure	218.6 atm
Triple point	0.0099°C and 4.579 mm Hg
Heat of fusion at 0°C	79.71 cal/g; 1.436 kcal/mol
Heat of vaporization at 100°C	538.7 cal/g; 9.705 kcal/mol
Heat of sublimation at 0°C	12.16 kcal/mol

A simple phase diagram indicating the several physical states of water at equilibrium is shown in Figure 10.1. A "phase" can be defined as a homogeneous and physically differing part of a system separated from the other parts of fixed borders (Deshpande et al. 1984). The equilibrium conditions between the solid, liquid, and gaseous states of water are temperature and pressure dependent. A state of equilibrium of ice, liquid water, and water vapor exists only at the "triple point" with a pressure of 4.579 mm Hg and a temperature of 0.0099°C (Figure 10.1, Table 10.1). Below the triple point, the ice is directly sublimed into a gaseous state. This forms the basis of the freeze-drying technology. As shown in Figure 10.1, water can exist, under proper conditions of temperature and pressure, in 11 different physical states. These include one each for liquid and vapor phases, and nine different crystalline solid phases. The latter are the different high pressure polymorphs of ice.

The phase transitions that occur in true solutions can be analyzed fairly accurately based on the regularities of ice formation in nonstructural systems. However, for multicomponent or pseudosystems such as diagnostic reagents, many generalizations have to be made. The most important phase transitions that take place at low temperatures in multicomponent reagent systems include the following:

1. Crystallization and melting (including the eutectic phenomena),
2. Vitrification,
3. Glass (vitreous) transformation,
4. Devitrification (crystallization) following glass transformation,
5. Recrystallization preceding melting, and
6. Antemelting.

The formation of ice in a simple, diluted binary (two-phase) system is accompanied by the following.

1. The separation of water as pure ice crystals
2. A progressive increase in the concentrations of the dissolved substance, and
3. A gradual decrease in the melting point of the concentrated solution.

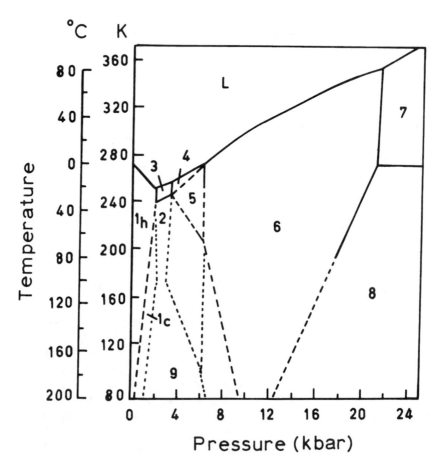

Figure 10.1. Phase diagram of water
From: de Quervain (1975)

At eutectic temperature, both the unfrozen water phase and the substance dissolved in it crystallize simultaneously in a fixed relationship, forming a mixed conglomerate (hydrate). In multicomponent solutions, the dissolved solids crystallize out successively according to their eutectic temperatures. Unfrozen water is retained in the system until the lowest eutectic temperature becomes saturated.

Diagnostic reagents are not "true" solutions, because in addition to water and the primary component (e.g., enzyme-hapten or enzyme-antibody conjugate, standard or calibrator compound), they also contain a large number of other soluble components ("excipients") added to enhance the stability of the primary component as well as the elegance of the final dried product. However, for reasons of simplicity, diagnostic reagents in the following discussion are considered as pseudobinary systems, where all substances dissolved in water are considered as one component.

A simple phase diagram of a binary mixture is shown in Figure 10.2. If a binary mixture is cooled from its initial temperature T_A under conditions allowing equilibrium to be attained, then a solid crystalline phase (pure ice crystals) begins to appear in the liquid phase. This corresponds to composition W_A and the freezing point T_{A^1} in the phase diagram. Usually only one component in the binary mixture crystallizes out in a pure form.

At the initial freezing temperature (T_{A^1}), where the crystallization process begins, only small amounts of crystals are generated. As crystallization proceeds, the concentration of liquid water in the solution will decrease. The dissolved substances will therefore be concentrated progressively in the liquid phase. As will

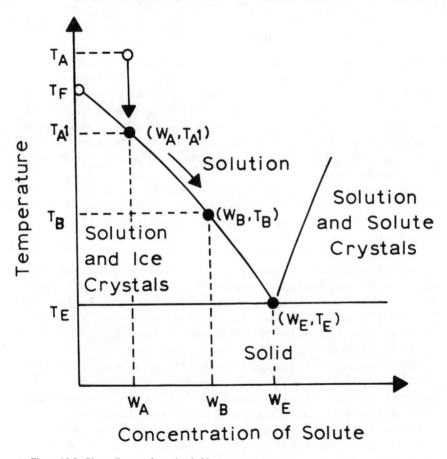

Figure 10.2. Phase diagram for a simple binary system

be described later, the formulation of diagnostic reagents to be lyophilized assumes greater importance, as certain dissolved concentrated substances may adversely affect the quality of the primary component, i.e., either the enzyme or the antibody.

The progressive concentration of the dissolved solids in the liquid phase decreases the crystallization temperature of the remaining liquid; therefore, a lower operating temperature is required to effect further crystallization. At this stage, the liquid composition follows the line $[(W_A T_{A^1}) (W_B, T_B)]$ in the phase diagram (Figure 10.2). It is apparent that for higher conversions to the crystal phase, i.e., for effective ice separation, successively lower temperatures are required. At eutectic concentration (W_E) and temperature (T_E), the crystallizing solid has the same composition as the supernatant liquid. At this stage, removal of heat from the system converts more of the liquid phase to solid, but at a constant temperature.

Proteins at Low Temperatures

Because the primary reagents, enzymes and antibodies, used in the diagnostic test kits are proteins, it is essential to understand the behavior of proteins at low temperatures. Most proteins tend to adjust their conformations during a temperature change in accord with the temperature dependence of those forces that maintain the protein's conformation. The principal contributor to the stability of the "native" conformation of a protein is the tendency of its nonpolar groups to avoid contact with water, i.e., to engage in hydrophobic interactions. This process is largely entropy driven.

The unfavorable low entropy of "ordered" water, which would surround nonpolar groups when these are exposed to the aqueous medium, is avoided because the nonpolar groups fold inward. This is accompanied by an entropy-increasing release of ordered water to the relatively disordered state of the bulk solvent. Because the bulk water itself becomes more extensively hydrogen bonded and thus more highly ordered as the temperature is lowered, the entropic advantage of folding nonpolar groups inward is lessened at low temperatures. In this case, stabilization by hydrophobic interactions would be expected to decrease. It is widely recognized that hydrophobic interactions are of major importance with regard to the conformational integrity of proteins and that these interactions suffer when a protein is "cold-denatured."

Thus, low temperatures are expected to alter protein conformation by favoring the formation of and strengthening hydrogen bonds and diminishing the importance of hydrophobic interactions. In contrast, little is known about electrostatic forces in connection with low temperature effects on protein conformation. Such effects may be significant but their magnitude cannot be assessed, nor is it possible to determine whether such effects would strengthen, weaken, or alter the protein's conformation in a temperature-dependent manner.

A clear distinction must be drawn between "cold" and "freezing." Freezing is the removal of water as ice, and is therefore accompanied by dramatic increases in the concentrations of all water-soluble substances in the residual liquid phase. Most of the observed damage to proteins during freezing, which is a prerequisite for the freeze-drying process employed for the preparation of solid formulations, is in fact due to concentration of one type or another, rather than the freezing as such. This so-called freeze denaturation is largely irreversible and often takes the form of rapid aggregation, following an initial unfolding and/or subunit dissociation. The kinetics of freeze denaturation of proteins are often complex and not yet thoroughly understood. They are, however, very much dependent on possible eutectic phase separations of pH buffer salts and/or the occurrence of glass transitions in the freeze concentrate. Some specific examples in this regard are described later in this section.

For a detailed treatment of this subject, the readers are referred to several excellent reviews on the properties and behavior of proteins at low temperatures (Fennema 1979; Taborsky 1979; Kuntz 1979; Eisenberg and Richards 1995; Franks 1995).

Freeze-Drying Process

The freeze-drying process consists of three interdependent stages: freezing, primary drying, and secondary drying. In the diagnostics industry, the aqueous reagents are poured into suitable containers (usually vials) with stoppers partially inserted in the vial necks. The semistoppered position permits vapor flow from the container. The vials are then loaded onto temperature-controlled shelves in a large drying chamber. The shelf temperature is lowered to about −30°C to −44°C to transform the aqueous solution into a solid, i.e., ice and solid solute. The temperature is held at these levels for sufficient time to ensure complete freezing of the product.

After the product is completely solidified, vacuum is applied to the chamber and the shelf temperature is increased to initiate the ice sublimation stage of the process, called "primary drying" (Pikal 1990a). Water vapor created by sublimation is transported through the partially dried product cake and is condensed on low temperature surfaces (about −60°C) in the condenser chamber. Even after all ice has been removed by sublimation, the product still contains large amounts of water, typically ranging from 20–50, that is "dissolved" in the amorphous portion of the solid. Most or essentially all of this dissolved water is removed during the "secondary drying'" stage. This final step is normally carried out at elevated product temperatures to achieve efficient water removal.

The freeze-drying system itself comprises six integrated subsystems (Hull Corporation 1994). Their functions are briefly described below.

Product chamber: The product chamber is a vacuum vessel constructed to withstand the pressures and temperatures used in freeze-drying. The product shelves in the chamber also act as heat exchangers to remove heat during the initial freezing process and to provide heat during the final drying steps of the process.

Condenser: The condenser functions as a heat sink to collect the water vapor removed during the drying of the product. In most modern freeze-drying units, the condenser is commonly chilled by direct expansion of a refrigerant. The condenser must have a sufficient surface area to efficiently condense all the water vapor generated during primary and secondary drying. It must also effectively condense all the water vapor removed so that it does not reach and thus damage the vacuum pumps.

Heat transfer system: The heat transfer system provides shelf temperature control including chilling of the shelves during product freezing and heating during the drying process. Chilling is accomplished by passing a heat transfer fluid through a heat exchanger coupled with a refrigeration system. Heating is often accomplished by an immersion heater in a heat transfer fluid circuit (Hull Corporation 1994).

Refrigeration system: Refrigeration is required in several different steps of the freeze-drying process. Initially, all the refrigeration capacity is used to cool the shelves during the freezing step. For primary and secondary drying, the main system is used for chilling the condenser, while a second unit is available for shelf temperature control. For the initial freezing step, generally there is enough capacity to cool the shelves to temperatures of $-55°C$ or below. Often , depending upon the type of compressor and refrigerant used, a dedicated secondary system is available that has the capacity for maintaining cooling for shelf control to at least $-30°C$ to $-40°C$.

Vacuum system: Vacuum pumping units are required to evacuate the system and to reduce the pressure within the chamber and condenser at the beginning of the primary drying cycle in order for sublimation to occur. Their configuration may vary with the manufacturer. Single-stage units may achieve pressures down to 20–40 microns, while the more expensive two- and three-stage systems may achieve 10–20 microns. The vacuum pumps normally used in a freeze-drying unit are oil-sealed pumps. They must be protected from water vapor in order to maintain their pumping speed and operating pressures.

Instrumentation for process control: Depending on its sophistication, the freeze-drying unit may be provided with a variety of instrument controls. Four principal functions of the instrumentation include shelf temperature control, pressure control, process parameter recording, and process sequencing. The control and monitoring may be completed by standalone instruments or by an integrated process automation system (Hull Corporation 1994).

The characteristics of the three stages of the freeze-drying process are briefly described below. This discussion is primarily based on several excellent review articles available in the literature (Pikal 1990a,b; Hull Corporation 1994; MacKenzie 1977, 1985a).

1. Freezing. The initial freezing step in the lyophilization process significantly influences the process of sublimation and desorption as well as the final appearance of the product. Furthermore, several important properties such as the physical strength of the cake (friability) and reconstitution of the final dried product are controlled by the rate of freezing of the reagents.

In practice, ice does not form at the thermodynamic or equilibrium freezing point of the solution. The solute usually nucleates and crystallizes only after supercooling at about 10–15°C below the equilibrium freezing point. This degree of supercooling depends on the nature of the solutes, the freezing procedure or shelf temperature-time protocol, the container, and the presence of particulate matter to serve as heterogeneous nucleation sites for ice formation (Pikal 1990a). A higher degree of supercooling generally produces smaller ice crystals. The size of the ice crystals in turn determines the size of the pores or channels created during ice sublimation and the surface area of the porous solid produced by the sublimation process. The degree of supercooling, therefore, affects the rate of sublimation (e.g., large ice crystals create large pores, leading to rapid sublimation) as well as the rate of secondary drying (e.g., large ice crystals create a small surface area, leading to slow secondary drying) (Pikal et al. 1983, 1990). A moderate degree of supercooling (10–15°C) is generally beneficial. To obtain a quality product, the degree of supercooling must be uniform, both within a given vial and within the entire batch of vials. The optimal shelf temperature-time regimes must be determined for each individual product to be lyophilized.

The freezing process greatly influences the quality of the diagnostic reagents. Once freezing begins, the proteins are exposed to an environment quite different from that of the starting aqueous solution. As water is being converted to ice, all solutes begin to concentrate in the remaining liquid phase. The highly concentrated multicomponent system is potentially a hostile environment to enzymes and antibodies until the time the system is completely frozen. Thus, buffer components may selectively crystallize, producing massive pH shifts; high electrolyte concentration will be produced even when the initial salt concentration is low (Taborsky 1979; Pikal 1990a,b).

The concentration effects of solutes are illustrated in Figure 10.3. In this experiment, the freeze concentration effect for a normal saline solution (0.9% NaCl) was investigated by placing 5 ml of solution into 20-ml glass vials that were then placed on a –30°C shelf. Nucleation and freezing initiated at about 10 min into the freezing process at a solution temperature of –15°C. The percentage of liquid remaining decreased roughly linearly with time until most of the water was frozen. However, at the same time, the concentration of NaCl increased from the starting 0.15 molality to approximately 6 molality just before the supercooled eutectic system crystallized at –26°C. Thus, the concentration of NaCl increased approximately 40-fold during the freezing process. Thus, protein reagents formulated in a dilute NaCl solution will be exposed to very high concentrations of salt during the latter part of the freezing process, with possible adverse consequences for product stability as well as quality.

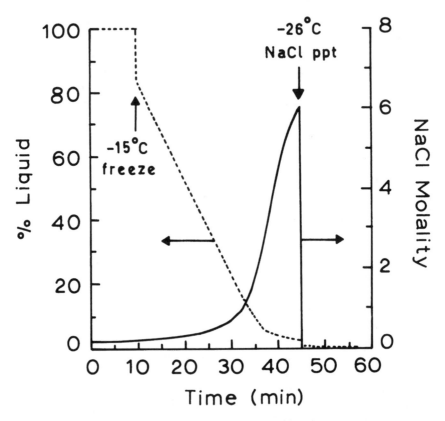

Figure 10.3. Concentration of solute during freezing, calculated from freezing 5 ml of 0.9% NaCl in 20-ml glass tubing vials
From: Pikal (1990b)

In addition to possible adverse effects that may arise from high electrolyte concentration, buffer components may crystallize, resulting in massive pH shifts during freezing. For example, even the relatively simple aqueous system composed only of sodium and potassium salts of phosphoric acid were shown to have 11 eutectic points associated with them (Taborsky 1979). The pH values and ionic strengths of these eutectic solutions range widely within their freezing range between 0°C and about −17°C. Thus, a dissolved protein's structural integrity can be expected to be challenged by these variations in pH and ionic strength.

The effect of freeze-thaw on the pH of a citric acid-disodium phosphate buffer system is illustrated in Figure 10.4. The system pH was measured using special low temperature electrodes (Larsen 1973). The arrows show the direction of increasing time. Ice crystallization was initiated by seeding, so no effects of water supercooling were present. The differences between the freezing and thawing

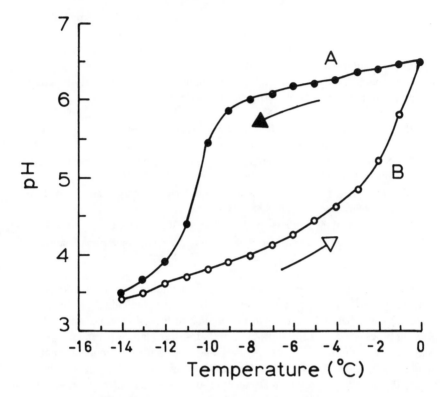

Figure 10.4. The effect of freezing on the pH of a citric acid-disodium phosphate buffer system during cooling (A) and thawing (B)
From: Pikal (1990b)

curves reflect supersaturation of solute during cooling. The large decrease in pH as the temperature decreases is due to crystallization of the basic buffer component (disodium phosphate). Obviously, this buffer system would cause problems if used with a protein that undergoes irreversible conformational changes at pH 5 and if the disodium phosphate crystallized from the mixture. Other buffer salts, such as potassium phosphate, produce much smaller pH shifts when one of the buffer components crystallizes (Larsen 1973). Such freezing-induced pH effects can be counteracted by the addition of "cryoprotectants" such as glycerol or dimethylsulfoxide, which are effective because the selective precipitation of buffer salts will occur only at lower temperatures in their presence.

2. Primary Drying. Primary drying consists principally of sublimation of ice formed during the freezing step. It occurs when the pressure of the environment is lower than the vapor pressure of ice. The process primarily takes place at the surface of the ice; the so-called sublimation front or solid-vapor interface. Because heat is lost during the process of sublimation, the product temperature will be lowered. To maintain a constant product temperature, approximately 620–650 calories of heat must be supplied from the shelves to the frozen product for each gram of ice sublimed.

Conceptually, the following things happen during the primary drying phase of lyophilization process (Hull Corp. 1994).

1. The chamber pressure (partial pressure of water) drops below the vapor pressure of the ice at the specific product temperature.
2. At this stage, the vapor pressure of ice is greater than the environmental pressure. The ice thus sublimes, evolving water vapor.
3. As the phase change from solid to vapor occurs, a quantity of heat (heat of fusion and vaporization) is removed from the product.
4. As a result of the heat loss, the product temperature is lowered.
5. When the product temperature falls below the shelf temperature, heat from the shelf will be transferred to the product.

Because the primary drying step involves heat and mass transfer, both parameters need to be optimized during process design for individual products. Pikal (1990a) has discussed the relative importance of these two issues the in freeze-drying process. Factors that influence mass transfer include parameters, such as partially dried product, stopper openings, and the chamber-to-condenser pathway, that resist the flow of water vapor. Generally, the greater the thickness of the dried layer, the greater the product resistance. Product resistance also decreases as vial diameter increases and the product thickness decreases. Thus, the larger the vial, the less the product thickens and the faster a given fill volume will freeze dry (Pikal et al. 1984). Because the maximum number of vials that can be loaded into the dryer is lower for larger vials, an optimum vial size must be determined for a given product.

Dried product resistance also generally increases with increasing solute concentration and frequently decreases as the temperature of the frozen product approaches the eutectic temperature. Production of larger ice crystals by an annealing process during freezing, in which the product temperature is increased to at least several degrees above the glass transition temperature and held there for several hours to allow the solute to crystallize and ice crystals to grow, may also decrease the product resistance (Pikal et al. 1983).

The rate of primary drying increases with increase in the product temperature. Therefore, it should be carried out at the highest temperature possible. Generally, there is a product-dependent upper temperature limit for primary drying. This "maximum allowable temperature" is the eutectic temperature for a solute system that crystallizes or the 'collapse temperature' for a solute system that remains amorphous (Pikal 1990a; Hull Corp. 1994; Mackenzie 1975, 1985b). Collapse is essentially the amorphous system analog of a eutectic melt. If the product temperature rises above the collapse temperature, the amorphous solute-water system gains enough fluidity to undergo viscous flow once the ice in that region has sublimed. Thus, the dried region adjacent to the ice will "flow" and lose structure. The collapse temperature and the glass transition temperature of the maximally concentrated solute are closely related and for most practical purposes are identical (Pikal and Shah 1990). Drying above the maximum allowable temperature normally produces no well-defined cake but results in a mass of ill-defined geometry that is unacceptable even if the product activity is unchanged. Exceeding the glass transition temperature may decrease product activity for most proteins. When a severe collapse occurs in a system of low solids content, the vial often appears empty.

Because eutectic and collapse temperatures vary over an enormous range (approximately −1°C to −50°C, depending on the nature of the formulation), determining the maximum allowable product temperature is extremely important and is the first step in formulation and process development (Pikal 1990a; Hull Corp. 1994). Collapse and eutectic temperatures may be determined by thermal analysis methods, electrical resistance measurement, and by direct microscopic observation of freeze-drying as a function of temperature.

Similar to mass transfer, several barriers also exist for heat transfer during the primary drying step. These may be described in terms of four barriers or resistances to heat flow: the shelf, the pan or tray, the vial, and the product. The thermal resistances (proportional to temperature differences) of both the tray and the vials are almost entirely due to the vapor boundary that exists between two solid surfaces that are not in perfect contact. In contrast, temperature differences across the pan bottom and across the glass in the vial bottom are generally very small.

Generally for a given formulation, fill volume, container, and freezing process, the chamber pressure and shelf temperature sequence with time will determine the product temperature during primary drying. The objective of process design is to maintain the product temperature essentially constant during primary drying at a safe level below the eutectic or collapse temperature (Pikal 1990a; Mackenzie 1985a). Depending on the situation, the safety margin is usually 2–5°C. If there is no real need to speed up the process, a larger safety margin can be employed.

As mentioned earlier, the resistance of the dried product increases with increasing dried product thickness, and therefore increases with time, the self-cooling rate from ice sublimation decreases with time throughout the primary drying process. Similarly, because the objective is to maintain constant product temperature, the heat input must also decrease with time. Thus, either the shelf temperature or the chamber pressure (or both) must decrease during the primary drying cycle. Depending on the product, the magnitude of the decrease needed may be significant or negligible. It essentially depends on how much dried product resistance increases with increased thickness of the dried cake.

A major problem in designing the primary drying portion of the process is selecting the appropriate chamber pressure and shelf temperature settings. The general rule is that the chamber pressure should be significantly lower than the vapor pressure of ice at the target product temperature (Pikal 1990a). At fixed product temperature, a nonzero chamber pressure decreases the driving force for sublimation and therefore prolongs primary drying of a given sample. However, uniform heat input from sample to sample is usually easier to achieve at chamber pressures in the range of 0.1–0.2 mm Hg (Pikal et al. 1984. Furthermore, operation at modest chamber pressures allows one to adjust the heat input by adjusting the chamber pressure. This is often advantageous in several process control strategies.

3. Secondary Drying.   Primary drying ends when all ice in all product containers has been removed. Most of the water vapor is thus removed in bulk by sublimation during this phase of the lyophilization process. The water remaining after the primary drying cycle is often referred to as "bound" water. It may be a combination of absorbed water, adsorbed water, or water associated with a hydrated salt.

It is important to understand the differences of the different types of bound water in order to evaluate their significance during the drying process. Absorbed water exists as a part of the "solid solution" of amorphous matrix and can permeate through the bulk solid phase. Adsorbed water, in contrast, is present on the surface of the dried material. It can be further classified as physically adsorbed (physisorption) and chemically adsorbed (chemisorption) water. Physisorption involves loose association of water with the solid material via van der Waals forces of weak intermolecular attraction. It may exist as a monolayer or as several different layers. Chemisorption involves a stronger association, including covalent and hydrogen bonding. Water is also associated with crystalline hydrates, where it is generally tightly held within the crystal lattice. Removing this water of hydration is often difficult depending on the characteristics of the material or the number of water molecules associated with the hydrate.

At the end of the primary drying cycle, the shelf temperature is normally increased to facilitate removal of most or all of the unfrozen bound water. Premature increase of the shelf temperature before all ice has been removed risks collapse or eutectic melt in those containers with residual ice; therefore, an indicator of the end of primary drying cycle is needed for process control. The most commonly method in this regard is product temperature response (Hull Corp. 1994; Pikal 1990a). Here, the shelf temperature is increased for secondary drying when all product temperatures are approximately equal to the shelf temperature. The validity of this method depends on the assumption that the product near the temperature sensors, which are limited in number, is typical of the batch as a whole. However, to avoid collapse, an empirically determined delay or "soak" period is placed in the production protocol. Thus, the increase in shelf temperature is often delayed until several hours after the product temperature response indicates nominal absence of ice.

Pikal (1990a) suggests that contrary to conventional wisdom and common practice, secondary drying should not be carried out at the lowest pressure attainable. The secondary drying rate is independent of chamber pressure up to at least 0.2 mm Hg, and lower pressures can cause problems in some process designs (Pikal et al. 1990). Thus, secondary drying should be carried out at around 0.2 mm of Hg. However, the temperature and time of secondary drying must be established by development studies. As a general rule, to avoid loss of product elegance, as well as to minimize damage to the protein, the temperature-time profile should ensure that the product never rises above the glass transition temperatures during secondary drying.

The desired level of residual moisture in the lyophilized product varies with the type of product. With nonprotein products, the lower the residual moisture, the more stable the product. Some moderate levels of residual moisture (up to 1%) are often beneficial for protein products (Pikal et al. 1992). The concept that overdrying a protein will cause loss of activity, although theoretically plausible, has little experimental support in the literature. If the protein product of interest requires an intermediate level of water to maintain activity, proper design of the secondary drying process becomes difficult. One approach that might prove satisfactory, however, is to carry out secondary drying at a low (approximately 0°C) shelf temperature for an extended time (Pikal et al. 1990). It is not unusual to observe increased formation of aggregates as the secondary drying temperature is increased from 10°C to 50°C.

4. Excipients.   Stability problems during freeze-drying and of the freeze-dried products are often addressed by varying the formulation. Thus, in addition to the active component (e.g., enzyme, antibody, standard, or calibrator), other components (excipients, cryo- or lyoprotectants) may be added for specific purposes. The stabilizing mechanisms may involve the effect of the additives on water structure. Generally, stabilizers are water structure promoters and destabilizers are water structure breakers.

Bulking agents such as mannitol or glycine are added to enhance product elegance and to prevent product "blowout" (Pikal 1990b). Blowout occurs with formulations very low in total solids (i.e., ~ 1%); the streaming water vapor can disrupt cake structure and carry some or all of the dried material out of the container with the water vapor stream.

Buffers are frequently added for pH control, while salts such as NaCl may be added to yield an isotonic solution. However, because of salt concentration and possible pH shifts during freezing, it is advisable to minimize the amount of buffer and/or salt relative to the amount of protein. This is particularly important with proteins that are known to be susceptible to damage by either pH shifts or high salt concentrations.

Excipients can also be used to increase solubility of the primary component in the formulation (Pearlman and Nguyen 1989). They can also be added to raise the collapse temperature of the formulation, which usually falls between the collapse temperatures of the individual components. However, collapse temperatures of a specific formulation must be determined, as they generally cannot be predicted with useful accuracy. Some useful collapse temperature modifiers and their respective collapse temperatures are: dextran ($-10°C$), Ficoll ($-20°C$), gelatin ($-8°C$), human serum albumin ($-9.5°C$), and hydroxyethyl starch ($> -5°C$) (Pikal 1990b; Mackenzie 1975). However, it should be noted that the composition of the amorphous phase formed during freezing determines the collapse temperature. Thus, trace organic solvents, buffers, and other salts that do not crystallize also affect (usually decrease) the collapse temperature.

Excipients or lyoprotectants are also frequently added to protein products to enhance stability during the freeze-drying process and/or to enhance stability of the freeze-dried solid during storage. Sugars such as trehalose and sucrose and proteins such as serum albumins are commonly used for this purpose (Pikal 1990b).

It is almost impossible to provide a set of rules for formulating a given protein product. However, it must be noted that formulation often has a dramatic effect on the degradation of proteins during the freeze-drying process as well as impacting on the shelf life stability of the freeze-dried product. The mechanisms responsible for the effects of additives on protein stability are rarely well understood. This difficulty in part stems from incomplete physical and chemical characterization of the formulations studied. For elucidating the mechanism of lyoprotection, it is helpful to separate the effect of freezing from the effect of drying by comparing freeze-thaw stability with freeze-dried stability.

In spite of the paucity of literature in this field, Pikal (1990b) has provided some general guidelines for formulating protein products for freeze-drying. If an additive is to be effective in stabilizing a protein, either as a lyoprotectant or to enhance stability of the dried solid, it must remain at least partially amorphous. Only then is there potential for stability enhancement by one or more mechanisms. Diluting the protein in the excipient phase, ultimately providing molecular dispersion of the protein in a rigid glass, may promote resistance to aggregation by minimizing bimolecular protein-protein interactions. Diluting the protein with excipients also dilutes salts, thereby minimizing the effects of high salt concentration. An amorphous excipient, however, is not a sufficient condition for stability enhancement. In fact, molecular interaction between the excipient and the protein may actually destabilize the protein.

Ideally, product formulations should also have a high glass transition temperature (Pikal et al. 1991). Any potentially reactive chemical system, including a protein, is much less stable above the glass transition temperature of the amorphous phase. Thus, instability might be expected if the temperature of the system exceeds the glass transition temperature for an extended period of time during processing or storage of the dried solid. Glass transition temperatures vary dramatically with the nature of the formulation and always decrease as the residual water content of the amorphous phase increases (Pikal 1990b).

Similarly, whenever possible, one should avoid the use of high levels of buffers and other salts. If the solute system consists largely of buffer, crystallization of one of the buffer components and a subsequent pH shift during freezing are more likely. Moreover, even if the buffer does not crystallize, thereby circumventing a pH shift problem, a high level of buffer remaining amorphous normally lowers the collapse temperature and the glass transition temperature of the dried material. According to Pikal (1990b), a low level of buffer tends to remain amorphous and without dramatically affecting the collapse temperature or glass transition temperature of the lyoprotectant-protein system.

Thus, the selection of potential lyoprotectants or enhancers of solid state stability is subject to several physical criteria as described above. However, selecting the optimum formulation still requires considerable development screening. Given the historical use of sugars and sugar alcohols as lyoprotectants and stability enhancers, they must be included in such a screening program. Ultimately, knowledge of the solution stability characteristics unique to the protein of interest must be coupled with empirical freeze-thaw and freeze-drying development studies using these physical criteria.

Because of the importance of solid reagent formulations in diagnostic kit development, it is therefore essential that one is thoroughly familiar with the freeze-drying technology. Hull Corporation, Hatboro, Pennsylvania,, offers excellent courses detailing the fundamentals of freeze-drying technology. In addition, there are several excellent review articles describing the use and importance of this technology (Pikal 1990a,b; Mackenzie 1975, 1977, 1982, 1985a,b; Goldblith et al. 1975; Williams and Polli 1984). Although his work is essentially geared for pharmaceutical process validation, Trappler (1993) has recently reviewed the validation aspects of lyophilized products and takes the readers through every single aspect of this technology from a process design and validation viewpoint.

PREPARATION OF REFERENCE STANDARDS AND CALIBRATORS

Immunoassays are used either to estimate the concentration of an analyte in a test specimen or to compare two unrelated substances in a particular test sample. The latter involves identifying cross-reacting substances that influence the specificity of an assay. Most immunoassays, however, are designed to measure specific analytes. Such assays, therefore, depend on comparisons between the unknown samples and standards for quantitation of the assay data, and require a preparation of the pure analyte that can serve as an appropriate standard for this purpose.

Thus, a prerequisite to evaluation of any immunoassay system is the calibration of standards.

Types of Standards

An immunoassay may be developed for a substance that has not been measured previously by this method. A reference preparation therefore may not exist. Moreover, the material may not be characterized by a single chemical structure. Under these circumstances, the first publication often serves as the reference point for the assays developed subsequently (Malan et al. 1983).

Well-defined compounds of high quality with established high purity and stability, however, are available for several analytes, and are often adequate as standards. In contrast, many biologically active substances are available only in crude form. For such analytes, special reference preparations have been established to improve accuracy of quantitation. Three types of such reference preparations are commonly used for standardization of immunoassay kits (Bangham 1971,1988). These are as follows.

1. International Standards (IS).   To calibrate a new method for biological analytes, an international standard (IS) or international reference preparation (IRP) must be used. Materials for these standards are collected, tested, and aliquoted by responsibility of the World Health Organization International Laboratory for Biological Standards. These compounds are extensively tested for their potency and stability. An international unit (IU) for activity has been assigned to these preparations after extensive collaborative effort by several different laboratories. These reference preparations, therefore, are to be regarded as the most reliable standards.

The term "IRP" is used to designate preparations that do not meet the demanding criteria for an IS, but nonetheless, are useful for method-to-method standardization.

International standards are available in limited quantity for a nominal charge for calibration of national or laboratory standards or reference preparations. However, they are not available in sufficient quantity for routine use as standards. A list of the international standards available and extensive discussion of their application to immunoassays are available (WHO 1979). Many of these are prepared in the United Kingdom by the National Institute for Biological Standards and Control (NIBSC, Blanche Lane, South Mimms, Potters Bar, Herts, EN6 3QG).

Extremely rigorous protocols are followed for the preparation and ampuling of international standards. These include avoidance of contaminants with enzymes such as peptidases from the source material, prevention of adsorption onto surfaces by adding a carrier substance, avoidance of oxidation by containing the preparation in an atmosphere of inert gas such as pure dry nitrogen in neutral glass ampules sealed by fusion of the glass, and limiting of moisture by desiccation so that the water content is below about 1–2%. These ampules are then stored in the dark at –20°C.

2. Reference Materials.   These are not as extensively tested as the established international standards, but do have certain potency and purity data that are primarily provided by the producer. These are distributed by such institutions as the Division of Biological Standards, National Institute of Medical Research, London; and the National Institute of Arthritis, Metabolic and Digestive Diseases, Bethesda, Maryland. These preparations constitute a valuable tool for calibration of the assays.

Reference preparations are particularly useful for substances that are (1) unable to be completely characterized by chemical and physical means alone, (2) heterogeneous (e.g., glycoproteins), (3) difficult to isolate in pure form (e.g., synthetic peptides), (4) scarce or expensive (e.g., several hormone preparations), (5) unstable or easily altered during isolation (e.g., human growth hormone), and (6) difficult or expensive to be assayed or characterized (e.g., prolactin, thyrotrophin).

3. In-House or "Working" Reference Preparations.   These are preparations that may have been produced by the laboratory performing the assays or acquired without any reliable potency estimates. Frequently, these materials are "calibrated" by reference to an international standard.

4. "Working" Standards.   In many respects, the working standard is the most important form of standard, as it constitutes the basis for accuracy of the routine assay. The extensive testing and validation work required for the introduction of reference standards is not normally necessary in the preparation of working standards. The laboratory, however, must assume the responsibility for maintaining the appropriate quality. These standards are normally made in larger volumes than the reference standards.

Preparation

Preparation of international standard solutions from ampules requires great care and should not involve inexperienced personnel. Ideally, two ampules should be reconstituted by different personnel on different occasions, with a check that the results from standard solutions made from the two ampules are in good agreement.

The accuracy of measurement of solutions to be dispensed is extremely important. Hence, newly calibrated and precise volumetric equipment and glassware should be used. A set of reference standards of different concentrations is most easily prepared by first making up a dilution of the reference preparation in the supporting matrix at a concentration greater than the maximum concentration to be used in the immunoassay. Individual standards then can be prepared by mixing aliquots of this standard with the analyte-free matrix. Ideally, volumes greater than about 1 ml should be used in an attempt to achieve maximum precision of measurement. Double dilutions of the solutions should be avoided. These standard solutions should be stored in aliquots at −70°C.

In order to quantitate the analyte from test sample, immunoassays often require the use of calibrators. Typically 4–6 calibrators are used to construct the calibration curve. The calibrators or the "working standards" can be made in bulk volumes by technically less demanding methods than those required for the preparation of reference standards.

Assays in which a new set of working standards is to be calibrated must be as representative of routine operations as possible. Thus, a variety of different production lots of labeled analyte, antisera, and other reagents should be used (Malan et al. 1983). To minimize the risks of miscalibration arising from between-assay variation, which could result in bias, the assays should be performed by a number of different operators and the assay runs (at least 20) spread over a period of time.

Cross-checks of the working standards or calibrators against established standards should be performed at least once a year. Standard calibration should also be checked when new antisera are introduced in an assay, because small differences in antisera specificity and affinity may alter the apparent potency of the working standard.

Matrix Considerations

The choice of solvent for calibrators often presents a major problem in the diagnostic reagent formulation. The accuracy of immunoassays relies on comparisons of the test sample with standards containing a predetermined amount of the analyte to be measured. The validity for such comparisons presumes that samples and standards are identical except with respect to the variation in the concentration of the analyte. This implies that standards should be prepared in a matrix that either is identical to that of the test sample, or does not influence the analyte-antibody binding reaction differently.

As a general rule, all antigen-antibody binding reactions are influenced by the matrix in which they are performed. Such influence, however, varies from assay to assay. It affects not only the binding reaction as such, but also other assay steps, such as the separation of the bound and free analyte. Therefore, each individual assay must be investigated to determine the extent to which dissimilarities between standard and sample solvents will influence the results.

When the results are affected by the type of solvent used, it is necessary to prepare the standards in something similar to the test sample. In the clinical field, various matrix materials are used for different analytes. The most commonly used matrix includes defibrinated, delipidated pooled human sera or plasma containing physiologically or pathologically low concentrations of the analyte. Such samples should be tested for the presence of anti-HIV and hepatitis B surface antigen.

A major difficulty in the preparation of a set of calibrators for the commercial immunodiagnostic kits is that it must include a 'zero' analyte calibrator. Several methods, including the use of antibodies, charcoal, silica and ion-exchange resins, have been used to physically "strip" the analytes from the serum to prepare "analyte-free" serum. Some of these procedures or materials are difficult to reproduce. Most methods developed by manufacturers to remove analytes from serum remain confidential commercial practices.

Alternatively, animal serum or buffer solutions containing protein can be used to prepare calibrators. However, both are prone to matrix effects.

Storage and Stability

Ideally, the reference standards should be first divided into small aliquots for ease of use. They can either be stored at $-70°C$, or lyophilized if the analyte is less stable in frozen solution. In the latter case, however, one must ensure that the freeze-drying process itself does not cause changes to either the analyte or the matrix.

As a precaution, stocks of reference standards should be stored at different temperatures (e.g., at $-70°C$ and at $-20°C$) and at different locations to safeguard against loss due to mechanical failures of deep-freezers. Given the economics and the amount of work involved in preparing the reference standards, they should be used only for the calibration of secondary or working standards for routine use.

The reference standards used in clinical diagnostics generally have adequate stability. Most international standards change by less (usually much less) than 1% per year at the recommended storage temperature (generally $-20°C$). Thus, they are stable enough to last virtually unchanged for about 10–20 years, and, if required, to withstand the rigors of transit in the mail for several days even under tropical conditions (Bangham 1988).

The degradation of a wide variety of reference materials in the dry form has been shown to follow first order reaction rates. The instability is primarily attributed to contamination with enzymes during extraction from biologic materials or those present in other reagents such as carrier proteins; chemical reactions involving oxidation, deamidation, and alkylation; conformational changes including dimerization and β-aspartyl shift; or adsorption onto surfaces or other molecules such as denatured proteins.

In contrast, the stability of reference standards in solutions is influenced by other conditions. Hence, their degradation in solution form is less likely to follow first order kinetics (WHO 1978). In particular, the stability of protein standards in solution can be enhanced by using a relatively concentrated solution of the standard protein. Similarly, they should be prepared in buffers that have high eutectic freezing point and that do not show pH shifts during freezing. In this regard, potassium phosphate buffers are much better than sodium phosphate (Bangham 1971). Furthermore, the use of a carrier protein at about 0.5% concentration greatly aids in preventing the loss of the standard material by adsorption onto glass and plastic surfaces. However, such carrier protein must be free of endogenous proteolytic activity (Caygill 1977).

Because they are not necessarily sterile, the use of preservatives is advisable for reference standards prepared in solution forms. The liquid standard also needs to be quickly frozen to a low temperature and then stored at a temperature below the eutectic freezing points of its constituents. A temperature of $-40°C$ or lower is generally adequate. It must be remembered that assay problems such as bias drifts can often be traced back to the use of inappropriately prepared reference standards.

Excellent discussions on several aspects of the importance and preparation of reference standards in the calibration and standardization of immunoassays, especially for use in the clinical field, are available in the literature (Bangham 1971, 1976, 1982, 1988; Bangham and Cotes 1971, 1974; WHO 1975, 1978, 1979, 1981; Jerne and Perry 1956).

SHELF LIFE EVALUATION

The shelf life (or stability) of a product may be defined as the time that essential performance characteristics are maintained under specific handling conditions. The change of quality of a product over time is a function of, among other things, storage temperature, humidity, package protection, and product formulation. Product expiration dating is the ultimate practical result of determining stability. The need to know how long a material will be stable affects not only commercial products but also a laboratory's own in-house prepared materials.

Government regulations are specific about the requirement for expiration dates on all materials. Section 211.166 of CGMPs regulations published by the Food and Drug Administration (FDA) requires that there be a written testing program designed to assess the stability characteristics of all clinical products. It lists five criteria for determining the product stability as follows:

1. Sample size and test intervals for each attribute examined,
2. Storage conditions to which retained samples are subjected,
3. Reliable, meaningful, and specific test methods to assess quality,
4. Testing to be carried out in the same container or closure system as that intended for marketing, and
5. Testing for products for reconstitution be performed at the time of dispensing as well as after they are reconstituted.

Generally, too long an expiration date produces compromised test results and would mean loss of repeat sales because of disappointed customers. In contrast, too short a date generates wastage and unnecessary work.

Three types of stability are important for diagnostic reagents: physical, bacteriostatic, and chemical or functional stability. Physical changes such as discoloration or appearance of precipitate are undesirable in diagnostic reagents. Most diagnostic products are formulated to contain a bacteriostatic or bactericidal agent to prevent the deterioration of the product due to microbial growth. The active components of such an agent must remain at sufficiently high concentration in the product to prevent microbial growth during the course of its projected shelf life. In contrast, chemical or functional stability is essential for better assay performance. Functional changes such as loss of binding affinity due to antibody or label degradation, or chemical changes, for example, in chromophore quality due to hydrolysis, oxidation, reduction, etc., will adversely influence the assay results.

A manufacturer's quality assurance criteria typically require that a product must recover at least 90% of the initial value throughout its life (Anderson and Scott 1991). Applying this performance criterion to the results of stability testing thereby determines shelf life. However, alternate criteria are sometimes used for evaluating the shelf life of a given product.

It is imperative that reagents used in diagnostic test kits be designed to have long stability. The desirable length of reagent stability is summarized in Table 10.2. Most reagents have adequate shelf life before opened. A long shelf life is desirable so that a single lot of reagent can be used for a long period of time, preferably longer than 1 year. This will minimize the reagent lot check-in process, which could be time consuming and expensive. The stability of the opened reagent stored at 4°C will allow the same calibration curve to be used. Most reagents are capable of achieving this step. A desirable characteristic of reagent stability of one month is more difficult to achieve on the automated instruments. The ideal automation, as described in Chapter 8, is random access with testing on a first-in-first-out basis. If the system is to be used on a 24-hr basis, the reagent will be expected to be stable and available for testing at all times (Chan 1992). Calibration curve stability is also important in the efficient operation of the system. A one-month stability will allow the clinical laboratory to perform less frequent tests without recalibration. Quality control performed on a daily basis will ensure the readiness and the adequacy of testing.

TABLE 10.2. Desirable Length of Reagent Stability[a]

Reagent condition	Length
Shelf life of unopened reagent	2 years
Opened reagent at 4°C	3 months
Opened reagent on the instrument	1 month
Calibration curve stability	1 month
Quality control frequency	1 day

[a]From Chan (1992).

In this section, various approaches used to evaluate shelf life of diagnostic reagents are briefly described. Similar approaches can also be used to test other kit components such as the stability of antigen- or antibody-coated solid-phase matrices using appropriate test criteria.

Stability is the inverse of degradation. All materials undergo degradation at a rate defined by the laws of chemistry and physics. Fortunately, for a given formulation of pH, ionic strength, composition, etc., the dominant variable affecting the degradation rate is temperature. Thus, degradation of the product can be monitored at each of several high temperatures and the information extrapolated to the anticipated storage temperature to determine the usable product shelf life. Two basic approaches used for shelf life determination are described below.

Real-Time Stability Testing

Real-time stability studies are the "gold standard" in determining expiration testing. Unfortunately, they are not practical in most situations. In this approach, the duration of the test period should be long enough to allow sufficient product degradation under recommended storage conditions (Anderson and Scott 1991). At the least, the testing protocol must permit one to distinguish percent degradation from interassay variation. For example, data may be collected at an appropriate frequency such that a trend analysis may discern instability from day-to-day imprecision.

The reliability of data interpretation can be increased by including in each assay a single lot of reference material with established stability characteristics. Sample recovery between assays then can be normalized to this reference, minimizing the impact of systematic drift and interassay imprecision. Unfortunately, an appropriate reference material is often not available for use as a control.

When one measures the stability of a reference material, imprecision may be introduced by changes in both reagents and instrumentation (Zakowski 1991; Anderson and Scott 1991). Ideally, reagents should be sufficiently stable that a single lot provides unchanging performance throughout the stability testing and instrument performance should remain constant. However, one must monitor the system performance and control for drift and discontinuity resulting from changes in both reagents and instrumentation.

Thus, real-time data are frequently confounded by drift or changes in the testing method used over a period of time. Use of extrapolations and approximation methods based on data acquired over a short period of time can resolve these issues (Connors et al. 1986). Nonetheless, real-time product stability testing is necessary to validate stability claims for clinical chemistry reagents and reference materials.

Accelerated Stability Testing

Accelerated stability testing is often used in the development of clinical reagents to provide an early indication of product shelf life and thereby shorten the development schedule. In this approach, a product is stressed at several high (warmer than ambient) temperatures and the amount of heat input required to cause product failure is determined. This information is then extrapolated to predict product shelf life or used to compare the relative stability of alternative formulations.

A reasonable statistical treatment in accelerated stability projections requires that at least four stress temperatures be used (Anderson and Scott 1991; Connors et al. 1986). In addition, more nearly accurate stability projections are obtained when denaturing stress temperatures are avoided. This is particularly true for reagents containing labile, proteinaceous components such as enzymes and antibodies.

Accelerated stability testing protocols allow one to stress samples, refrigerate them after the stressing, and then assay them simultaneously. Because the duration of the analysis is short, the likelihood of instability in the measurement system is reduced. Such protocols additionally permit one to compare the unstressed product with stressed material within the same assay, the stressed sample recovery being expressed as a percent of unstressed recovery. This utilization of product as an internal control is especially valuable when no suitably stable reference material is available (Kirkwood 1977).

Accelerated testing can be performed using several different approaches. The most commonly used are described below.

1. Arrhenius Relationship.   The Arrhenius relationship states that the functional relationship between time and stability of a product stored under constant conditions is dependent upon the order of reaction and a rate constant that determines the speed of reaction. The logarithm of the reaction rate is a linear function of the reciprocal of absolute temperature. Mathematically, the relationship can be expressed as follows:

$$\log (k_2 / k_1) = (-E_a/2.303R) [(1/T_2 - 1/T_1)] \qquad (1)$$

where k_2 and k_1 are, respectively, rate constants at temperatures T_2 and T_1, respectivel in kelvins; E_a is the activation energy; and R is the gas constant.

It is this relationship that allows data taken at elevated temperatures to be used to predict the degradation rate at ambient or any other desired temperatures, and hence, estimate shelf life of the product. The rate constants obtained at the elevated temperatures can be used to estimate parameters in the Arrhenius equation, which in turn can be used to estimate reaction rate at other temperatures. Activation energy, the independent variable in the above equation, is equal to the energy barrier that must be exceeded for the degradation reaction to occur.

An example of an Arrhenius plot in which three systems are theoretically represented is shown in Figure 10.5. The slope of each line is equal to the activation energy divided by R. A steeper slope means the reaction is more temperature de-

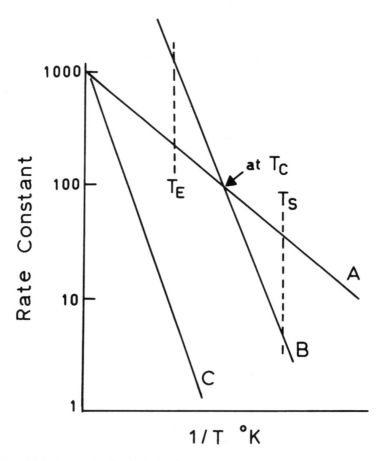

Figure 10.5. An example of an Arrhenius plot

pendent, i.e., as the temperature increases, the reaction increases at a faster rate. Thus, systems B and C in Figure 10.5 have the same temperature dependence, and increase in rate more rapidly than A as temperature increases ($1/T$ becomes less).

This figure also illustrates that, for A and B at some possible temperature, the rates are the same; however, above and below this level their relative rates are different. Thus, at temperatures below T_C (i.e., larger $1/T$) the rate of loss of A is faster than B; above T_C, B is lost more rapidly than A. Thus, for a reagent formulation with two modes of deterioration, the mode of deterioration may change with temperature, which could create a problem in predicting shelf life. For example, if a single study of shelf life were done at temperature T_E and the quality loss in the reagent could be measured by both mechanisms A and B, B would be faster and would result in the end of shelf life. If the slope of line B were known, one could then project to lower temperature to get the rate at T_S (the temperature of storage) and thus get the shelf life. However, as seen, at T_S the rate of deterioration for mode A is faster than for B, so the true shelf life could be much less than estimated. Unfortunately, not enough data are available on the stability of the diagnostic reagents in the literature to determine whether this is a likely occurrence.

Anderson and Scott (1991) have described a simple first order relationship for evaluating the stability of diagnostic reagents. The relationship between a product's quality criteria, such as residual standard concentration [D] after time (t) for a first order equation is

$$-d\,[\text{D}]/dt = K_1\,[\text{D}] \tag{2}$$

where K_1 is the reaction rate constant at a given temperature. Integrating from $t = 0$ to $t = \text{t}$, one obtains the following relationship.

$$\ln\,[\text{D}] = \ln\,[\text{D}]_0 - K_1 t \tag{3}$$

$$\text{or} \qquad \log\,[\text{D}] = \log\,[\text{D}]_0 - (K_1 t/2.303) \tag{4}$$

where $[\text{D}]_0$ is the standard drug concentration at time zero. By substituting $0.90\,[\text{D}]_0 = [\text{D}]$ in equation (4), the time required to reach 90% of the original drug concentration can be shown to be

$$t_{90} = 0.105/K_1 \tag{5}$$

A common practice of manufacturers in the pharmaceutical and diagnostic reagent industries is to utilize various shortcuts, e.g., bracket tables and the Q rule to estimate product shelf life. These techniques, briefly described below, have the advantage that decisions may be made by analyzing only a few stressed samples (Connors et al. 1986; Anderson and Scott 1991). However, they are based on assumptions about the activation energy of product components and are valid only as long as these assumptions are accurate. Irrespective of the method chosen, the validity of product stability projections depends on analytical precision, the use of appropriate controls within the experimental design, the assumptions embodied in the mathematical model, and the assumed or measured activation energy of product components.

The reaction rate is also influenced by pH, osmolarity, the presence of stabilizers, etc. Similarly, as mentioned above, key product components may degrade or otherwise become unavailable through multiple mechanisms (Connors et al. 1986; Lachman and DeLuca 1976). In complex chemical systems, therefore, a minor variation in formulation can profoundly affect lot-to-lot stability and, indeed, the activation energy of product degradation. Consequently, shelf life projected from accelerated studies must be validated by appropriate real-time stability testing.

2. The Q Rule.   Using the Arrhenius equation, the end point data can be transformed into a shelf life plot as shown in Figure 10.6, which is usually a straight line. Theoretically, if only a small temperature range from the Arrhenius plot is used (no more than a 20–40°C change) then the same data will give a fairly straight line on the shelf life plot.

Two examples are shown in Figure 10.6, both of which pass through 200 days at 25°C. What is needed to construct this plot is: (1) some measure of loss of quality, (2) some end point value for unacceptability, (3) the order of the reaction so as to predict the time to reach this end point, and (4) experiments to measure this loss at least at two temperatures so the line can be constructed.

It is evident from Figure 10.6 that the steeper the slope, the more sensitive the reagent formulation (or the reaction) to temperature change. A measure of this sensitivity is called the Q_{10} of the reaction. The Q rule states that a product degra-

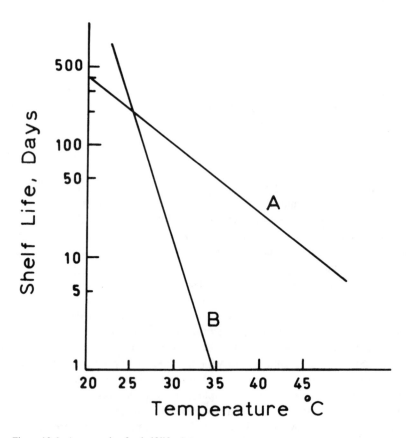

Figure 10.6. An example of a shelf life plot

dation rate decreases by a constant factor (Q_{10}) when the storage temperature is lowered by 10°C. Mathematically, it is defined as

$$Q_{10} = [\text{rate at temperature } (T + 10°C)] / \text{rate at temperature } T°C \qquad (6)$$

The Q_{10} can also be calculated from the shelf life plot as:

$$Q_{10} = \text{shelf life at } T°C / \text{shelf life at } (T + 10°C) = \theta_{ST} / (\theta_{ST} + 10) \qquad (7)$$

where θ_{ST} is the shelf life at any given temperature. Equation (7) assumes that the rate is inversely proportional to the shelf life. For a zero order reaction this is exactly true because:

$$\text{rate} = k = \text{amount lost at end point} / \theta_S \qquad (8)$$

If the amount lost at θ_S is defined, then the Q_{10} can be calculated as in equation (7) above. A similar relationship can be derived for first order reactions to also show that Q_{10} can be calculated by equation (7). Thus, the fit of data to Figure 10.6 gives no clue as to reaction order.

From Figure 10.6, it can be calculated that for reagent formulation A, the Q_{10} is (for example)

$$Q_{10} = \theta_S \text{ at } 25°C / \theta_S \text{ at } 35°C = 200 / 50 = 4 \qquad (9)$$

and for reagent formulation B, the Q_{10} is

$$Q_{10} = 800 / 3 = 266 \qquad (10)$$

Thus, formulation B or reaction B is much more sensitive to temperature change than A is. Theoretically, the Q_{10} and the activation energy can be related by:

$$\log 10 \, Q_{10} = (2.19 \, E_a) / (T) (T + 10) \qquad (11)$$

Figure 10.6 illustrates the usefulness of a shelf life plot when applied to studying the loss of quality, as it shows that holding a reagent at a higher temperature causes it to degrade faster. Thus, if studies at two different temperatures are made, the shelf life at some lower temperature can be predicted if the line is assumed to be straight. One cannot, however, study the deterioration at only one temperature, because it is not possible to predict beforehand the slope of the line or the Q_{10} exactly. It should also be remembered that different reactions may occur at different temperatures to cause end of product acceptability, so the projected line might be incorrect. Therefore, in Figure 10.6, reaction B would cause end of shelf life in

12 days at 30°C, but below 25°C, reaction is faster and would cause the end of shelf life.

For most diagnostic reagents, a Q_{10} value of 2 provides a conservative estimate of shelf life and results calculated with this value are considered "probable." A Q_{10} value of 4 is less conservative and yields results considered "possible" (Anderson and Scott 1991).

3. Bracket Tables.   The bracket table technique assumes that, for a given analyte, the activation energy is between two limits, e.g., between 10 and 20 kcal. As a result, a table may be constructed showing "days of stress" at various temperatures (Table 10.3). The use of a 10 to 20 kcal bracket table is reasonable, because broad experience indicates that most analytes and reagents of interest in pharmaceutical and clinical laboratories have activation energies in this range (Anderson and Scott 1991; Porterfield and Capone 1984; Kennon 1964). Protein denaturation and enzyme inactivation, both likely causes of deterioration in diagnostic reagents, may typically show activation energies in this region (Jerne and Perry 1956; Silver et al. 1986).

Because the bracket table provided in Table 10.3 does not specify stability requirements at a stress temperature of 55°C, a conservative use of the table requires that projections be taken from the 47.5°C data.

Prudent use of either of these rapid techniques would dictate that data at three or four higher temperatures be incorporated into the projection of refrigerated shelf life. To evaluate the usefulness of the Q rule and the bracket table, one must

TABLE 10.3.   Bracket Table[a,b]

| Storage temp., °C | *Days of stress to predict stability at 5°C for* | | | | | | | |
| | *6 months* | | *1 year* | | *2 years* | | *3 years* | |
	20 kcal	10 kcal	20 kcal	10 kcal	20 kcal	10 kcal	20 kcal	10 kcal
14.5	55.3	100	111	201	221	402	332	603
25	16.1	54	32	108	64	217	97	326
35.5	5.1	30.6	10	61	20	122	31	183
47.5	1.5	16.6	3	32	6	66	9	100
60	0.5	9.2	0.9	18	1.9	37	2.8	55

[a]From Anderson and Scott (1991).

[b]To use this table, select the column appropriate to the stability claim to be evaluated, e.g., 1 year. Acceptable performance after the number of days of stress in the 20-kcal column predicts that it is "possible" the selected claim will be observed (e.g., stressed for 32 days at 25°C). Acceptable performance indicated in the 10-kcal column predicts that it is "probable" the selected claim will be observed.

determine the activation energy for the compound of interest and then project the refrigerated shelf life by using the Arrhenius equation (Anderson and Scott 1991).

4. General Protocol for Stability Testing.   As a guide, the following protocol is suggested to perform accelerated stability testing studies. To design such studies, one must consider four parameters: preparation of samples, storage conditions, analytical procedure, and analysis and interpretation of data.

PREPARATION OF SAMPLE.   Although information on the stability of a product is frequently required at an early stage of its development and complete details of the sales container are not available, nevertheless, it is important that containers, especially for liquids, be similar to those in which the product will be sold. For example, if plastic containers will be used, then accelerated stability studies of the product in sealed glass ampules will give false data.

STORAGE CONDITIONS.   Samples should be stored at at least three or four elevated temperatures. The selection of storage conditions can be based more on the practical considerations involved. The highest temperature chosen is limited by the nature of the product (e.g., denaturation of antibody or enzyme), and the lowest temperature by the length of time it is proposed to devote to experimentation. When plastic containers are used, lower storage temperatures (determined by the polymer employed) are necessary to avoid physical changes in the container walls, which would not take place under normal storage conditions.

Ideally, the storage temperature should be constant. However, a maximum variation of ±1°C is generally permissible. During the shelf life of the product, nothing greater than a 10% decomposition is tolerable by the FDA guidelines. In accelerated stability testing, however, it is advisable to follow the stability up to 50% decomposition because this ensures great accuracy in determining the rate constant. Furthermore, a reliable result can be expected, especially if the method of analysis is not particularly reliable.

ANALYTICAL PROCEDURES.   The analytical method used must be meaningful for the active reagent of the product. Most chemical assays are reliable up to ± 2%, while biological assays can only be accurate up to ± 5%. Irrespective of the method chosen, it is important to determine the initial assay very accurately in multiple replicates. Otherwise, 5% or even 10% decomposition at room temperature, which are of interest, cannot be determined accurately. Similarly, the starting material should be analyzed alongside the stored samples.

CALCULATION AND INTERPRETATION OF RESULTS.   The data can be analyzed by any of the methods described above. If one is interested in screening different formulations of a product or a reagent, then it is generally not necessary to determine the rate of degradation. For example, one can determine the time for 10% decomposition at different temperatures and using these values can predict the time needed for this degradation to occur at the desired shelf life temperature.

The following is a general guideline for designing an accelerated stability experiment.

1. Select three or four temperatures at which to study degradation. Degradation can be defined as a loss in total function or a critical parameter such as active enzyme concentration, percent binding of a reference material to the antibody at a predetermined concentration, etc. The reagents are placed in incubators set at these predetermined temperatures. If the degradation of the reagent is not observed at these temperatures within a reasonable period of time (e.g., 1–2 months), then higher temperatures must be selected. One must achieve degradation in order to use accelerated stability studies for determining the shelf life of the reagents.

2. Place a sufficient amount of reagent in the final containers at the selected temperatures for specified periods of time depending on the estimated rate of degradation. The amount placed at different temperatures can be determined from the test method chosen. One must have sufficient reagent to run duplicate assays for each point on at least a three point standard curve. Higher temperatures should be sampled more frequently than lower temperatures. The following protocol is generally found acceptable for testing of diagnostic reagents: once every 5 days at 37°C, once every 4 days at 45°C, once every 3 days at 50°C, once every 2 days at 60°C; and daily at 80°C. Preferably, the high and low standards should be run in duplicate.

3. The samples should be cooled upon removal from the incubator and promptly analyzed.

4. The measurements can be made on any critical component, e.g., absorbance of the enzyme label in the presence and absence of a given reference standard.

5. When the data are plotted against the number of days, the estimated values should give an approximate straight line on semilogarithmic graph paper. Conventionally the slope of this line is measured. However, as discussed earlier, the length of time required for the potency to drop to 90% of the original value (t_{90}), or any other appropriate criterion selected, can be determined from the graph. In reality, when using this technique, one must estimate the maximum allowable drop in activity that would not affect the functioning of the product and thus assay performance.

6. From such a set of data, one can then predict the length of time required to reach 90% potency at the desired storage temperature. The logarithms of the t_{90} values are linearly related to the reciprocal of the absolute temperature.

7. A semilog plot of t_{90} values as logarithmic ordinate and $1/T$ in kelvins as the abscissa should yield a straight line. From this plot, the predicted shelf life at the desired storage temperature can be calculated.

This procedure yields fairly reliable results. However, often the data points do not lie on a straight line. In such instances, regression techniques must be employed to determine the best fit. If there is more than one component, each one should be tested separately. Ideally, either a critical component should be chosen or the total function determined.

There are also certain instances when extrapolations to desired temperatures cannot be made. These include parallel reactions, excipient degradation and impurity interactions in the formulations, and any reactions that occur only at elevated temperatures.

Based on the author's own experiences in the diagnostics industry, the following guidelines can be used to approximate the shelf life of a product at the desired temperature:

1 month at 50°C is equivalent to 1 year at room temperature

2 months at 50°C is equivalent to 2 years at room temperature

3 months at 37°C is equivalent to 1 year at room temperature

6 months at 37°C is equivalent to 2 years at room temperature

7 days at 37°C is equivalent to 1 year at 4°C

3 days at 37°C is equivalent to 3 to 6 months at 4°C

Thus, accelerated stability testing and the various models available for interpreting these data can provide valuable information for evaluating reagents and control products. By optimizing the analytical precision and other aspects of test protocol design, one can expect both real-time and accelerated stability studies to provide more valid information. For analytes with high activation energies, both bracket tables and the Q rule provide useful information when they are applied conservatively. Furthermore, use of published or experimentally derived activation energy values can significantly lower the risks inherent in projecting product shelf life.

REFERENCES

ANDERSON, G., and SCOTT, M. 1991. Determination of product shelf life and activation energy for five drugs of abuse. *Clin. Chem.* 37:398–402.

BANGHAM, D. R. 1971. Reference materials and standardization. In *Principles of Competitive Protein-Binding Assays,* eds. W. D. O'Dell and W. H. Doughaday, pp. 85–105, J.B. Lippincott Co., Philadelphia, PA.

BANGHAM, D. R. 1976. Standardization in peptide hormone immunoassay. Principle and practice. *Clin. Chem.* 22:957–961.

BANGHAM, D. R. 1982. Biological standards in clinical endocrinology. Some soluble and insoluble problems. In *Quality Control in Clinical Endocrinology,* eds. D. W. Wilson, S. J. Gaskell, and K. W. Kemp, pp. 43–49, Alpha Omega Publ. Ltd, Cardiff, Wales.

BANGHAM, D. R. 1988. Reference preparations and matrix effects. In *Complementary Immunoassays,* ed. W. P. Collins, pp. 13–25, John Wiley & Sons, New York.

BANGHAM, D. R., and COTES, P. M. 1971. Reference standards for radioimmunoassay. In *Radioimmunoassay Methods,* eds. K. E. Kirkham and W. M. Hunter, pp. 345–368, Churchill-Livingstone, London.

BANGHAM, D. R., and COTES, P. M. 1974. Standardization and standards. *Br. Med. Bull.* 30:12–18, 1974.

BLOCKX, P., and MARTIN, M. 1994. Laboratory quality assurance. In *The Immunoassay Handbook,* ed. D. Wild, pp. 263–276, Stockton Press, New York.

CAYGILL, C. P. J. 1977. Detection of peptidase activity in albumin preparations. *Clin. Chim. Acta* 78:507–515.

CHAN, D. W. 1992. *Immunoassay Automation. A Practical Guide.* Academic Press, San Diego, CA

CHAPMAN, J. S. 1994. An effective biocide for in vitro diagnostic reagents and other products. *Amer. Clin. Lab.* 13(8):13–14.

CONNORS, K. A.; AMIDON, G. L.; and STELLA, V. J. 1986. *Chemical Stability of Pharmaceuticals.* 2d ed., John Wiley & Sons, New York.

DE QUERVAIN, M. R. 1975. In *Freeze Drying and Advanced Food Technology,* eds. S. A. Goldblith, L. Rey, and W. W. Rothmayr, pp. 3–16, Academic Press, London.

DESHPANDE, S. S.; CHERYAN, M.; SATHE, S. K.; and SALUNKHE, D. K. 1984. Freeze concentration of fruit juices. *CRC Crit. Rev. Food Sci. Nutr.* 20:173–248.

EISENBERG, D. S., and RICHARDS, F. M. 1995. *Advances in Protein Chemistry. Protein Stability.* Vol. 46, Academic Press, San Diego, CA.

FENNEMA, O. 1979. *Proteins at Low Temperatures.* Amer. Chem. Soc., Washington, D.C.

FRANKS, F. 1995. Protein destabilization at low temperatures. In *Advances in Protein Chemistry. Protein Stability.* Vol. 46, eds. D. S. Eisenberg and F. M. Richards, pp. 105–139, Academic Press, San Diego, CA.

GOLDBLITH, S. A.; REY, L.; and ROTHMAYR, W. W. 1975. *Freeze Drying and Advanced Food Technology.* Academic Press, London.

HULL CORPORATION. 1994. *Fundamentals of Freeze Drying. Proceedings of a Workshop,* May 5–6, San Diego, CA.

JERNE, N. K., and PERRY, W. L. M. 1956. The stability of biological standards. *Bull. World Health Org.* 14:167–182.

KENNON, L. 1964. Use of models in determining chemical pharmaceutical stability. *J. Pharm. Sci.* 53:815–818.

KIRKWOOD, T. B. L. 1977. Predicting the stability of biological standards and products. *Biometrics* 33:736–742.

KUNTZ, I. D. 1979. Properties of protein-water at subzero temperatures. In *Proteins at Low Temperatures,* ed. O. Fennema, pp. 27–34, Amer. Chem. Soc., Washington, D.C.

LACHMAN, L., and DELUCA, P. 1976. *Kinetic Principles and Stability Testing. The Theory and Practice of Industrial Pharmacy.* 2d ed., Lea and Febiger, Philadelphia, PA.

LARSEN, S. 1973. *Arch. Pharm. Chem. Sci. Ed.* 1:41–53 (cited by Pikal, 1990b).

MACKENZIE, A. P. 1975. Collapse during feeze drying. Qualitative and quantitative aspects. In *Freeze Drying and Advanced Food Technology,* eds. S. A. Goldblith, L. Rey, and W. W. Rothmayr, pp. 277–307, Academic Press, London.

MACKENZIE, A. P. 1977. Solvent exchange and removal: Lyophilization. In *Proc. Intl. Workshop on Technology for Protein Separation and Improvement of Blood Plasma Fractionation,* pp. 185–201, Sept. 7–9, Reston, VA.

MACKENZIE, A. P. 1982. Freeze drying of three component systems of pharmaceutical interest. Paper presented at the Annual Meeting of the Parenteral Drug Association, Nov. 8–19, Philadelphia.

MACKENZIE, A. P. 1985a. A current understanding of the freeze drying of representative aqueous solutions. In *Fundamentals and Applications of Freeze Drying to Biological Materials, Drugs and Foodstuffs.* pp. 21–34, International Institute of Refrigeration, Paris, France.

MACKENZIE, A. P. 1985b. Changes in electrical resistance during freezing and their application to the control of the freeze drying process. In *Fundamentals and Applications of Freeze Drying to Biological Materials, Drugs and Foodstuffs.* pp. 155–163, International Institute of Refrigeration, Paris, France.

MALAN, P. G.; SIMPSON, J. G.; STEVENS, R. A. J.; GROVE-WHITE, J. F.; and WHITWORTH, A. S. 1983. Calibration of standards for immunoassay, including an appendix on spectrophotometric standardization of solutions of thyroxine and triiodothyronine. In *Immunoassays for Clinical Chemistry,* eds. W. M. Hunter and J. E. T. Corrie, pp. 48–57, Churchill-Livingstone, London.

PEARLMAN, R., and NGUYEN, T. H. 1989. Formulation strategies for recombinant proteins. Human growth hormone and tissue plasminogen activator. In *Therapeutic Peptides and Proteins. Formulations, Delivery and Targeting,* eds. D. Marshak and D. Liu, pp. 23–30, Cold Spring Harbor Laboratory, New York.

PIKAL, M. J. 1990a. Freeze drying of proteins. Part I. Process design. *Biopharmacology* 3(9):18–27.

PIKAL, M. J. 1990b. Freeze drying of proteins. Part II. Formulation selection. *Biopharmacology* 3(10):26–30.

PIKAL, M. J.; DELLERMAN, K. M.; and ROY, M. L. 1992. Formulation and stability of freeze dried proteins. Effects of moisture and oxygen on the stability of freeze dried formulations of human growth hormone. *Dev. Biol. Stand.* 74:21–38.

PIKAL, M. J.; DELLERMAN, K. M.; ROY, M. L.; and RIGGIN, R. M. 1991. The effects of formulation variables on the stability of freeze dried human growth hormone. *Pharm. Res.* 8:427–436.

PIKAL, M. J.; ROY, M. L.; and SHAH, S. 1984. Mass and heat transfer in vial freeze drying of pharmaceuticals: Role of vial. *J. Pharm. Sci.* 73:1224–1237.

PIKAL, M. J., and SHAH, S. 1990. The collapse temperature in freeze drying. Dependence on measurement methodology and rate of water removal from the glassy phase. *Int. J. Pharm.* 62:165–186.

PIKAL, M. J.; SHAH, S.; ROY, M. L.; and PUTMAN, R. 1990. The secondary drying stage of freeze drying. Drying kinetics as a function of temperature and chamber pressure. *Int. J. Pharm.* 60:203–217.

PIKAL, M. J.; SHAH, S.; SENIOR, D.; and LANG, J. E. 1983. Physical chemistry of freeze drying. Measurement of sublimation rates for frozen aqueous solutions by a microbalance technique. *J. Pharm. Sci.* 72:635–650.

PORTERFIELD, R. I., and CAPONE, J. J. 1984. Application of kinetic models and Arrhenius methods to product stability evaluation. *Med. Device Diagnostic Ind.* April 1984: 45–50.

SILVER, A.; DAWNAY, A.; LANDON, J.; and CATTELL, W. L. 1986. Immunoassays for low concentrations of albumin in urine. *Clin. Chem.* 32:1303–1306.

TABORSKY, G. 1979. Protein alterations at low temperatures. An overview. In *Proteins at Low Temperatures,* ed. O. Fennema, pp. 1–26, Amer. Chem. Soc., Washington, D.C.

TRAPPLER, E. H. 1993. Validation of lyophilized products. In *Pharmaceutical Process Validation,* eds. I. R. Berry and R. A. Nash, pp. 445–477, Marcel Dekker, New York.

WHO. 1975. Expert Committee on Biological Standardization. 26th report, WHO Tech. Rep. Ser. No. 565, World Health Organization, Geneva.

WHO. 1978. Expert Committee on Biological Standardization. 29th report. Annex 4: Guidelines for the preparation and establishment of reference materials and reference reagents for biological substances. WHO Tech. Rep. Ser. No. 626, World Health Organization, Geneva.

WHO. 1979. Biological Substances. International Standards, Reference Preparations, and Reference Reagents. World Health Organization, Geneva.

WHO. 1981. Expert Committee on Biological Standardization. 31st report. Annex 10: Requirements for immunoassay kits and reagents. WHO Tech. Rep. Ser. No. 658, World Health Organization, Geneva.

WILLIAMS, N. A., and POLLI, G. P. 1984. The lyophilization iof pharmaceuticals: A literature review. *J. Parenteral Sci. Technol.* 38:48–59.

ZAKOWSKI, J. 1991. Determination of stability. *Clin. Chem.* 37:313–314.

11
Data Analysis

INTRODUCTION

Data analysis is an important aspect of using solid-phase immunoassays. The basic principles of obtaining dose-response curves using immunoassay techniques were described earlier in this book. In quantitative immunoassays, the two fundamental variables involved are the analyte concentration (the "dose") and the signal level (the "response"). Depending on the label used, there are many different types of signals. The most commonly used ones include radioactivity in radioimmunoassays (RIAs), color intensity in conventional enzyme immunoassays (EIAs), and fluorescence intensity in fluoroimmunoassays (FIAs).

Because the dose-response curves plotted in conventional ways are often nonlinear, it is sometimes useful to work with functions of these variables. For example, analyte concentration can be plotted against the percentage bound, which is the signal level expressed as a percentage of the total signal originally added to the assay tubes. Generally, a response variable such as change in bound fraction is plotted on the y-axis against the total analyte (x-axis) present in a series of standard or calibration tubes, and the concentration of the unknown is determined from the resulting curve. In practice, with the advent of computers, a variety of methods of plotting the data have been developed, with different ways of expressing both the response variable (y-axis) and the dose variable (x-axis).

The fundamentals of mathematical and statistical analyses of immunoassay data, and in general of ligand-binding assays, were first established using response variables used in RIAs (Table 11.1). However, they are equally applicable to assay systems involving nonisotopic labels for the antigen. In a standard RIA,

TABLE 11.1. Response Variable Options Used in Immunoassay Data Analysis[a]

• B	• B_o / B
• F	• $(B_o - NSB) / (B - NSB)$
• B / T	• $B / (T - B)$
• $(B - NSB) / T$	• $(B - NSB) / (T - B)$
• $(B - NSB) / (T - NSB)$	• $(B - NSB) / [T - (B - NSB)]$
• B / B_o	• $(T - B) / B$
• $(B - NSB) / (B_o - NSB)$	• $(T - B) / (B - NSB)$
• T / B	• $[T - (B - NSB)] / (B - NSB)$
• $T / (B - NSB)$	• $(T - F) / (F - NSB)$
• $(T - NSB) / (B - NSB)$	

[a]Abbreviations used are: B = bound fraction, F = free fraction, T = total counts or absorbance, B_o = counts or absorbance at zero analyte concentration, and NSB = nonspecific binding.

the bound fraction (B) is measured as the counts obtained in the precipitate after separation of the bound and free fractions. In EIAs, the bound fraction is measured as the change in absorbance due to the reaction of the bound enzyme label with its substrate. The free fraction (F) in the RIA is determined by subtraction from total counts (T) after correction for nonspecific binding (NSB). In EIAs, the total change in absorbance is affected by the reaction of the total amount of added enzyme and its substrate. NSB is measured as the change in absorbance due to the reaction of the substrate and the enzyme label that is bound nonspecifically to the matrix. In EIAs, both these parameters are determined from controls included in every experiment. In EIA, the free fraction is determined by subtraction from the total change in absorbance after correction for NSB. It is thus obvious that the data processing methods and models of data reduction that have been developed for RIA analysis are equally applicable to EIAs. There is no need for independent development of mathematical analyses for EIA data reduction.

In this chapter, the evaluation of standard or calibration curves, data reduction and curve-fitting techniques, and some of the more commonly used statistical parameters in data analysis are briefly described.

STANDARD OR CALIBRATION CURVES

Basic Considerations

The primary objective of a quantitative immunoassay is to measure the analyte concentration in the test sample and to estimate the error associated with its measurement. The analyte concentration in the test sample can be estimated using a two step procedure involving calibration and interpolation.

Calibration is the process of establishing a clear relationship between what the assay actually determines (i.e., the response corresponding to a given standard concentration) and what is actually measured, viz., the analyte concentration. Calibration may be accomplished graphically by plotting the response versus the standard analyte concentration. Such a plot is referred to as a "calibration curve." The calibration curve then can be used to estimate the analyte concentration corresponding to the response obtained. This process is known as "interpolation."

Because standard solutions are used to generate a calibration curve, such a curve is also referred to as a "standard curve." Similarly, because dose is related to the analyte concentration, the term "dose-response" curve is also sometimes used, particularly in describing the results of a bioassay procedure. However, the term "calibration curve" is preferable to both standard curve and dose-response curve as it more accurately describes what the curve actually does: it calibrates the assay measurement method (Channing Rodgers 1981).

Ideally, the calibration curve should be monotonic; i.e., the slope of the curve does not change sign. Depending on the assay format, the slope may be either positive or negative. A monotonic calibration curve does not have any local maximum or minimum values (peaks or valleys). Thus, each value of the response corresponds to one and only one value of analyte concentration. In nonmonotonic calibration curves, a response may correspond to multiple values of analyte concentration, thereby necessitating additional analysis of an unknown test sample to determine the appropriate analyte estimate.

The possible responses (Table 11.1) used in immunoassay data analysis can be divided into two groups: the direct responses and the indirect responses (Channing Rodgers 1981). The direct group of responses consists of B, F, 1/B, 1/F, B/F, F/B, B/T, F/T, T/B, and T/F, where B, F, and T represent, respectively, the bound, free, and the total labeled ligand in a quantitative immunoassay. The value of T is generally assumed to be constant for all assay reaction mixtures. The term "direct" is used to indicate that these responses are direct reflections of the extent of chemical interaction of the antibody with the labeled antigen. More complex functions of B, F, and T can also be used as responses in constructing the calibration curves. These are known as "indirect" responses, and are confined to use in automated data processing packages.

Calibration curves based on direct responses can be further classified into several groups. Based on the fraction, these can be classified as bound-based methods (e.g., B, B/F, B/T, and 1/B) or free-based methods (e.g., F, F/B, F/T, and 1/F). Based on the slope of the calibration curve which they produce, they can also be classified as positive slope methods (e.g., F, F/B, F/T, and 1/B), and negative slope methods (e.g., B, B/F, B/T, and 1/F). These categories are schematically shown in Figure 11.1. Furthermore, the calibration curves produced by the direct responses can be described by two different types of mathematical equations: cubic equations describing curves resulting from using F, F/T, B/T, and B; and hyperbolic equations resulting from employing F/B, 1/F, 1/B, and B/F (Channing Rodgers 1981).

The hyperbolic curves are smooth curves that are concave upward or downward (Figure 11.1). The cubic forms also appear similar, but can also assume a sigmoid (S-shaped) form. This categorization assumes that the analyte concentration is expressed on a linear scale.

Several responses thus can be used for plotting an immunoassay calibration curve. The selection of an appropriate response variable can dramatically influence the shape of the calibration curve as well as the variance distribution (Fackrell et al. 1985; Ekins 1974; Rodbard et al. 1970, 1978; Lo 1981; Feldman and Rodbard 1971; Rodbard 1981; Walker 1977; Rodbard and Frazier 1975). However, if a proper mathematical relationship is used, the results of an assay should be independent of the choice of response used in the calculation. The difficulty of the correct procedure will vary greatly according to the choice; hence, the response selected should allow the simplest possible approach to be taken.

The response variable selected should ideally have random errors that follow a Gaussian distribution. When preparing the calibration curves, one cannot assume that a given response will work for a given assay. Therefore, the distribution of errors in measuring a response must be tested independently for each assay system.

Most immunoassay calibration curves exhibit certain general characteristics that often make routine manual analysis and interpretation of data difficult (Raggatt 1991). These include the following.

1. The calibration curve exhibits a typical nonlinear relationship between the response or signal measured and the analyte concentration. This precludes simple mathematical analysis of resulting data and often necessitates the use of more sophisticated techniques for data reduction and interpretation.

2. Generally, more than one curved line can be drawn through a set of calibrator points. Thus, a choice has to be made regarding the appropriate curve, which, in turn, can introduce the risk of bias.

3. Immunoassay data are characterized by large assay errors that occur relative to the analyte levels measured. Thus, there is a significant uncertainty in just where the line should be drawn even though the general shape is often well defined. Furthermore, these errors tend to be different in magnitude in almost every region of the assay's working range. They also tend to be quite variable from batch to batch, and hence, necessitate a fresh calibration curve for every new batch.

Some of these drawbacks can be minimized by increasing the number of calibrants used. In an ideal situation, the calibrators can be more closely spaced in the working range of the assay with several replicates being used at each calibrator level. In practice, this is often difficult to follow because of economics and longer turnaround times. Fortunately, several mathematical relationships have been well established for analyzing immunoassay data.

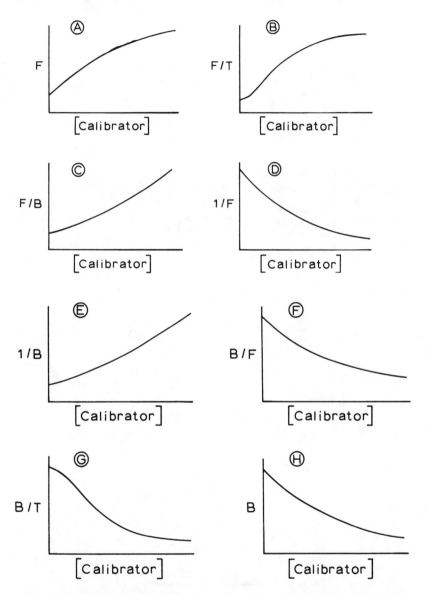

Figure 11.1. General shapes of immunoassay calibration curves obtained by using the direct responses: F (free fraction), B (bound fraction), and T (total counts or absorbance). These can be categorized as either free-based (curves A–D) or bound-based (curves E–H) methods. Curves in A, B, C, and E have a positive slope; the remaining calibration curves are characterized by a negative slope. Two different types of mathematical equations, cubic (curves A, B, G, and H) equations and hyperbolic (curves C, D, E, and F) equations are shown. The x-axis in all these examples is assumed to be linear.

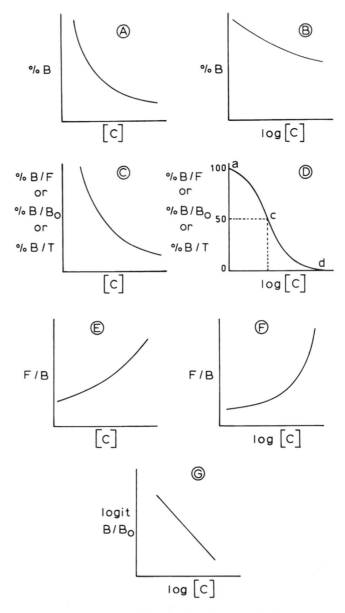

Figure 11.2. Examples of commonly used approaches for constructing immunoassay calibration curves. [C] on the *x*-axis represents the calibrator concentration on either linear (curves A, C, and E) or log (curves B, D, F, and G) scales. B, B_o, F, and T represent the bound fraction, counts (RIA) or absorbance (EIA) at zero calibrator concentration, free fraction, and total counts (RIA) or absorbance (EIA), respectively. The parameters *a* and *d* in curve D represent assay responses at zero

Graphic Representation

Some of the common approaches for graphic representation of the calibration curve are illustrated in Figure 11.2. In competitive immunoassays, as the total concentration of unlabeled analyte increases in the sample, that of the labeled analyte bound decreases, resulting in a corresponding decrease in signal intensity. The simplest way is to plot the absorbance reading versus the calibrator concentration in Cartesian coordinates (Figure 11.2A), or to use a logarithmic scale for calibrator concentration (Figure 11.2B). However, as the reagents age (especially the tracer), the assay tends to lose sensitivity.

Therefore, to normalize this effect and to minimize the variance distribution associated with the measurement of response variable, a ratio such as bound (B) to free (F), total (T) or absorbance at zero dose (B_0) as response variable is plotted against analyte concentration on a linear scale (Figure 11.2C). This technique of plotting shows the same relationship as simply plotting absorbance values (or raw counts in RIA and fluorescence intensity in FIA) versus calibrator concentration (Figure 11.2A). The ratio B/F is primarily useful when the primary calibration curve is used as a basis for calculating the Scatchard plot.

Usually, the relationship described in Figure 11.2C appears to be hyperbolic or curvilinear in shape. Ideally, as the analyte concentration becomes infinite, the B/B_0 ratio should decrease asymptotically to zero. However, because of the inherent characteristics of immunoassays or small systematic errors in the separation of the bound and free forms, the lower asymptote is usually not zero, but rather some small value related to nonspecific binding of the label (Rodbard 1981).

Further mathematical manipulations, such as using a logarithmic scale for calibrator concentration, can transform the curve from a hyperbola into an S-shape (Figure 11.2D). Such a curve can be described using four parameters. The parameters a and d represent the responses at zero and infinite dose, respectively. The parameter c describes the concentration of the calibrant, which results in a decrease of B/B_0 to a value at midpoint between a and d. This is often described in terms of ED_{50} or IC_{50} (Figure 11.2D). The parameter b is an exponent that determines the steepness of the curve at its center or, for that matter, at any other point. Using these parameters, the response R at any given calibrator concentration X can be determined using the following relationship:

$$R = (a - d)/[1 + (X/c)^b] + d \qquad (1)$$

Caption to Figure 11.2. (*cont.*)
and infinite dose, while c describes the calibrator concentration that results in a decrease of B/B_0 to a value at midpoint between a and d. Along with b (not shown here), which is an exponent function that determines the steepness of the curve at its center or at any other point on the curve, these parameters are used in logit-log (curve G) and four-parameter logistic (4-PL) curve-fitting techniques for immunoassay calibration curves. See text for details.

To construct a calibration curve, the inverse ratio, free to bound, can also be used on both linear (Figure 11.2E) and logarithmic (Figure 11.2F) scales for the calibrator concentration. Such a plot has a completely different appearance. It also suffers from a major drawback in that the slope at various parts of the curve is quite unrelated to the experimental findings. Thus, the apparent slope of the curve, if equated with precision, does not distinguish the experimental difference between two successive data points.

One must be aware of pitfalls in using a linear scale for calibrant concentration for data analysis and interpolation. The calculation of small concentrations of analyte by interpolation may be awkward, because the part of the scale covering these concentrations is relatively cramped. However, such an approach may offer some practical advantages. For example, if the assay is directed to the determination in a narrow physiological range with a correspondingly narrow range of standards, manual extrapolation is possible. It also permits plotting of the zero standard, which is not possible with a logarithmic plot.

In addition to the direct responses described above for constructing immunoassay calibration curves, several indirect methods can also be used. These involve more complex calculations, and are often confined to use in automated data processing packages. A notable exception is the well-known "two-parameter logit" method, also known as the "two-parameter logistic" or "logit-log" method. This curve-fitting procedure was first introduced in the early 1970s by Rodbard and his colleagues in linearizing immunoassay calibration curves (Rodbard and Cooper 1970; Rodbard and Lewald 1970; Feldman and Rodbard 1971). This technique can be used manually with the aid of a calculator or special logit graph paper.

The logit-log method is a popular way to treat immunoassay data because in many assays it produces a linear calibration curve, thereby making interpolations easy (Figure 11.2G). However, linearization of sigmoidal curves using the logit transformation may lead to severe heteroscedasticity, i.e, the error in the signal level varies greatly with calibrator concentration.

The logit is a general mathematical transformation named by Berkson (1944,1951), who had used it for various bioassays since the 1920s. The logit function is a continuous sigmoidal function with a single point of inflection. It is expressed as

$$\text{logit } B/B_0 = \log_e[(B/B_0)/(1 - B/B_0)] \tag{2}$$

or
$$\text{logit } Y = \log_e[1/(1 - Y)] \tag{3}$$

where
$$Y = (B - NSB)/(B_0 - NSB) \tag{4}$$

where NSB represents the nonspecific binding of the label. When logit Y (same as logit B/B_0) is plotted against the log dose, a linear curve is usually obtained (Figure 11.2G). The logit-log relationship is theoretically exact under conditions of saturation of the antibody. An implicit assumption of this system is that most assays will be operating under near saturation and that in any event the logit transformation will result in a useful approximation in most assay systems.

Several practical considerations must be taken into account to use logit transformation of the immunoassay data. These include:

1. The NSB must be determined and subtracted from each data point. A failure to subtract NSB values yields a nonlinear response. Similarly, assays with a high and variable NSB value are often poorly suited for logit transformation.

2. The variance is not uniform along the entire logit-log curve, being the least in the midrange and greatest at the ends. This drawback can often be minimized by eliminating the upper and lower 10% of a calibration curve prior to curve fitting using a regression technique. Alternatively, a weighted least-squares regression must be applied to the data.

3. In many systems, the logit transformation does not yield a straight line. This is because both the log function and logit function are symmetrical. Thus, logit transformation may not fit very well in an immunoassay that produces an asymmetrical calibration curve. This is often the case when polyclonal antiserum is optimized for use at very low analyte concentration. Such a situation may also arise in assays in which the labeled analyte is chemically different from the test analyte and in immunometric assays.

Ekins (1974) has proposed a variation of the logit-log method that avoids the need for logit paper. In this approach, one plots the logarithm of the free-to-bound ratio minus the free-to-bound ratio for zero dose, i.e. $\log [(F/B)—(F_o/B_o)]$ against the logarithm of the concentration of calibrators. This will result in linearization of the calibration curve exactly as in the logit-log method. Ekins (1974) has further pointed out that one could also utilize logit $[(B/F) / (B/F)_o]$ rather than logit B/B_o. This results in a shift in the calibration curve. However, because this shift is the same for both calibrators and unknowns, one still has a valid method for constructing the calibration curve.

All these approaches are mathematically similar to the logit transformation, and therefore, have similar advantages and drawbacks.

CURVE-FITTING TECHNIQUES

Once assay response data are obtained for a set of calibrants, two possible approaches can be used in fitting a curve to this set of data points: the empirical or interpolatory methods and the theoretical or regression methods (Box and Hunter 1962; Nix 1994; Raggatt 1991). The first approach has the line described by the function pass exactly through the data points. These empirical or interpolatory methods therefore assume that the function connecting the points is continuous and smooth and that the calibrator signal levels are correct, even though their absolute accuracy is not possible in practice (Figure 11.2A). Such an approach often yields undesirable results with immunoassay data that have inherent scatter. These weaknesses in the interpolatory approach are highlighted when using linear or curvilinear (spline functions) interpolation techniques because these forms of empirical curve fitting always pass through the mean value of the signal for each calibrator.

In contrast, the theoretical or regression methods fit a given functional form or model to the data, thereby partially correcting the errors in the calibration points and making the curve more robust. These methods are generally based on the underlying physical-chemical mechanisms of immunoassays. Their only drawback is that each is based on a simplified model of immunoassay, which does not take all the possible factors into account (Nix 1994). However, in practice, this rarely affects substantially the quality of the curve fitting, and for this reason these methods are generally considered to be superior in performance.

Two statistical techniques are commonly used to estimate the parameters in any regression method: the least-squares and weighted least-squares procedures. The first is the most commonly used approach. In it the squares of the vertical distance of each data point from the curve are added together to determine the total sum of squares. The curve-fitting procedure then selects the curve that gives the smallest sum of squares.

In the second approach, the weighted least-squares procedure, the precision of measurement at each calibrator point is taken into consideration, the more precise points influencing the curve fitting the most. The curve-fitting routine is thus forced to concentrate on those data points on the calibration curve having the greatest precision. Assuming there are no correlations between the signal levels from different calibrators, the weighted least-squares procedure involves minimizing the weighted sum of squares. However, if the calibration curve is fitted to the means of replicate signal levels rather than to the individual points, a modified weighting function is required if the orders of replication for the calibrators differ. This is because the variance of the mean is dependent on the number of replicates.

Detailed mathematical treatments of these two approaches are commonly found in any standard statistics book.

Several reviews describe a variety of curve-fitting techniques available for immunoassay data analysis (Nix 1994; Raggatt 1991; Channing Rodgers 1981, 1984). No one algorithm has proven itself appropriate for all immunoassays. Considering the complex nature of antigen-antibody reactions, the inherent variability in immunoglobulins and hence heterogeneity of these reactions, and the many variables that must be controlled in immunoassays, it is reasonable to expect that no single mathematical model could prove itself adequate for all assays. Thus, recognition of the inherent limitations of each of the models described below is important. A choice of model should incorporate that information that is deemed desirable and suitable for a particular set of data. The most "correct" mathematical model is not necessarily required; only a model that gives reliable and clinically useful results.

Prior to using the appropriate curve-fitting algorithm, it is often useful to plot the data manually in a variety of ways. Visual inspection of manual plots can frequently reveal areas where the assay may lose sensitivity, and under these circumstances, it may be wise to use a less extensive calibration curve. For example, the highest or the lowest calibrator may be eliminated from data analysis. These plots can be prepared from laboratory data generated in-house or from package insert data. After inspecting the manual plots, the data can then be tested using the appropriate curve-fitting technique and computer software. Some commonly used curve-fitting methods are briefly described below.

Hand Plots

Hand plots using a simple or special graph paper are easier to draw when the dose-response relationship is reasonably linear. The curve is carefully drawn through the set of data points. The method is reasonably precise and unlikely to be biased. It is often used as a reference method for checking computerized curve-fitting techniques (Nix 1994). However, the quality of the results depends on the skill of the user, especially when the assay errors tend to be heteroscedastic. It is also time-consuming and impractical for high-throughput laboratories.

Linear Interpolation

Linear interpolation involves drawing straight lines between the mean signal or response for neighboring calibrator points. However, if the curve appears to be linear, it is often preferable to fit all the points by linear regression than by linear interpolation. This method is simple to use and normally provides reasonably unbiased curve fitting. However, bias can occur between the calibrator concentrations when the curvature is high, causing consistent bias in controls. Furthermore, because of lack of smoothing, such point-to-point plots are significantly influenced by outliers and poor replicates. In such instances, assay-to-assay reproducibility is often poor.

Linear Regression

A linear curve fit can be obtained by fitting the assay data to the equation

$$Y = a + b * X \tag{5}$$

where a is the y-intercept of the line, b the slope of the curve, and Y and X, respectively, the assay response and calibrator concentrations. A linear fit should be used whenever the standard values appear to lie on or scattered around a straight line.

Similarly, semi-log and log-log curve-fitting techniques can be utilized using the following equations:

$$Y = a + [b * \log_{10}(X)] \tag{6}$$

$$\log_{10}(Y) = a + [b * \log_{10}(X)] \tag{7}$$

The semi-log or log-log fit should be used for a curve that has strong curvature up or down when a linear curve-fitting technique is used.

Spline Fits

The spline fitting technique is interpolatory. There is no physical-chemical structure to the model, and it can be used for any combination of dose and response variables (Nix 1994). The inherent lack of versatility of low order polynomials can be overcome if the calibration curve is split into several small sections, each of which is fitted by a polynomial. Each section then can be made to join neighboring sections smoothly, creating one smooth curve. This technique is commonly known as the spline method. In its simplest form, the equation used for each segment is usually a cubic polynomial:

$$Y = a + bX + cX^2 + dX^3 \tag{8}$$

In order to calculate a cubic polynomial, at least three independent data points are required. The equation is used to fit different cubic polynomials through the mean values of the response variable for each neighboring calibrator value, imposing connectivity conditions making the curve value, slope, and curvature match at each point. Because of the need for each segment to be recalculated until the curve is smooth, i.e., because of the iterative nature of the calculations, a computer is essential. A quadratic spline can also be used if a second order polynomial is used.

More sophisticated spline fits have a smoothing parameter, which relaxes the condition that the curve has to pass through the mean responses of the calibrators, so that a straighter line can be produced (Nix 1994; Raggatt 1991). The most sophisticated spline programs allow for the number of curve turning or inflection points to be restricted, typically to no more than one. Smoothed spline functions were first described by Reinsch (1967, 1971) and then applied to immunoassay by Marschner et al. (1974). Malan et al. (1978a,b) further refined the use of this technique in immunoassay data analysis.

Spline-fitting techniques are applicable to a wide range of assays as they are the mathematical equivalent of a flexicurve. Smoothing gives some protection against outliers and poor duplicates. Unsophisticated spline functions, however, are quite capable of producing a calibration curve with a wiggle in it, i.e., one containing more than one inflection point. Furthermore, in regions where the calibration curve flattens out at asymptotes, the spline may oscillate badly. This severely restricts the range of interpolation for which it can be used reliably. Extrapolation beyond the calibration curve should also be avoided, as spline fits do not have any built-in biochemical restrictions on the curve shape (Nix 1994; Raggatt 1991).

Polynomial Regression

Calibration curves can also be fitted to a curve described by the following equation:

$$Y = a + bX + cX^2 + dX^3 + \ldots + pX^q \tag{9}$$

where the order of the polynomial is q. Generally, the order is ≤ 3 to avoid oscillations. Higher powers can also be used; however, they require proportionately more data in order to get an unambiguous solution. The curve fitting uses a weighted least-squares approach to account for the different levels of precision across the calibrator concentration range.

Polynomial curve fitting is very flexible with a wide application. Similar to the spline-fitting technique, extrapolation beyond the calibration curve should be avoided, as polynomial regression also does not have any built-in biochemical restrictions on the curve shape.

Logit-Log

This technique was described earlier in this chapter in conjunction with graphic representation of immunoassay calibration curves. This mathematical conversion comes close to linearizing the calibration curves of both competitive and immunometric assays because it simulates the physicochemical situation. In competitive assays, the logarithm of the calibrator concentration is plotted against the logit of the signal, after correcting for the B_o and NSB; whereas in immunometric assays, the NSB is replaced by the maximum replicate signal of the highest concentration of the calibrator.

The advantages and drawbacks of this technique were described earlier.

Four-Parameter Logistic (4-PL)

This model was developed by Healy (1972) to improve the logit-log method. The logit-log model produces two parameters: the slope and the intercept of the linear regression line. Healy added two more parameters to deal with the data for the assay zero (B_0) and the assay blank (NSB), and expressed the model as described in equation (1) earlier. The algorithm for the 4-PL curve fit begins with the linear regression results from the logit-log computation. The assay blank and the assay zero are regarded as fixed but not error free, and are determined afresh for each assay. The fitting of a 4-PL is an iterative procedure done on a computer.

The 4-PL model yields results that are more accurate and less susceptible to variation, as compared to the logit-log methods. The fit at the extremes of the calibration curve is also much better. The curve, however, is still essentially symmetrical about the ED_{50} point like the logit-log curve. Nonetheless, the 4-PL method is widely preferred over the logit-log technique.

More complex functions, such as a five-parameter logistic model, wherein a fifth parameter is allowed to introduce some asymmetry in the curve, are also available. However, because of their complexity, they are seldom used in routine immunoassay data analysis and curve-fitting techniques.

Four-Parameter Law of Mass Action

Primarily based on the chemistry of radioimmunoassays, the four-parameter law of mass action method was specifically developed for competitive assays. The four parameters estimated are equilibrium dissociation constant of the antibody, antibody concentration, labeled antigen concentration, and the fraction of the total signal attributed to NSB (Nix 1994). The fitting is again achieved by an iterative computer algorithm.

The quality of the fit produced by a curve-fitting program needs to be monitored routinely. An inappropriate combination of assay characteristics and software can introduce bias into the results. The simplest indicator of how well the data and software are matched is found in the comparison of the actual and calculated dose of the calibrators. As a rule, this difference should be within the expected precision of the assay at various dose levels. If the two values coincide, the data fit is acceptable. If, however, there is a wide variance between the values throughout the concentration range, another data fit should be sought because the results, or unknowns, may be skewed. This distortion can have an impact on the normal range and is sometimes the cause of poor performance in proficiency surveys. Once the laboratory is confident that the fit and assay are well suited, monitoring the actual versus calculated dose can be used as a way to signal contaminated calibrators, pipetting and other errors, and deterioration of assay performance. Such a quality check will also reveal a concentration-dependent bias and the imprecision in terms of concentration, as these parameters are directly related to the particular assay for which the curve-fitting method is being used.

Several software packages are available for analyzing data from immunoassays on personal computers. The characteristics of some of the popular software programs are summarized in Table 11.2. The preferred system should ideally be modular, allow data input from any assay, be flexible enough to allow for different assay designs and protocols, allow computation of data by a variety of different algorithms, permanently store data, allow either manual or automatic data input, operate on an inexpensive computer, and be operable by technical staff with very little training.

STATISTICAL DESCRIPTORS

All approaches to calculation, analysis, and interpretation of immunoassay data are statistical in nature because a sample population is used to estimate the range end points of a much larger population. An implicit assumption is that the values of the analyte in the population do in fact reflect some single central tendency. Furthermore, because random errors (see Chapter 9 for more details) are inherent in any laboratory measurement, no chemical measurement is exact. Statistical descriptors of data sets therefore are useful for several purposes in analyzing data from immunoassays. Some of these include the following (Garber and Carey 1984):

1. To assess random variation in a population of data,
2. To compare the amounts of random variation among populations of data,
3. To test for systematic differences between populations of data, and
4. To assess the degree of correlation between populations of data.

Some statistical descriptors commonly used in immunoassay data analysis and interpretation are briefly described below. It is also essential that the reader be familiar with sound experimental designs to yield data amenable to unambiguous statistical treatment and interpretation. Such information can be found in any basic statistics textbook.

Population Distribution

To be able to use the appropriate statistical parameters for data analysis and interpretation, it is of foremost importance to identify the population distribution of data measurement values. The simplest way to describe a population of data is to construct a histogram (also referred to as a "frequency diagram" or "frequency distribution"). If enough data are represented in the histogram and the data are truly random, i.e., each result was affected by random errors only, one can then use the histogram to predict the probability of obtaining future results above or below a given value.

Three terms are commonly used to define the values about which a population is centered. The arithmetic average of a set of data is defined in terms of a "mean." The

TABLE 11.2. Characteristics of Some Commercially Available Software Programs for Immunoassay Data Analysis[a]

Parameter	BioCalc	DeltaSoft	ELISAAID	ELISAPLUS	KinetiCalc	Microplate Manager	MicroTek	SOFTmax	Titersoft II
Type of Plot									
Linear	Y[b]	Y	Y	Y	Y	Y	Y	Y	Y
Semi-log	Y	Y	Y	Y	Y	Y	Y	Y	Y
Log-log	Y	Y	N	Y	Y	Y	Y	Y	Y
Logit-log	Y	Y	Y	Y[c]	Y	Y	Y	Y	Y
Curve-Fitting Technique									
Linear regression	Y	Y	Y	Y	Y	Y	Y	Y	Y
Polynomial	Y	N	N	Y	Y	Y	N	Y	Y
4-PL	Y	Y	Y	Y	Y	Y	Y	Y	Y
Spline	Y	N	Y	N	Y	N	N	N	N
Linear interpolation	Y	Y	Y	Y	Y	N	Y	N	N

[a]Software programs are provided by the following companies: BioCalc, Dynatech Laboratories, Inc.; DeltaSoft, BioMetalics, Inc.; ELISAAID, Robert Maciel Associates, Inc.; ELISAPLUS, Meddata, Inc.; KinetiCalc, Bio-Tek Instruments, Inc.; Microplate Manager, BioRad, Inc.; MicroTek, Denly Instruments, Inc.; SOFTmax, Molecular Devices, Inc.; and Titersoft II, ICN Biomedicals/Flow Laboratories, Inc.

[b]Y = has function, N = does not have function.

[c]The software uses a logit-log transformation internally, but this is plotted in a semi-log manner.

"median" defines a value or interval of a population occurring in the middle of a population, half the total values fall above and half below this value. Finally, the "mode" defines the value or interval of a population occurring with the greatest frequency.

If the frequency distribution of data is Gaussian (generally referred to as "normal" in standard statistics textbooks) or symmetric about the mean value of the population, it will be represented by the well-known bell-shaped curve (Figure 11.3A).

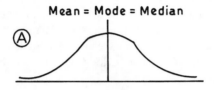

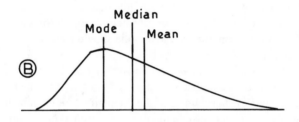

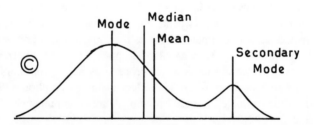

Figure 11.3. Examples of population distribution. Gaussian or "normal" distribution, defined by curve A, is symmetric about the mean and also has the same value for median and mode. Gaussian populations can be analyzed using parametric statistical tests. Curve B, with a long tail of high values, represents a nonsymmetric distribution characterized by different values for mean, median, and mode. Curve C, characterized by two values for mode (bimodal distribution), is an extreme example of a nonsymmetric distribution. Nonparametric statistical tests are used to analyze nonsymmetric distributions. For most immunoassays, the response variable in terms of B/B_o behaves according to Gaussian parametric statistics.

In a Gaussian distribution, the central value represents all the three parameters, i.e., the mean, median, and mode values are the same. Such a frequency distribution is generally obtained when repeated measurements are made on the same sample with only the small inherent random errors of the method affecting the values.

Many statistical tests assume that the data are "normally" distributed. If this assumption is true, then the sample population can be described by the mean $(\bar{X})$ and standard deviation (SD) of the observed values, which in turn, estimate the true mean (μ) and variance (S^2) of the entire population from such a subset of data. A Gaussian distribution of data also allows one to draw certain statistical inferences about where a value may lie. A generally accepted normal range has been the 95% confidence limits nominally defined as the mean ± 2 SD. Statistical methods used to describe and characterize Gaussian or log-Gaussian distributions are also also known as parametric methods.

As compared to Gaussian distributions, in nonsymmetric distributions (Figure 11.3B,C), the mean value alone cannot describe the center of the distribution satisfactorily. Different values for mean, mode, and median indicate that the distribution is skewed with a long tail of higher values (Figure 11.3B). A bimodal distribution (Figure 11.3C) is an extreme example of a nonsymmetric distribution. It is, however, generally not expected for repeated testing of the same sample. Differences between the mean, median, and mode should alert one to the presence of a nonsymmetric distribution.

Statistical tests used for analyzing nonsymmetric distributions are also known as nonparametric methods. These estimates do not depend on an a priori assumption of the frequency distribution of the population. Nonparametric statistics can be used to analyze ranked or semiquantitative as well as quantitative data when the population distribution is unknown. The computational procedures are simple and can be performed quickly. The advantage of these approaches is that even if the underlying distribution is Gaussian, the methods are efficient in establishing the confidence limits. However, when using nonparametric methods, more observations are necessary to estimate the limits for a given proportion of the population with a given probability. They are also less powerful than the corresponding parametric tests. Nonparametric statistical procedures are not commonly used for data analysis in clinical diagnostics. The effect of the assumption and the statistical method on the estimated normal range of data is comprehensively described by Reed et al. (1971).

Measures of Variation

The mean $(\bar{X})$, standard deviation (SD), and percent coefficient of variation (CV) are commonly used as indicators of the variation of immunoassay measurements. These are calculated as follows:

$$\bar{X} = \Sigma \, X_i/N \tag{10}$$

$$SD = \Sigma \, (X_i - \bar{X})^2/(N-1) \tag{11}$$

$$\% \, CV = [SD/\bar{X}] * 100 \tag{12}$$

where X_i is an individual measurement and N the sample size.

CVs are commonly used to express random variation of analytical methods in units independent of analytical methodology.

Confidence Interval or Limit of Mean

A Gaussian population distribution can be completely described by two statistics: the mean and the SD. It has the unique property that if vertical lines are drawn one SD to either side of the mean (or the peak), they will contain 68% of the area under the curve (Figure 11.4). This is equivalent to saying that 68% of the values used to create the distribution will fall within one SD of the mean. Limits may be found that contain any given proportion of the values. Such limits relate to the frequency of occurrence of individual values and are known as "statistical confidence limits" or "intervals." The 95% confidence limits are located 1.96 SD to either side of the mean, and the 99% limits fall 2.58 SD to either side (Figure 11.4). In practice, these values are usually rounded to 2 and 3, respectively. Generally, confidence in the validity of measurement increases as the sample size or N increases.

The confidence interval for a mean essentially describes where the population mean (μ) lies with respect to a sample mean ($\bar{X}$) with a given probability. If several means are obtained from different groups of measurements of the same sample, the individual means will be distributed about the grand mean. This random variation in the population of means is described by the standard error of the mean ($S_{\bar{X}}$) as follows:

$$S_{\bar{X}} = SD/\sqrt{N} \tag{13}$$

By using the standard error of the experimentally determined mean, one can thus develop a confidence interval that has a known probability of including the true population mean, μ. This value depends on the number of measurements (N), the SD, and the level of confidence desired, and is calculated as follows:

$$\mu = \bar{X} \pm \left(t \, SD/\sqrt{N} \right) \tag{14}$$

or
$$\mu = \bar{X} \pm \left(t \, S_{\bar{X}} \right) \tag{15}$$

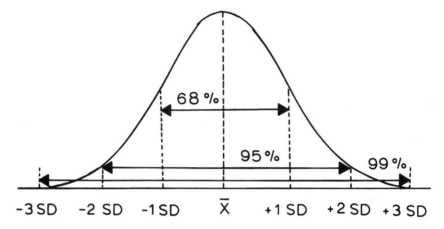

Figure 11.4. Probability distribution for a population defined by a Guassian curve. Mean ($\bar{X}$) ±1, 2, and 3 standard deviations (*SD*), respectively, encompass 68%, 95%, and 99% measurements from a population.

The value for *t,* taken from a *t* table, depends on the level of confidence or the desired probability (*p)* that μ is outside the confidence interval because of chance alone, and the number of degrees of freedom (ν, one less than the sample size, or $N-1$) associated with the estimation of *SD*. Thus, p = 0.05 implies a 95% confidence [100 $(1-p)$%] that the interval includes the true mean.

The *t* values describe the same Gaussian probability distribution as shown in Figure 11.4, the only difference being that in the latter case one assumes that the true population mean and *SD* are known. Because a confidence interval is being constructed to include the true mean, the 0.05 probability that μ is beyond the calculated limits must be spread over both ends or tails of the distribution (Figure 11.5). Thus a two-sided, *p* = 0.05, *t* value is used. If one is only interested in stating that the μ is greater than some limit, a one-sided *t* value would be used. The *t* value for a two-sided interval at *p* = 0.05 is the same as the *t* value for a one-sided limit at *p* = 0.025.

The following points should be kept in mind concerning Gaussian confidence limits for means (Channing Rodgers 1981).

1. No information about random error can be obtained from one sample alone. Replicates are thus absolutely necessary to determine the *SD* of the mean.
2. The *SD* of the mean, analogous to $S_{\bar{X}}$, is smaller than the sample *SD*. It decreases as the square root of the number of samples averaged. Thus, the mean is a better approximation to the true mean (μ) than any single sample. As the number of samples increases, the confidence interval for the mean gets increasingly smaller, because of the decreasing *SD* of the mean and also because of the decreasing value of *t*.

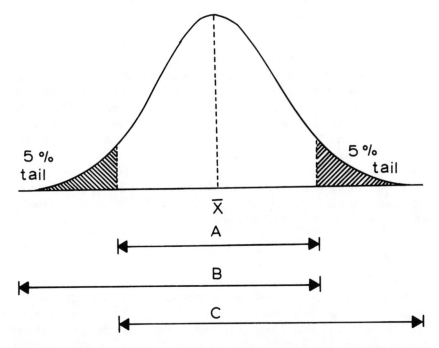

Figure 11.5. Determination of confidence interval or limit for a Gaussian population or distribution using *t* test. $\bar{X}$ defines the mean of the population. A, B, and C define, respectively, the 90% confidence interval, 95% one-sided (upper) limit, and 95% one-sided (lower) limit of confidence. Curve A is an example of a two-tailed t test, while B and C represent one-tailed *t* tests. Confidence intervals can be defined for any limits of population. For example, a 95% confidence interval for a Guassian population can be defined by using a 2.5% tail on either side of the mean.

3. Any two limits that contain 95% of the area under a distribution define 95% confidence or tolerance limits. Therefore, there are an infinite number of ways of drawing such limits. However, the symmetrical limits, described by two-tailed and one-tailed *t* tests, define the narrowest interval obtainable.

4. The definition of confidence limits requires the subjective choice of a level of confidence on the part of the analyst. The higher the level of confidence, the wider the limits.

For the calculation of the confidence limits, the random errors in the measurement must follow a Gaussian distribution. This can be tested by applying the chi-square test or the Kolmogorov-Smirnov tests, described in standard statistics textbooks. Fortunately, for most immunoassays, the response variable in terms of B/B_o behaves according to Gaussian parametric statistics. Although nonparametric statistics can be used, they are not practical for most assays as they require larger numbers of replicates than parametric methods.

Statistical Tolerance Interval or Limit

Statistical tolerance intervals or limits are not to be confused with confidence intervals or limits for a population mean. A tolerance interval represents the limits within which a specified percentage of the population is expected to lie with a given probability. Tolerance limits are of use in establishing "normal ranges" of physiological values for a given analyte (see Chapter 9). Statistical tolerance limits are therefore especially useful for specifying the limits of variability in a composition of samples. If the *SD* of the population of samples were known, the limits for a given percentage of the population could be calculated with certainty. Because only an estimate of the *SD* is usually known, based on a limited sampling of the population, a tolerance interval based on inclusion of a percentage of the population with a specific probability of inclusion is all that can be calculated. It is described by the following relationship.

$$\text{tolerance interval} = \bar{X} \pm k\, SD \tag{16}$$

where k is a factor based on the percentage of population to be included, the probability p of inclusion, and the number of measurements used to calculate $\bar{X}$ and *SD*. The width of the tolerance interval depends on the required confidence in the result. The estimated upper and lower limits come closer to the true upper and lower limits as the number of observations (N) increases. Similar to confidence intervals of mean, this method does not guarantee which 95% of the population is included.

Outliers

Occasionally, one or more data points of an immunoassay calibration curve will have an unusually large error and will deviate excessively from its corresponding true value. A pair of duplicates may not agree well, or one point may lie well off the line that goes through all the other data points. Such values are often termed "outliers" and may be rejected from the calibration curve. Healy (1979) defined an outlier as "an observation which departs from expectation to an improbable extent."

Outliers in data sets can result from such causes as blunders or malfunctions of the methodology, or from unusual losses or contamination. If outliers occur too often, this could indicate deficiencies in the quality control programs, which can be corrected and thus improve the measurement process. One should always search diligently for causes before data are rejected.

Possible outliers can be identified when data are plotted, when results are ranked, and when control limits are exceeded. Only when a measurement system is well understood and the variance is well established, or when a large body of data is available, is it possible to distinguish between extreme values and true outliers with a high degree of confidence.

Whenever an outlier is suspected, the analyst should look for a reason for its occurrence, i.e., an assignable cause. Miscalculations, use of wrong units, system malfunctions, misidentification of a sample, transcription errors, and contamination are some examples. In some cases, it may be possible to take corrective action and salvage what would otherwise be a bad measurement. Records should be kept of such problems and the corrective action taken, for guidance in possibly modifying quality control procedures.

Several statistical tests, such as the Dixon test, Grubbs test, Youden test, and Cochran test, are available for detecting outliers. However, these tests are often not useful for borderline cases in assays using small numbers of replicates, such as triplicate determinations of a data point. Outliers in calibration curves should always be a cause of concern, especially if samples are tested as singletons, as it is extremely difficult to detect outliers in singleton values.

If data are Gaussian, then outliers can be defined as those data points outside mean $\pm m$ SD where m is some multiplier. Mean ± 3 SD is commonly used, although the value of m ought to be varied depending on the size of the sample and the degree of confidence required (Burnett 1975).

Immunoassay data analysis, especially the selection of a suitable mathematical model to describe the calibration curve, thus can pose several perplexing problems. Calibration curves should always be analyzed first by the simplest model available. More complex models should be used only when the simple ones prove inadequate. Although an experienced analyst can manually plot immunoassay calibration curves using any mathematical model, this method is slow and its quality is totally dependent on the skill of the analyst. For these reasons, the use of manual interpretation is generally not recommended. Instead, the objectivity gained through the use of statistics can be usefully employed. This and the use of computers have resulted in considerable change in the techniques for the gathering, processing, and reporting of immunoassay data.

REFERENCES

BERKSON, J. 1944. Application of the logistic function in bioassay. *J. Amer. Statis. Assoc.* 39:357–365.

BERKSON, J. 1951. Why I prefer logits to probits. *Biometrics* 7:327–339.

BOX, F. E. P., and HUNTER, W. G. 1962. A useful method for model-building. *Technometrics* 4:301–318.

BURNETT, R. W. 1975. Accurate estimation of standard deviations for quantitative methods used in clinical chemistry. *Clin. Chem.* 21:1935–1938.

CHANNING RODGERS, R. P. 1981. *Quality Control and Data Analysis in Binder-Ligand Assay.* VolS. 1 and 2. Scientific Newsletters, Inc., Anaheim, CA.

CHANNING RODGERS, R. P. 1984. Data analysis and quality control of assays. A practical primer. In *Practical Immunoassay,* ed. W. R. Butt, pp. 187–211, Marcel Dekker, New York.

EKINS, R. P. 1974. Basic principles and theory. Radioimmunoassay and saturation analysis. *Br. Med. Bull.* 30:3.

FACKRELL, H. B.; SURUJBALLI, O. P.; and BURGESS, S. R. 1985. ELISA data reduction. A review. *J. Clin. Immunoassay* 8:213–219.

FELDMAN, H., and RODBARD, D. 1971. Mathematical theory of radioimmunoassay. In *Principles of Competitive Protein-Binding Assays,* eds. W. H. Daughaday and W. D. O'Dell, pp. 158–203, J.B. Lippincott, Philadelphia, PA.

GARBER, C. C., and CAREY, R. N. 1984. Laboratory statistics. In *Clinical Chemistry. Theory, Analysis and Correlation,* eds. L. A. Kaplan and A. J. Pesce, pp. 287–300, The C.V. Mosby Co., St. Louis, MO.

HEALY, M. J. R. 1972. Statistical analysis of radioimmunoassay data. *Biochem. J.* 130:207–210.

HEALY, M. J. R. 1979. Outliers in clinical chemistry and quality control schemes. *Clin. Chem.* 25:675–677.

LO, D. H. 1981. A quality control system in RIA. *Ligand Quarterly* 4:45–50.

MALAN, P. G.; COX, M. G.; LONG, E. M. R.; and EKINS, R. P. 1978a. Development in curve fitting procedures for radioimmunoassay data using a multiple binding site model. In *Computing in Clinical Laboratories,* ed. F. Siemaszko, pp. 282–290, Pitman, Medical, Tunbridge, Wells.

MALAN, P. G.; COX, M. G.; LONG, E. M. R.; and EKINS, R. P. 1978b. Curve-fitting to radioimmunoassay standard curves: Spline and multiple binding-site models. *Ann. Clin. Biochem.* 15:132–134.

MARSCHNER, I.; ERHARDT, F.; and SCRIBA, P. C. 1974. Calculation of the radioimmunoassay standard curve by spline function. In *Radioimmunoassay and Related Procedures in Medicine,* pp. 111–117, International Atomic Energy Agency, Vienna, Austria.

NIX, B. 1994. Calibration curve-fitting. In *The Immunoassay Handbook,* ed. D. Wild, pp. 117–123, Stockton Press, New York.

RAGGATT, P. 1991. Data processing. In *Principles and Practice of Immunoassay,* eds. C. P. Price and D. J. Newman, pp. 190–218, Stockton Press, New York.

REED, A. H.; HENRY, R. J.; and MASON, W. B. 1971. Influence of statistical method used on the resulting estimate of normal range. *Clin. Chem.* 17:275–284.

REINSCH, C.H. 1967. Smoothing by spline functions. *Numer. Math.* 10:177–183.

REINSCH, C.H. 1971. Smoothing by spline functions. II. *Numer. Math.* 16:451–454.

RODBARD, D. 1981. Mathematics and statistics of ligand assays. An illustrated guide. In *Ligand Assay,* eds. J. Langan and J. J. Clapp, pp. 45–99, Masson Publishing, New York.

RODBARD, D., and COOPER, J. A. 1970. A model for the prediction of confidence limits in radioimmunoassay and competitive protein binding assays. In *In Vitro Procedures with Radioisotopes in Medicine,* pp. 659–673, International Atomic Energy Agency, Vienna, Austria.

RODBARD, D., and FRAZIER, G. R. 1975. Statistical analysis of radioligand assay data. *Methods Enzymol.* 37:3–22.

RODBARD, D., and LEWALD, J. E. 1970. Computer analysis of radioligand assay and radioimmunoassay data. *Acta Endocrinol. Suppl.* 147:7–103.

RODBARD, D.; MUNSON, P. J.; and DE LEAN, A. 1978. Improved curve-fitting, parallelism testing, characterization of sensitivity and specificity, validation, and optimization for radioligand assay. In *Radioimmunoassay and Related Procedures in Medicine,* pp. 469–504, International Atomic Energy Agency, Vienna, Austria.

RODBARD, D.; RAYFORD, P. L.; and ROSS, G. T. 1970. Statistical quality control of radioimmunoassays. In *Statistics in Endocrinology,* eds. J. W. McArthur and T. Colton, pp. 411–429, MIT Press, Cambridge, MA.

WALKER, W. H. C. 1977. An approach to immunoassay. *Clin. Chem.* 23:384–387.

12

Documentation, Registration, and Diagnostic Start-Ups

INTRODUCTION

Commercial production of immunodiagnostic kits requires that the various components be manufactured consistently over a long period of time. The process involves availability of a consistent source of antibody, either through the pooling of large quantities of polyclonal antibodies or the use of monoclonal antibodies. In the former case, immunogen should also be readily available to ensure and enable continued production of polyclonal antibodies. Procedures for other kit components, such as enzyme conjugates, various buffers and reagents used, enzyme substrates, and stop solutions, also need to be scaled up and their reproducibility documented. The standardization of the entire process thus requires a combination of reagent stabilization, development of a comprehensive documentation system, and production of large lot sizes.

It is essential that the product have a reasonably good shelf life, preferably at least up to 9–12 months at 4°C. Longer shelf life allows larger lot sizes to be manufactured on a less frequent basis. This helps in minimizing lot-to-lot variation in the product.

A detailed documentation system describing how each component is prepared and how the quality assurance testing should be performed, as well as the limits of acceptable performance ensures the production of a reliable and reproducible assay system over the long term (Rittenburg and Grothaus 1992).

Criteria for documentation systems are provided in the *Code of Federal Regulations, Title 21*. These regulations, available to various regulatory agencies, such as the U.S. FDA and U.S. EPA, are critical in assuring the quality and safety of products. They were enacted to assure that companies involved in the development of new products follow established protocols in evaluating the performance, quality, and safety of these products. The manufacturers must rigorously document that these protocols have been followed. Furthermore, the records and experimental designs used in validation processes must be available for review by personnel from the regulatory agencies. These records must stand up to scrutiny when examined by the agency and when the company files an application to market such products.

The documentation of a quality assurance system, as has been so dramatically emphasized in the ISO 9000 guidelines described later in this chapter, may soon be a requirement for business in the global marketplace. Therefore, it is not only instructive but imperative to understand how to design these systems so that they support routine operations.

A comprehensive documentation system is also essential for registration under ISO 9000 guidelines. This internationally accepted quality assurance certification is almost mandatory for global marketing of products and ensures that they meet minimum quality standards. The importance of an appropriate and comprehensive documentation system in this regard is quite evident from a survey indicating that the most consistent barrier to ISO 9000 registration revolves around documentation issues (Morrow 1993). This survey included 1700 ISO registered firms in the United States and Canada. Procedure creation was cited by 19.7% of the participants as a barrier to registration, while 18.7% cited document development as the most significant barrier to registration. In spite of such hurdles, nearly all participants (about 96%) indicated a desire to use or plan to use ISO 9000 registration status for public relations purposes.

The immunodiagnostics business also offers many opportunities for entrepreneurs. In addition to the core business, several opportunities exist in providing ancillary services, such as reagent supply for clinical laboratories and supply of custom-made antibodies and/or labeled conjugates to both basic research laboratories and the diagnostic industry. Similar opportunities also exist for scientists working in universities and other academic institutions to do contract work or license their products, from antibodies to novel compounds, to various diagnostics companies.

In this section, the basics of documentation systems, the ISO 9000 registration system, and some tips for raising venture capital for diagnostic start-ups are briefly covered.

DOCUMENTATION SYSTEMS

A comprehensive documentation system is essential to a company's internal and external interactions. It must be a proactive vehicle of communication, and must be flexible enough to grow and change with the company. A well-designed documentation system ensures that quality standards are met routinely. It minimizes the potential for error, reduces downtime when deviations or failures occur by providing immediate access to well-organized data, and serves as a consistent training tool for technical and managerial personnel alike (DeSain 1993a; Stier et al. 1993). A poorly designed system can create several pitfalls, from the assay development stage through large scale manufacturing and the QA/QC process.

A comprehensive documentation system must take into consideration both good laboratory practices (GLPs) and good manufacturing practices (GMPs). A detailed description of the steps and procedures involved in this process is beyond the scope of this chapter; only the salient features are described here.

Companies involved in the immunodiagnostics business should employ GLPs and GMPs. There are several reasons to develop and maintain good practices. These include legal requirements, protocols to assure safety and reduce liability, control of processes in large scale manufacturing of the product, and product quality, even building design and maintenance. Perhaps the most important reason is economics. Without a thorough understanding of what is being done, how it is done, and where responsibilities lie, a commercial manufacturer of immunodiagnostic products is simply asking for trouble in the form of product recalls, adverse publicity, worker injury, environmental problems, or product quality.

The ideal way to develop and establish a documentation system begins with the management. A simple organization chart illustrates this point (Figure 12.1). Different groups, although working toward a common goal, must be separated. Production and Quality Control groups must adhere to GMPs to satisfy federal regulations. The Research and Development (R&D) group as well as the Quality Control group need to follow GLPs to satisfy internal needs and regulatory requirements when applicable.

The Quality Assurance (QA) group is held accountable for monitoring compliance to each of these programs. The *Code of Federal Regulations, Title 21,* mandates that this group should be entirely separate from and independent of the personnel engaged in the direction and conduct of all other aspects of a research and development process. The overall responsibilities of the QA group should include the following (Stier et al. 1993; DeSain 1993a):

1. Maintaining a master schedule of all studies/work being conducted,
2. Maintaining copies of all protocols,
3. Inspecting each phase of the work to assure compliance with requirements,
4. Submitting status reports to management,
5. Determining that there are no deviations from established protocols without authorization and approval,
6. Reviewing the final reports to assure that they correctly reflect the data,
7. Preparing a report describing inspections and findings, and
8. Maintaining the records.

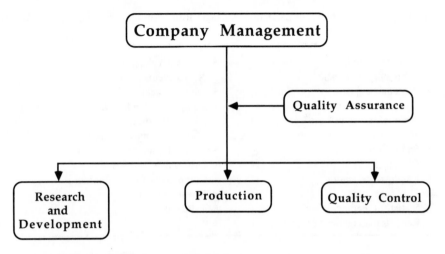

Figure 12.1. An example of an organizational chart

A good QA manager with his/her group can generate huge cost savings to the company if the job is done competently. The QA person must also focus on how the group benefits the company by preventing problems, eliminating waste, and contributing to more efficient operations. The various aspects of GLPs and GMPs are briefly described below.

Good Laboratory Practices (GLPs)

Standardization of methods and operating procedures is essential to good laboratory operations. In a new facility, writing protocols and establishing a formal manual for laboratory staff to follow may be the most important tasks of the laboratory or quality operations manager (Stier et al. 1993). One of the critical points involved in standardizing procedures is accurate record keeping. Both the procedures and the record keeping must be kept as simple and direct as possible. Various protocols which the laboratory manager is responsible for developing are summarized in Table 12.1. This task has been greatly simplified with the advent of computer technology. Several software programs, from the large and expensive LIMS (Laboratory Information Management Systems) to a wide range of sample log programs, are available for this purpose (Blumenthal 1992).

Analytical procedures or methods must be established in each laboratory as part of the standard operating procedures (SOPs). Many laboratories, including most commercial testing laboratories, use "official" and/or "recommended" methods provided by professional organizations and societies in a given field of application. Procedures may also be developed in-house, drawn from the technical literature, or supplied by manufacturers of equipment or testing materials. In

TABLE 12.1.   Examples of Laboratory Manager Responsibilities in Preparing Protocols to Implement GLPs in the R&D and QC Departments

• Analytical methods	• Forms development
• Laboratory notebooks	• New systems evaluation
• Procedures update	• Sample tracking
• Equipment usage	• Equipment standardization
• Materials inventory	• Materials storage
• Material safety data sheets (MSDSs)	• Materials supply and aging
• Safety systems	• First aid training
• Cleanup systems	• Disposal techniques
• Employee education	• Computer software training
• Staff schedules	• Downtime analysis
• Statistical methods	• Validation
• Records management (audit trail)	

any event, validation of such methods and their approval for use in laboratory operations is essential. Although official methods or collaborative procedures can sometimes be used directly, their suitability for a given need must be established. Tested and approved procedures and appropriate references must be indexed and filed in a laboratory manual in a way that ensures that only the current approved procedure is used.

Laboratory procedures sometimes can be rewritten to reflect the needs of the technical personnel and for ease of use. By clearly delineating the materials required, adding step-by-step instructions for the method, and writing in an easy style to ensure that every user of the method will clearly know exactly what to do at each step without making any assumptions, such procedures can become easier to use (Stier et al. 1993). An independent evaluation of the first draft often helps. The final draft must be validated by actual laboratory performance by a typical user of the procedure as written, without consultation with the writer.

Laboratory instrument calibration is often neglected or overlooked. Each instrument used in the laboratory for analytical purposes or safety monitoring must be maintained, inspected, and calibrated at frequent intervals, and records must be properly kept. As suggested by the manufacturer, appropriate standards or calibrators must be used to monitor instrument performance. These protocols should also form a part of the laboratory manual. It is also desirable that instruments should be labeled as to their calibration status. Ideally, they should not be used beyond their next scheduled calibration date. Calibration records are often helpful in identifying errors in subsequent measurements.

The data from all the experiments should be recorded in appropriate laboratory notebooks or standardized worksheet forms. If a mistake is made, a single line is drawn through the error, and the correct value is inserted with the appropriate comments and initialed by the person to denote the change. If a record pertaining to product quality, validation, or safety that a regulatory agency is entitled to review appears to have been altered without following the above procedure, complications may arise. The laboratory data notebooks or forms should also be reviewed and signed off by the manager.

In recent years, computerized data collection has greatly enhanced the concept of GLPs. It allows one to quickly and easily examine and evaluate data generated by the staff without going through notebooks or files. It has also facilitated easy analysis of the data by using a variety of statistical techniques.

The chemicals and other supplies used routinely in day-to-day laboratory work also need to be monitored for their quality and expiration dates, if applicable. Chemical deterioration over a period of time can affect reagent quality and lead to undesirable side reactions affecting the performance and quality of the finished product. Developing a set of guidelines for ordering, tagging, and properly storing chemicals and reagents is an effective means to manage resources and reduce costs.

The importance of good record keeping and appropriate documentation of analytical laboratory work is thus an integral part of the GLPs. An excellent monograph detailing the importance of good record keeping is available (OFR/NARA 1989). It summarizes all the records that are required by the *Code of Federal Regulations*. Programs to address these issues and assure compliances should be included in every company's SOPs.

Detailed record keeping and documentation also provide the management with a history of how the R&D department is performing. Such records can be used as part of an information base to upgrade systems, understand trends, and streamline operations. They are a vital part of a healthy and well-run company, a help in troubleshooting, and a mainstay of a GLP program.

Good Manufacturing Practices (GMPs)

Within the working environment of a company, the production and QC group must adhere to GMPs (QC group adheres to GLPs as well) for several reasons, the foremost being regulatory compliance. A thorough discussion of this topic, given its nature and scope, is beyond the limits of this chapter. However, the readers are referred to several articles and books that describe in detail the implementation of GMPs in a manufacturing process (DeSain 1993a,b; CFR 1990; CDER 1987a,b; Guerra 1986; Finkelson 1986; CEC 1989,1990; Weiss 1990; DeSain and Vercimak 1993a,b,c; DeSain and Sutton 1994,1995a,b,c). The following information is largely based on these articles.

Documenting the rationale for responsible decision making is integral to the philosophy of GMPs. This is obvious when one considers the need to document the rationale for changing a specification or changing a manufacturing process; it is, however, not so obvious when one considers the need to document the rationale behind the design of a documentation system. Without the rationale and without guidelines for decision making, the system would be vulnerable to major inconsistencies, and the work would proceed very slowly. Either possibility costs time and money, and is a source of tremendous frustration.

There are two ways to develop a documentation system to implement GMPs. The first is a top-down approach, which presents the general commitments of the system and then describes how the system is implemented. The second is a bottom-up approach, which first describes the construction of the most fundamental building blocks of the system, then builds on this structure. Ultimately, these two differing viewpoints reflect the fact that there are two types of users: manufacturing personnel who need to know what to do and how to do it, and administrative personnel (regulatory, development, finance, and marketing) who must understand and interact with the system by reference.

Several basic components that interrelate to serve a common system function are involved in preparing a documentation system describing the GMPs. These include the following:

Descriptive documents: These documents describe how to perform certain tasks, such as SOPs, data collection documents, specifications, and master production records.

Data collection documents: Data collection documents facilitate the timely and accurate documentation of tasks and events, such as forms, reports, production batch records, and logbooks.

Numbering systems: These documents account for and track information and documents, such as part numbers, lot numbers, equipment numbers, form numbers, SOP numbers, receiving codes, etc.

Data files: These files organize the data into useful categories of concern for review and support accountability and traceability requirements. Data files include specification files, equipment history files, product files, and facility qualification files.

The primary function of a documentation system in a GMP operation is to establish, monitor, and document quality (DeSain 1993a). The descriptive documents define and establish the quality of the raw materials, environment, production process, and final product. The data collection documents confirm that the materials, environment, production process, or product routinely meet the established quality characteristics. The numbering systems serve to control and track the use of descriptive documents and data collection documents. Data files organize the data into useful categories to facilitate review and retrieval. The preparation and importance of the various documents are briefly described below.

1. Part Numbering System.   A part numbering system is a basic building block of GMP documentation. A part number is a simple code used to identify one of a variety of items, its identity is defined in a separate part number specification document. The latter describes the item in detail and can include a way to test the quality of that item. In a GMP operation, the quality of an item differentiates it from similar items. Usually, part numbers, used primarily because of convenience, can also identify items purchased from outside vendors as well as those formulated or produced within the facility (e.g., solutions, components, and products).

All the critical items in a facility that directly affect the quality of the finished product should be given a designated part number. Similarly, any item whose control can provide convenience or ultimately assist in troubleshooting analytical assays or production processes needs a part number. When performing assays, whether for research or to support production of a product, it is far more convenient to cite a part number when preparing a solution than to write it out longhand. Also, the ability to track reagent use is invaluable when troubleshooting an assay.

Part numbers should be assigned by the QC or QA department. They can be simply random numbers; however, it is convenient to organize these numbers into identifying categories. There should also be only one part number for each unique item. Assigned and yet-to-be-assigned part numbers can be kept in an electronic data base to facilitate their use. This data base should be supported by two lists: a numerical list of part numbers by category and part names, and an alphabetical list of parts and part numbers. All departments in the manufacturing facility should have access to updated lists.

2. Part Numbering Specifications. These documents summarize the description of each part numbered item, containing purchasing information, chemical formulas, dimensions, sampling information, handling precautions, storage conditions, and when required, testing methods and acceptance criteria. Each page of a specification should contain the name of the company, a title or narrative description of the item, the part number of the item, the edition number of the specification, pagination, and approval signatures. In addition, each specification should contain a list of approved vendors for that item with catalog numbers, an expiration date for the item, sample size (when testing is required), file sample size when required for critical items, handling precautions, and testing or acceptance criteria.

Each category of part numbered items can have its own specification form to facilitate documentation of its characteristics. Some examples include a chemical specification form listing a physical appearance description of the chemical, chemical grade, empirical formula, and formula weight; a component specification form containing sections on material composition, size, dimensions, and color; a cell-line specification form indicating whether the cell line is suspension or anchorage dependent and detailed cell line history; a printed materials or labeling specification form containing an actual approved master copy of the printed item or label signed appropriately; and a solution specification form containing information such as formulation instructions, final solution appearance, pH, viscosity, and specific gravity.

The format of a specification should be simple and ideally one page long. Because testing methods are usually documented in SOPs, when a specification document cites a testing method, a simple reference to the SOP number is sufficient. A specification notebook should be available in the QC laboratory, and the specification files should contain the entire history of material receipt and testing. SOPs should be written on specification writing, approval, use, and change. There should also be a written commitment to audit the specifications annually.

3. Lot Numbers.   The accountability and traceability fundamental to GMPs cannot be achieved with part numbers alone. To identify an item completely, both a part number and a lot number are required. The part number tells "what" item it is, and the lot number tells "which" item it is (DeSain 1993a). Unless both numbers appear on an item, its identity is questionable.

Lot numbers are assigned when an item is received into the facility (receiving codes) or when an item is produced within the facility (in-house lot numbers). The assignment of these numbers must be documented and controlled. SOPs should be written to support policies on lot numbering, receiving, labeling, and handling of all materials. A lot number should only differentiate identical items from one another by parameters of time or location.

4. Standard Operating Procedures (SOPs).   Standard operating procedures (SOPs) are documents that describe how to perform various routine operations in a GMP manufacturing facility. They contain step-by-step instructions that technicians in QC, production, maintenance, and material handling consult daily in order to complete their tasks reliably and consistently. SOPs are, in essence, written commitments to the FDA describing the performance of routine tasks. An FDA inspection evaluates how well these written commitments are fulfilled.

An SOP can be written in several ways. However, the following elements must be considered while writing such a document: title, purpose, scope, responsibility, references/applicable documents, safety considerations, procedural principles, preliminary operations, procedures, calculations, and documentation requirements. The document should be written by or with the individuals who perform the operations. It should be written in a proactive and simple language. An SOP must be specific enough to be clear and accurate, yet flexible enough to be useful.

SOP numbers should be assigned and controlled by one individual in QA. The document should be reviewed and approved by at least two qualified people. SOPs must be kept in the areas of the facility where the work they describe is performed. These documents may also need to be changed to accurately reflect the work as it actually is performed.

5. Data Collection Forms.   Data collection forms are documents completed by a technician while performing routine tasks. These forms often include the step-by-step instructions of the SOP and provide fill-in-the-blank spaces for the collection of raw data entries. In traditional research laboratories, laboratory notebooks are used to record data. However, in a GMP manufacturing environment, forms are created to ensure that all information that must be documented while performing a procedure is, in fact, recorded.

Upon completion of the work, these forms should be verified and signed by the supervisors. Blank forms should be available for use at all sites throughout the facility where they might be required. Easy access to data forms helps to ensure that they are used. The original forms must be kept in a known, secure location because FDA reviewers are likely to request originals for review.

6. Master Production Batch Records (MPBRs).   The *Code of Federal Regulations, Title 21* (21 CFR 210.3) defines a batch as "a specific quantity of a drug or other material that is intended to have uniform character and quality, within specified limits, and is produced according to a single manufacturing order during the same cycle of manufacturing." Once the batch is defined with a known beginning and end, and the separate processing events are identified, a master production batch record (MPBR) can be designed.

The MPBR is a detailed, step-by-step description of the production process. It explains exactly how the product is produced, indicating specific types and quantities of components and raw materials, processing parameters, in-process quality controls, processing intermediate specifications, environmental controls, etc. It should contain fill-in-the-blank spaces throughout the text to facilitate the documentation of events for each individual batch.

MPBRs are typically written by production and QC personnel in consultation with individuals who routinely perform the manufacturing operations. Similar to SOPs, a number should be assigned to the MPBR.

In addition to these elements, documentation is also required for equipment installation and identification, and equipment monitoring, repair, and preventive maintenance in a GMP environment. Protocols may also be written to describe the general conduct of operations or activities in GMP facilities. The validation process must also be documented to ensure that a specific process, method, or equipment system consistently produces a product that meets predetermined specifications and quality attributes.

ISO 9000 CERTIFICATION

ISO is an acronym for the International Organization for Standardization based in Geneva, Switzerland. It was created in 1947 to "promote the development of standardization and related activities in the world with a view to facilitating international exchange of goods and services, and to develop cooperation in the sphere of intellectual, scientific, technological, and economic activity." ISO is a federation of over 90 member bodies, each representing the primary organization that sets standards in the respective countries. The United States is represented on ISO by the American National Standards Institute (ANSI). The ANSI represents more than 250 domestic standards writing organizations, the foremost being the American Society for Quality Control (ASQC).

ISO 9000 equivalent quality management standards for selected countries are listed in Table 12.2. Rather than reinvent the wheel, authorities in many countries have accepted the ISO standards essentially in toto, sometimes only changing the designation numbers. For example, ANSI refers to ISO 9000 by Q90. As of January 1994, 74 countries worldwide had adopted ISO 9000 series standards (Thayer 1994).

In 1976, ISO started developing the ISO 9000 series of standards to describe a quality management system. These standards were largely modeled after British Standard (BS) 5750, with substantial input from the United States and Canada (Nadkarni 1993). These standards were first published in 1987, and were originally intended to be advisory in nature and aid in the development of contracts between customers and suppliers. Of late, these standards have been used as a basis for third-party audits and for registration of quality systems (Surak 1992; Nadkarni 1993; Thayer 1994).

TABLE 12.2.   ISO Equivalent Quality Management
Standards for Selected Countries

Country	Equivalent standard
Australia	AS 3900
Belgium	NBNX 50
Canada	CSA 2299
Denmark	DS/EN 29000
European Economic Community	EN 29000
France	NFX 50
Germany	DIN ISO 9000
Hungary	MI 18990
India	IS 10201
Ireland	IS 300
Italy	UNI/EN 29000
Netherlands	NEN ISO 9000
New Zealand	NZS 5600
Norway	NS 5801
Spain	UNE 66900
Sweden	SS ISO 9000
Switzerland	SN 9000
United Kingdom	BS 5750
United States	ANSI/ASQC Q90

Contrary to common belief, ISO 9000 is not a product standard, but a "quality system standard." It applies not to products or services, but to the process that creates them. It is designed and intended to apply to virtually any product or service made by any process anywhere in the world. To achieve this generic state, ISO 9000 refrains, to the greatest extent possible, from mandating specific methods, practices, and techniques. It emphasizes principles, goals, and objectives with the common and specific purpose of meeting customer expectations and requirements.

Some of the salient features of ISO 9000 standards and the certification process itself are described below.

Quality Systems Defined by ISO

The aim of a quality system is to ensure that a company's product or service meets the customer's quality requirements. The quality system incorporates both quality assurance and quality control. The ISO 9000 standard has meticulously defined these terms in its document ISO 8402, *International Standard Quality Vocabulary*. Some of the critical terms defined by this ISO document are as follows:

1. Quality.   Quality is an integration of the features and characteristics that determine the extent to which the product or service satisfies the customer's needs.

2. Quality Assurance (QA).   Collectively, quality assurance defines planned, formalized activities intended to provide confidence that the product or service will meet required quality levels. In addition to in-process activities, quality assurance includes an array of activities external to the process, including activities undertaken to determine customer needs.

3. Quality Control (QC).   Quality control is the collective term for in-process activities and techniques intended to create specific quality characteristics. These activities include monitoring, reduction of variation, elimination of known causes, and efforts to increase economic effectiveness.

A quality system, as ISO guidelines define, is thus a management-driven, facilitywide, and processwide program of plans, activities, resources, and events. This program is implemented and managed with the aim of ensuring that process output will meet customer quality requirements, logically ensuring that the return on investment in ISO certification goals is met.

ISO 9000 SERIES STANDARDS

The ISO 9000 standards comprise five distinct but interrelated standards (Table 12.3). Overall, there are 20 elements or criteria in the ISO 9000 series encom-

TABLE 12.3.   An Overview of ISO 9000 Standards for Quality Management

ISO standard	Category	Title
9000	Guidelines	Quality Management and Quality Assurance Standards: Guidelines for Selection and Use
9001	Model	Quality Assurance in Design/Development, Production, Installation, and Servicing
9002	Model	Quality Assurance in Production and Installation
9003	Model	Quality Assurance in Final Inspection and Testing
9004	Guidelines	Quality Management and Quality System Elements: Guidelines
10011	Guidelines	Auditing, Qualification Criteria for Quality Systems Auditors, Management of Audit Programs

passing various aspects of quality management. These are defined later in this chapter. In addition to these five documents, ISO publishes three documents under the ISO 10011 series, entitled *Guidelines for Auditing Quality Systems* that govern the conduct of audits required for the certification process.

1. ISO 9000.   This is a guidance standard, setting forth principal concepts and defining terms. It also introduces the other four standards and provides guidance on the selection and use of ISO 9001–9003 standards. These can be used for internal quality management purposes (ISO 9002) and/or for external quality assurance (ISO 9001–9003).

Quality management includes planning, resources, operations, and all "systematic activities for quality." Contracts and audits may mandate demonstration of a quality system. The QC aspects involve monitoring processes and eliminating causes of poor performance to reduce costs. The QA aspects include planned and systematic actions to provide adequate confidence that a product or service will satisfy quality requirements. QA is both a management tool and the customer's guarantee of quality.

2. ISO 9001.   This most comprehensive standard is the model for QA in design/development, installation, and service capabilities, and is used during several stages when the supplier must assure conformance to specified requirements. It calls for management to define, document, implement, and maintain a quality policy. A separate group of internal auditors and a management representative are responsible for evaluating the effectiveness of the quality system. ISO 9001 includes all of the requirements found in ISO 9002 and 9003.

Suppliers implement this standard through an effective, documented quality system that outlines procedures and instructions. They must also have procedures for contract review, coordinate the review with the purchaser, and document the review. The supplier must also control the design process itself and its testing; these need to be continually updated and fully documented. Both incoming and outgoing supplies need to be inspected. Test equipment must conform to adequate accuracy and precision, and appropriate procedures must be used to calibrate equipment with national reference standards. Such data must be recorded and made available for audits. Similarly, a nonconforming product should not be released or installed without documented procedures. In such instances, the supplier must establish, document, and maintain procedures for corrective actions. Other protocols for purchasing, handling, storage, record keeping, and training form part of the ISO 9001 standard.

3. ISO 9002.   ISO 9002 is similar to 9001 in that it covers production and installation, but it does not include development and service. It covers the following basic QA elements: management responsibility, quality system principles, internal auditing, contract review, purchasing, process control, production control, material control and traceability, inspection and test status, product inspection and testing, control of test equipment, control of nonconforming products, corrective action, handling and post-production functions, document control, quality records, and the use of statistical methods. ISO 9002 thus also contains QC elements contained in ISO 9003 as described below.

4. ISO 9003.   This is the least comprehensive standard; it provides a model for quality assurance when only final inspection and testing are required.

5. ISO 9004.   In contrast to the other standards, ISO 9004 offers guidelines for developing and applying internal quality management elements and activities. It primarily deals with such issues as product liability and safety. This standard describes a basic set of elements by which quality management systems can be developed and implemented. However, it is not intended as a checklist for compliance with a set of requirements. Many of its guidelines are similar to those of ISO 9001. The use of charts and statistical procedures is recommended for production control. Purchasing, maintenance, record keeping, and post-production handling criteria are also very similar for ISO 9001 and ISO 9004.

6. ISO 10011 Series Standards.   The ISO 9000 quality system mandates continuous gathering and evaluation of objective evidence about the performance of the system against specified requirements. This is carried out through audits, or assessments. There are three kinds of assessments as follows:

- *Internal assessments:* These are carried out by facility personnel in accordance with the facility's quality policy. They are required for contractual models.
- *Second-party assessments:* Here a customer audits the facility's quality systems.
- *Third-party assessments:* Also called "extrinsic audits," these are usually carried out by an accredited third party, often a registration body, to provide objective ev-

idence that the facility's quality system meets the published standards. Audit approval is essential for registration to the ISO 9000 quality system standard.

Introduced in 1990, the ISO 10011 series standards, entitled *Guidelines for Auditing Quality Systems,* include three documents that govern the conduct of audits. The Auditing document (ISO 10011–1) establishes basic audit principles, criteria, and practices. It spells out a system governing all aspects of quality system audits, from audit planning through followup on corrective action requests. The document Qualification Criteria for Quality Systems Auditors (ISO 10011–2) outlines educational, training, personnel, and experiential requirements for quality system auditors and lead auditors. ISO 10011–3, Management of Audit Programs, provides important guidelines for organizing, staffing, and carrying out audits. Although it is general enough to apply to all types of audits, its principles should be observed as facilities develop internal audit programs consistent with the ISO 9000 standard.

In addition to these standards, ISO is scheduled to release the ISO 14000 series of standards beginning in October 1996 (Gillespie 1995). Unlike the other standards, ISO 14000 deals only with one aspect, environmental management. It is concerned with controlled and uncontrolled emissions in atmosphere, discharges to water, land contamination, and disposal of solid and other wastes. In all, there will be five standards dealing with environmental management systems (EMSs), environmental performance evaluation, environmental labeling and life cycle assessment, a guide for environmental aspects in product standards, and environmental auditing. ISO 14000 leaves it to the organization to determine the environmental aspects it can control, depending on regulatory demands and its own operations.

Selection Criteria

It is not essential to apply for and meet the requirements for all the ISO 9000 series standards. In obtaining registration, a company must first determine the appropriate standard for registration (Table 12.3). For most R&D facilities, ISO 9001 and/or 9002 are the most appropriate standards when product development, design, and service are involved. ISO 9002 is more appropriate for chemical and process industry manufacturing plants involved in design and specification activities and in manufacturing products from supplied raw materials.

Companies do not need to register the quality system for the entire company. They can limit the scope of registration. For example, a company can obtain ISO 9002 standalone certification only for the QC/QA laboratories at the manufacturing sites. Whether or not a company obtains formal registration for the entire company is dependent on contractual requirements and the need to gain a competitive edge in the global market. Even if a company limits the scope of registration, it should still develop a quality system that meets or exceeds the ISO standard for the entire company.

Quality System Elements of ISO 9000 Series Standards

The ISO 9000 series standards involve 20 elements or criteria for defining a quality system. Not all are included or mandatory for obtaining registration under a particular standard. For example, all of the 20 criteria listed below are mandatory for ISO 9001 certification, whereas ISO 9002 and 9003 standards require only 18 and 12 of the following elements, respectively, for registration.

1. Management Responsibility.   This defines and documents the quality policy, quality organization, management representative, and management review of the quality system, and is a required element in all series of standards.

2. Quality System.   This is a system to ensure that products and services conform to specified requirements. This should be documented as a manual and properly implemented.

3. Contract Review.   This ensures that contractual requirements are met in selling products and services. This element is not present in the ISO 9003 standard.

4. Design Control.   This requires planning the design project; defining design input and output parameters, verifying the design, and controlling design changes. It is not present in ISO 9002 and 9003.

5. Document Control.   This element requires procedures for controlling generation, distribution, and changes in documents, and ensures that employees are using the proper procedures, instructions, and documents. This element is an integral part of all three series of standards.

6. Purchasing.   This is a system to ensure that purchased product conforms to specified requirements to meet ISO 9001 or 9002 standards only.

7. Control of Customer-Supplied Product.   This system controls the handling, storing, and maintenance of materials supplied by the purchaser or customer as a part of the ISO 9001 or 9002 standard.

8. Product Identification and Traceability.   This requires product identification by item, batch, or lot during all stages of production, delivery, and installation, and is mandatory in all three standards.

9. Process Control.   Process control procedures ensure that operations are controlled in a specified manner and sequence so that the final product will meet requirements specified in ISO 9001 or 9002.

10. Inspection and Testing.   This element requires procedures for inspecting and testing incoming materials, products as they are processed, and finished products. Records of inspection, test data, and disposition of products must be maintained.

11. Control of Inspection, Measuring, and Test Equipment.   Procedures are required for calibrating, checking, controlling, and maintaining test and measuring equipment to provide confidence in decisions and actions that are based on measurement data.

12. Inspection and Test Status.   This is a system to verify the status of materials and components throughout the entire manufacturing process. It requires proper markings, stamps, or labels be used on products as they move through various processing steps to indicate conformance or nonconformance to testing.

All the three elements requiring inspections are part of the ISO 9001–9003 standards.

13. Control of Nonconforming Products.   The manufacturing processes must be controlled so that product not conforming to standards is not inadvertently used. It requires formal procedures to review and dispose of such products. This element is also common to all three standards.

14. Corrective Action.   Corrective action describes a system to investigate, eliminate, or minimize the reoccurrence of problems, and forms part of the ISO 9001 and 9002 standards.

15. Handling, Storage, Packaging, and Delivery.   Common to all standards, this system is designed to control the handling, packaging, storage, and delivery of product up to the time the product is put into use.

16. Quality Records.   This system controls the proper handling and maintenance of records for the quality system, and is common to all standards.

17. Internal Quality Audits.   This element defines a system to conduct internal audits that will verify the effectiveness of the quality management system, and is a required part of ISO 9001 and 9002.

18. Training.   Training needs are identified and training must be provided to all necessary personnel who have assigned activities that affect the quality of products and services. This element is common to all standards.

19. Servicing Procedures.   Required only for ISO 9001, this element describes a system to meet the after sales servicing requirements of the contract.

20. Statistical Techniques.   Used in all series of standards, statistical techniques are used to verify acceptability of the process, product, and service.

Registration/Certification Process

Getting an ISO 9000 certification is essentially a team effort. Typically, the following steps are involved in obtaining registration to an ISO standard (Nadkarni 1993; Burr 1990; Craig 1991, 1992; Surak 1992; Johnson 1993; Clements 1994; Hassler and Yankowsky 1995).

1. Commitment by Top Management.   The top management of the company must commit itself to successful implementation of a quality system. Not only must they believe that a quality system will benefit the company, they must also be actively involved in the implementation process.

2. Formation of an ISO Steering Team.   This team is responsible for the development and implementation of the plan that will lead to registration. Their activities typically include a review of the ISO standard, an evaluation of how the current quality system complies to the ISO standard, and identification of gaps between the existing quality system and the standard. This team is also responsible for developing a plan for eliminating the deficiencies in the quality system.

3. Training in ISO.   During the initial phase of the registration process, all employees who are involved in processes that affect the quality of products or services should receive appropriate training in the ISO standard, registration process, and audit procedures. Sometimes additional training may be given to employees to eliminate technical deficiencies.

4. Development of a Quality Manual.   A documentation system is developed to describe the quality management system, procedures, and work instructions. In addition, appropriate records need to be kept to prove that the quality management system is operating in an effective manner. This documentation system must be developed in a way that satisfies the appropriate ISO standard.

5. Start-Up of Internal Audits.   The ISO standard mandates that the company carry out internal audits to determine the effectiveness of the quality management system. The company must also develop a corrective action system. This system must be able to identify and eliminate the root cause of problems.

6. Selection of a Registrar.   Caution must be exerted in selecting a registrar for the certification process. Registrations are not transferable between different registrars. If a company decides to change registrars, it must undergo a complete registration audit.

7. Preassessment.   A company may elect to have a registrar or a consultant conduct a preassessment of the quality system to determine if there are major problems in the quality system that could cause the company to fail the formal assessment.

8. System Improvement. Deficiencies discovered during the preassessment are corrected, and improved quality systems are implemented.

9. Assessment. Formal assessment is carried out at this stage. The audit team or registrar will first determine if the quality manual conforms to the appropriate ISO standard and then determine if the procedures and work instructions conform to the quality manual. Finally, the team will determine if employees are conducting their jobs in accordance with the procedure and work instructions.

Depending on whether major or minor findings are identified during the audit, the audit team may or may not recommend registration of a company. A delay in the registration can be recommended if there are significant audit findings. In such a case, the registrar may either require a followup audit or a complete audit of the quality system. This audit will be conducted after the company has implemented the improvements to the quality system.

10. Fixing of Minor Discrepancies. The company will fix any minor discrepancies. Some may be fixed during the audit, while others may need to be corrected at a later date. The audit team will determine how the audit closeout process will be conducted so that the registration process can be completed.

11. Obtaining Registration from the Registrar. Upon successful completion of the audit, the registrar will issue a registration certificate indicating that the company's quality system meets the criteria of the appropriate ISO standard.

Once the initial registration is achieved, the quality system is maintained by regular internal audits and third-party surveillance audits. The latter are done at least twice a year. Typically, the registrar will audit from one-third to one-half of the quality system during each surveillance audit. Some registrars will conduct only surveillance audits as long as the registration remains in effect. Others will conduct surveillance audits twice a year for two and one-half years and then conduct a complete audit of the quality system every third year. An awarded registration can be withdrawn if, in subsequent audits, the system no longer meets ISO standards.

Advantages and Drawbacks

ISO 9000 sets universally accepted, uniform standards of quality management for all industries. ISO certification assures that the right quality product will be consistently produced without preventing continuous improvement of a company's quality systems. ISO certification offers two main benefits: it has a positive impact on a company's bottom line, and it improves quality throughout the company. Additional advantages include the following:

1. It opens export markets,
2. ISO certification provides tangible evidence of the management's commitment to quality practices,

3. It enhances market credibility,

4. It provides economic benefits through better use of raw materials as well as reduced inspection costs and time spent on reworking, thus providing continuous improvement and eliminating waste,

5. Expensive errors are minimized, and

6. Overall cost of quality is reduced.

In spite of several obvious advantages, ISO standards do have certain drawbacks. Some of these include:

1. The certification process is expensive and time-consuming, which can be prohibitive to small companies,

2. It does not guarantee economic success and true total quality management,

3. ISO standards lack a requirement for either continuous improvement or customer satisfaction, and

4. The quality system covers only the activities of those employees that can affect the quality of goods and services.

These drawbacks notwithstanding, interest is growing in the ISO 9000 quality standards in the global marketplace. There is also an increasing pressure, especially from the European Economic Community (EEC) countries, to assure the customers that the products they purchase will meet minimum quality standards defined by ISO 9000. The U.S. businesses will be forced to adopt the ISO 9000 standards to export their products to EEC countries in the near future, and to the rest of the world in the long range. Achieving ISO 9000 certification also provides a solid foundation for total quality management. ISO 9000, therefore, can be looked upon as the lowest common denominator of quality systems on which to base enhanced practices to achieve total quality, lowered costs, increased customer satisfaction, and improved market share.

DIAGNOSTICS START-UPS

The diagnostics industry offers several opportunities and challenges to savvy entrepreneurs. The recent history also indicates a willingness on the part of venture capital firms to invest money (over $10 billion during the past 4 years) in diagnostics and medical start-ups because returns have been good. This section is focused on the needs for raising capital for the start-up through good planning. The information is presented under three categories: (1) how to conduct market research, (2) how to prepare a business plan, and (3) how to value market a start-up company to venture capitalists.

Market Research

Start-up companies spend a significant amount of money on product development but often spend almost nothing on understanding the marketplace. There are several reasons for inadequate market research. First, founders often come from science and engineering backgrounds and have had little exposure to this valuable tool. Second, market research conjures up images of spending large sums of money to obtain information that is "already known." Third, entrepreneurs are sometimes so convinced that their products are needed by prospective customers, that they don't see the value of asking anyone for feedback. In fact, nothing is more important for a start-up company than developing a product that customers want and will buy. A start-up company's survival usually depends on being able to sell the first product it develops.

Primary market research is any formal or informal process for eliciting information and feedback from prospective customers (Dotzler 1994a). For a start-up company, the initial objective of this research is to assess whether or not there is a need for a potential service or product, and to help determine its market potential. Later, feedback from prospective customers can provide important insights into what is needed to make the company's product more attractive to buyers. Market research also can help the company determine an appropriate price for its services, gain a better understanding of competitive products, identify the ideal selling channel, and even test promotional materials such as mailings, advertisements, and brochures.

The key to good market research is to avoid influencing the results. Company employees tend to lack the discipline and objectivity necessary for such research. Ideally, if the start-up has the resources, it should employ a professional market researcher. Both qualitative and quantitative market research need to be conducted.

The objective of the qualitative or exploratory research is to increase understanding of the factors that influence prospective customers to buy a given product. This type of research can be conducted using focus group meetings in which a general topic and a product concept are discussed, or by conducting face-to-face interviews with potential customers.

Quantitative market research is undertaken to measure or quantify the broader market's reaction to a product concept or procedure. It is useful for projecting market size and sales potential, ranking a product's attributes and identifying product voids, better understanding of the currently used products and procedures, and defining the market share of competing companies (Dotzler 1994a). This type of research is conducted using face-to-face surveys, mailed surveys, and phone surveys with numerous participants, so that the results can be projected to a group of customers.

Finally, product testing or test marketing research may be required to help measure product performance or the strength of marketing plans, before large sums of money are invested in the product launch. Using these techniques, a company can more effectively ensure that the product will be well received in the marketplace.

Business Plan

Before beginning a search for potential investors, most start-ups will write a business plan describing what they intend to do and how they will to do it. The objectives of such a plan are 2-fold: to set a direction for the company, and to capture the interest of potential capital investors (Dotzler 1994b). An example of a typical business plan is shown in Table 12.4. These are typically 20–50 pages long, with optional supporting data placed in an addendum. Various sections of a business plan are briefly described below.

TABLE 12.4.   An Example of a Business Plan for a Diagnostics Start-Up Company

   I. Executive Summary/Company Strategy
  II. Technology/Products
      A. Technology description
      B. Development plan
      C. Proprietary protection
 III. Markets/Marketing
      A. Markets
      B. Selling and distribution
      C. Competition
      D. Forecasts
  IV. Clinical/Regulatory Issues
   V. Manufacturing and Operations
      A. Facilities
      B. Equipment
      C. Process
      D. Cost of goods sold
  VI. Management Team
      A. Administration
      B. Research and Development
      C. Manufacturing
      D. Marketing
      E. Finance and Accounting
 VII. Financials
      A. Profit and loss
      B. Balance sheet
      C. Cash flow (sources and uses)
VIII. Risks
  IX. Appendices/Addendum

1. Executive Summary.   This section should highlight the business strategy emphasizing how and why it will succeed in the marketplace. It should explain the company's products, why they are proprietary, market needs for this type of product, a sales and earnings forecast, and an outline of the financing strategy.

2. Technology/Products.   This section highlights the basic technology and how it can be developed into marketable products or services. It should contain the following information: resources required to complete the development phase of the work, development milestones and when they will be achieved, significant obstacles to completing the work, protection of the technology, a discussion on the ratio of price/performance to competitive technologies, lead time over other start-ups in the field, the probability of obsolescence, regulatory hurdles, and the use of the technology in other products.

3. Markets/Marketing.   The markets for current and future products based on this technology or services should be defined along with a marketing strategy and a sales forecast. This section should also highlight primary customer prospects, their size, the competitive products they are using currently, creation of buyer interest in the product, possible sales strategies, and any customer training that might be required before the launch of the product.

4. Clinical/Regulatory Issues.   A strategy should be outlined to gain marketing approval from the regulatory agencies. The entire spectrum of required preclinical/field studies as well as clinical/field trials should be defined, particularly with respect to sample size, number of test centers, and the length of such studies.

5. Manufacturing and Operations.   This section of the business plan will include requirements for a manufacturing facility; information on how the product will be manufactured; identification of any single-source or difficult-to-purchase components; and estimated manufacturing costs including labor, materials, and overheads. Compliance with regulatory guidelines such as GMPs should also be described.

6. Management Team.   Biographies of the entire management team including research and development, marketing, manufacturing, finance, and accounting need to be included in the business plan. If the team is not yey in place, hiring plans must be spelled out. The strength of the management team is an important factor to potential investors. Venture capitalists, in particular, want to invest in people who have the ability and motivation to build a company. Every management team member should have relevant industry experience and functional expertise.

7. Financials.   The financial summary should be concise and should include a projected five-year profit-and-loss statement, balance sheet, and cash flow statement. The first-year financials should be shown monthly, years two and three can be shown quarterly, and yearly data are generally sufficient for the last two years. Backup data should be available so that potential investors can do a more in-depth financial analysis if they wish. The cashflow analysis should include when and how much capital will be required.

8. Risks. It is extremely important to disclose all risks and obstacles the company may face. Most investors are natural risk takers and are not afraid to invest money when the outcome is unpredictable. However, they must know what the risks are so that they can calculate the probability of success and the rate of return they require on their investment.

Valuation of a Start-Up Company

The valuation of a start-up company is usually negotiated by the company's founding team and the venture capitalists (Dotzler 1994c). Eleven factors influence these negotiations (Table 12.5). These include factors that can be controlled, such as the technology, target markets, management team, and financial potential, as well as external factors that affect the return on investment. Companies should strive to complete milestones that will eliminate risks associated with these variables.

Presently, nationwide more than 700 venture capital partnerships manage over $40 billion dollars. These firms specialize by industry, geography, and stage.

TABLE 12.5. Eleven Factors Influencing Valuation of a Diagnostics Start-Up Company

Variable	Lower valuation	Higher valuation
Technology		
Stage of development	Concept	Product
Patent status	None filed	Issued
Time to market	Long	Short
Market		
Demonstrable need	No	Yes
Size and growth	Small	Large
Market penetration	Slow	Rapid
Management Team Financial	None	Complete
Profit margins	Low	High
Total capital required	High	Low
Return on Investment		
Potential value in future	Low	High
Time to liquidity	Long	Short

Some invest small amounts, $50,000 to $100,000; others will not join a financing syndicate if they cannot invest $1.5 million dollars or more (Dotzler 1994d). The start-ups need to identify venture capital partnerships that have invested in their particular market and geographical region. Their experience with other companies in related but not competitive markets will often be helpful for the start-ups. It is also essential to find out whether the venture capitalists have competing investments that would preclude their consideration of the start-up company.

The amount of money a start-up company plans to raise should be enough to enable it to accomplish key milestones that will either make the company self sufficient or will enable it to raise additional funds at a higher valuation on a later date. One needs to be in close contact with those investors who can offer strategic advice based on their knowledge of the industry; guidance stemming from prior operating experience; vital network connections to potential management, board members, and professionals; and help in achieving liquidity through a merger, acquisition, or initial public offering.

REFERENCES

BLUMENTHAL, M. M. 1992. PCs in research and production laboratories. *Inform* 3:574–581.

BURR, J. T. 1990. The future is necessary. *Qual. Progress* 23(6):19–23.

CDER. 1987a. *Guideline on Sterile Drug Products Produced by Aseptic Processing.* Center for Drugs Evaluation and Research, Division of Manufacturing and Product Quality, Food and Drug Administration, Rockville, MD.

CDER. 1987b. *Guideline on General Principles of Process Validation.* Center for Drugs Evaluation and Research, Division of Manufacturing and Product Quality, Food and Drug Administration, Rockville, MD.

CEC. 1989. *Development Pharmaceutics and Process Validation. Rules Governing Medicinal Production in the European Community.* Vol. 3. Commission of the European Communities, Office of Official Publications of the European Community, Brussels, Belgium.

CEC. 1990. *Validation of the Purification Procedures. Rules Governing Medicinal Production in the European Community, Production and QC of Human Monoclonal Antibodies.* Vol. 3. Commission of the European Communities, Office of Official Publications of the European Community, Brussels, Belgium.

CFR. 1990. *Code of Federal Regulations, Title 21.* U.S. Govt. Printing Office, Washington, D.C.

CLEMENTS, R. B. 1994. *Quality Manager's Complete Guide to ISO 9000.* Prentice-Hall, Englewood Cliffs, NJ.

CRAIG, R. J. 1991. Road map to ISO registration. *ASQC Quality Cong. Trans.* 45:926–930.

CRAIG, R. J. 1992. Reflections: Life beyond ISO 9000 registration. *ASQC Quality Cong. Trans.* 46:1103–1109.

DeSain, C. V. 1993a. *Documentation Basics That Support Good Manufacturing Practices.* Aster Publ. Corp., Eugene, OR.

DeSain, C. V. 1993b. *Drug, Device and Diagnostic Manufacturing. The Ultimate Resource Handbook.* 2d ed., Interpharm Press, Buffalo Grove, IL.

DeSain, C. V., and Sutton, C. V. 1994. *Validation for Device and Diagnostic Manufacturers.* Interpharm Press, Buffalo Grove, IL.

DeSain, C. V., and Sutton, C. V. 1995a. Validation for device and diagnostic manufacturers. Part 2. Working with vendors and contractors. *BioPharm* 8(2):40–46.

DeSain, C. V., and Sutton, C. V. 1995b. Validation for device and diagnostic manufacturers. Part 3. Detecting change in validated systems and processes. *BioPharm* 8(4):42–43.

DeSain, C. V., and Sutton, C. V. 1995c. Validation for device and diagnostic manufacturers. Part 4. Decision making associated with change in validated systems and processes. *BioPharm* 8(5):73–74.

DeSain, C. V., and Vercimak, C. 1993a. Documentation basics: The master plan. *BioPharm* 6(5):27–30.

DeSain, C. V., and Vercimak, C. 1993b. Documentation basics: Clinical protocols and clinical SOPs. *BioPharm* 6(7):32–37.

DeSain, C. V., and Vercimak, C. 1993c. Documentation basics: Accountability and traceability. *BioPharm* 6(9):28–33.

Dotzler, F. 1994a. *Market research basics for the start-up biomedical company, or "you don't know what you don't know."* Medicus Venture Partners, Menlo Park, CA.

Dotzler, F. 1994b. *Raising venture capital: Tips for the medical start-up.* Medicus Venture Partners, Menlo Park, CA.

Dotzler, F. 1994c. *What percent of a medical company should founders sell for seed/start-up venture capital?* Medicus Venture Partners, Menlo Park, CA.

Dotzler, F. 1994d. *Raising venture capital: Managing the process for a medical start-up.* Medicus Venture Partners, Menlo Park, CA.

Finkelson, M. J. 1986. Validation of analytical methods by FDA laboratories II. *Pharm. Technol.* 10(3):75,78,80–84.

Gillespie, H. 1995. ISO 14000. What's driving the new environmental standard? *Research and Development* 37(9):29.

Guerra, J. 1986. Validation of analytical methods by FDA laboratories I. *Pharm. Technol.* 10(3):74,76,78.

Hassler, J., and Yankowsky, A. 1995. An overview of ISO 9001 certification. *Biopharm* 8(2):48–50.

Johnson, P. L. 1993. ISO 9000. *Meeting the New International Standards.* McGraw-Hill, New York.

Morrow, M. 1993. ISO 9000 survey finds real benefits in registration. *Chemical Week* 153(11):52.

Nadkarni, R. A. 1993. ISO 9000. Quality management standards for chemical and process industries. *Anal. Chem.* 65:387A–395A.

OFR/NARA. 1989. *Guide to Record Retention Requirements.* Office of the Federal Register/ National Archives and Record Administration, U.S. Govt. Printing Office, Washington, D.C.

RITTENBURG, J. H., and GROTHAUS, G. D. 1992. Immunoassays: Formats and applications. In *Food Safety and Quality Assurance. Applications of Immunoassay Systems,* eds. M. R. A. Morgan, C. J. Smith, and P. A. Williams, pp. 3–10, Elsevier Applied Science Publishers, London.

STIER, R. F.; BLUMENTHAL, T. K.; and BLUMENTHAL, M. M. 1993. GLPs: What are they? And how can they help food processors? *Dairy, Food and Environmental Sanitation* 13:272–275.

SURAK, J. G. 1992. The ISO 9000 standards. Establishing a foundation for quality. *Food Technol.* 46(11):74–80.

THAYER, A. M. 1994. Chemical companies see beneficial results from ISO 9000 registration. *Chem. Eng. News* 72(17):10–26.

WEISS, M. 1990. Advantages and validation issues for long term continuous production. *Genetic Eng. News* 10(7):8.

Index